METHODS IN MOLECULAR BIOLOGY

Series Editor
John M. Walker
School of Life and Medical Sciences
University of Hertfordshire
Hatfield, Hertfordshire, UK

For further volumes:
http://www.springer.com/series/7651

For over 35 years, biological scientists have come to rely on the research protocols and methodologies in the critically acclaimed *Methods in Molecular Biology* series. The series was the first to introduce the step-by-step protocols approach that has become the standard in all biomedical protocol publishing. Each protocol is provided in readily-reproducible step-by-step fashion, opening with an introductory overview, a list of the materials and reagents needed to complete the experiment, and followed by a detailed procedure that is supported with a helpful notes section offering tips and tricks of the trade as well as troubleshooting advice. These hallmark features were introduced by series editor Dr. John Walker and constitute the key ingredient in each and every volume of the *Methods in Molecular Biology* series. Tested and trusted, comprehensive and reliable, all protocols from the series are indexed in PubMed.

Doubled Haploid Technology

Volume 1: General Topics, Alliaceae, Cereals

Edited by

Jose M. Seguí-Simarro

Cell Biology Group, COMAV Institute - Universitat Politècnica de València, Valencia, Spain

 Humana Press

Editor
Jose M. Seguí-Simarro
Cell Biology Group
COMAV Institute - Universitat Politècnica de València
Valencia, Spain

ISSN 1064-3745   ISSN 1940-6029   (electronic)
Methods in Molecular Biology
ISBN 978-1-0716-1317-7   ISBN 978-1-0716-1315-3   (eBook)
https://doi.org/10.1007/978-1-0716-1315-3

This Humana imprint is published by the registered company Springer Science+Business Media, LLC, part of Springer Nature.
The registered company address is: 1 New York Plaza, New York, NY 10004, U.S.A.

Preface

Doubled haploid (DH) technology is a very convenient approach in plant breeding to save a considerable amount of time, and the corresponding costs, to produce pure, fully homozygous parental lines, essential for large-scale production of hybrid seed. This is the main reason why it is widely used in private companies and academia to produce homozygous individuals. In addition, DH plants are also useful in applied genetic research, including estimation of recombination fractions and linkage studies, genetic mapping of complex qualitative traits, fixation of transgenes in homozygosis, or unmasking of recessive mutants, among others.

In essence, DH technology relies on the use of haploid cells (basically the plant gametes or their precursors) to promote their development as haploid embryos. This is something that rarely happens in nature, and this is why such a process must be induced through experimental manipulation. Once the haploid embryo is obtained, it may become diploid (DH) on its own, or additional treatments for chromosome doubling may be needed. The methods to produce haploids and then DHs are remarkably diverse. A broad classification divides them into in vivo (methods that are carried on in planta) and in vitro methods (methods that require the in vitro culture of some parts of the plant). There are species amenable to one or some of these methods, and others (some of them of high agronomic interest) that do not respond to any of the available methods, which justifies the research in this field.

In this work, 62 chapters written by experts in the production of DHs in different species are compiled. These chapters cover the most important topics dealing with DH technology, as well as different methods to produce DHs in different species through different in vivo and in vitro approaches. This compilation is divided into three volumes. Each volume starts with a section of review-type chapters that cover general topics and different transversal methods of relevance in DH technology, and follows with parts covering practical protocols to produce DHs in a total of 44 different species, grouping each plant family in a separate part. In some of these species, different approaches such as anther culture, microspore culture, wide hybridization, hybridization with irradiated pollen, or in vitro culture of ovules or ovaries are covered in different chapters.

In *Doubled Haploid Technology, Volume 1: General Topics, Alliaceae, Cereals*, Part I deals with general topics and transversal methods in DH technology and includes five chapters. Chapter 1 is an overview of the different in vitro and in vivo technologies currently available to produce DHs in different species. Chapter 2 summarizes the different applications of DHs in plant breeding and applied research. Chapter 3 is a bibliographical compilation of selected references to produce haploid and/or DH plants in 383 different species. Chapter 4 reviews the most used methods to analyze ploidy in haploid and DH materials. Chapter 5 covers the different methods available for chromosome doubling of haploid individuals. Part II presents protocols for DH production in two alliaceae, onion and leek (Chapters 6 and 7). Part III includes 13 protocols of different methods for DH production in a number of cereals such as barley, durum, bread and spelt wheat, maize, triticale, oat, indica and japonica rice (Chapters 8–20).

In *Doubled Haploid Technology, Volume 2: Hot Topics, Apiaceae, Brassicaceae, Solanaceae*, Part I covers different hot topics in DH technology. Chapter 1 deals with a major problem of

DHs produced in some cereal species by in vitro methods: albinism. Chapter 2 focuses on DH production by in vivo methods in maize, a model species where these methods are widely used. Chapter 3 deals with the use and application of molecular markers for DH technology with two parallel perspectives, their application in plant breeding programs by private companies, and their use for research in academic institutions. Chapter 4 is a review of zygotic embryogenesis, useful as a reference of the development of the zygotic embryo to compare with the abnormal haploid/DH embryos frequently produced mostly through in vitro DH methods. Part II presents protocols to produce DHs in four apiaceae, caraway, fennel, dill, and carrot (Chapters 5–7). Part III includes six protocols for DH production in different brassicaceae including *Brassica napus*, *Brassica rapa*, *Brassica carinata*, *Brassica oleracea*, and *Raphanus sativus* (Chapters 8–13). Part IV includes eight protocols to produce DHs by different androgenesis-based methods in several members of the solanaceae family, including eggplant, pepper, tobacco, potato, *Physalis ixocarpa*, and *Datura metel* (Chapters 14–21).

In *Doubled Haploid Technology, Volume 3: Emerging Tools, Cucurbits, Trees, Other Species*, Part I includes three review-type chapters about emerging tools in DH technology. Chapter 1 deals with centromere engineering, an emerging tool being currently applied to an increasing number of crops. Chapters 2 and 3 focus on two underutilized methods not directly related to DH technology, but whose implementation to DH protocols would facilitate research and analysis of haploid/DH materials considerably: fractional factorial designs for rapid in vitro system development (Chapter 2) and impedance flow cytometry (Chapter 3). Part II presents six protocols based on anther culture, in vitro gynogenesis and hybridization with irradiated pollen for DH production in cucurbits, including cucumber, melon, watermelon, summer squash, pumpkin, and winter squash (Chapters 4–8). Part III presents protocols to produce haploids and DHs in different woody species and genera such as *Citrus*, coconut, walnut, almond, hazelnut, cork oak, and *Jatropha curcas* (Chapters 9–13). Finally, Part IV compiles a series of protocols for DH production in other seven species, not belonging to the previous sections. These species include borage, *Saintpaulia ionantha*, cow cockle, marigold, chickpea, red beet, and sugar beet (Chapters 14–20).

Some chapters describe the production of DHs in a same species but using different approaches (e.g., those focused on barley DHs by microspore culture and by the "bulbosum method"). There are also different chapters on variations of the same method (e.g., the conventional and double-layered method for sweet pepper). Despite the apparent redundancy, these chapters are useful to stress the importance of the genotype, a concept repeatedly pointed out throughout the book, to show the diversity of approaches available and principally to provide researchers with alternative tools to apply in case a standard protocol does not work well in a given material. Together, the chapters of this book provide a broad overview of the most important species (and varieties as for indica and japonica rice, for example) with available protocols to produce haploid and DH individuals. Thanks to the effort and contributions of 145 different authors from all over the world, these 62 chapters make this compilation the most comprehensive and ambitious book published up to now on this topic.

Valencia, Spain   *Jose M. Seguí-Simarro*

Contents

Contributors

BEHZAD AHMADI • *Department of Maize and Forage Crops Research, Agricultural Research, Education and Extension Organization (AREEO), Seed and Plant Improvement Institute (SPII), Karaj, Iran*

ALI RAMAZAN ALAN • *Plant Genetics and Agricultural Biotechnology Application and Research Center (PAU BIYOM), Pamukkale University, Denizli, Turkey; Department of Biology, Pamukkale University, Denizli, Turkey*

SANDRA ALLUÉ • *Department of Genetics and Plant Production. Aula Dei Experimental Station, Spanish National Research Council (EEAD-CSIC), Zaragoza, Spain*

OLFA AYED SLAMA • *Laboratory of Genetics and Cereal Breeding LR14AGR01, National Agronomic Institute of Tunisia, Department of Agronomy and Plant Biotechnology, University of Carthage, Tunis, Tunisie*

JAVIER BELINCHÓN MORENO • *Cell Biology Group – COMAV Institute, Universitat Politècnica de València, Valencia, Spain*

HEIKE BÜCHNER • *Leibniz Institute of Plant Genetics and Crop Plant Research (IPK) Gatersleben, Plant Reproductive Biology, Seeland, Germany*

FEVZIYE CELEBI-TOPRAK • *Plant Genetics and Agricultural Biotechnology Application and Research Center (PAU BIYOM), Pamukkale University, Denizli, Turkey; Department of Biology, Pamukkale University, Denizli, Turkey*

LUÍS CISTUÉ • *Departamento de Genética y Producción Vegetal, Estación Experimental de Aula Dei, Consejo Superior de Investigaciones Científicas, Zaragoza, Spain*

MARÍA ASUNCIÓN COSTAR • *Department of Genetics and Plant Production. Aula Dei Experimental Station, Spanish National Research Council (EEAD-CSIC), Zaragoza, Spain*

ANA MARÍA CASTILLO • *Department of Genetics and Plant Production. Aula Dei Experimental Station, Spanish National Research Council (EEAD-CSIC), Zaragoza, Spain*

ILONA CZYCZYŁO-MYSZA • *The Franciszek Górski Institute of Plant Physiology, Polish Academy of Sciences, Kraków, Poland*

ISIDRE D'HOOGHVORST • *Departament de Biologia Evolutiva, Ecologia i Ciencies Ambientals, Secció de Fisiologia Vegetal, Universitat de Barcelona, Barcelona, Spain; ROCALBA S.A., Girona, Spain*

DIAA ELDIN S. DAGHMA • *Leibniz Institute of Plant Genetics and Crop Plant Research (IPK) Gatersleben, Plant Reproductive Biology, Seeland, Germany*

SARA AUDIJE DE LA FUENTE • *Leibniz Institute of Plant Genetics and Crop Plant Research (IPK) Gatersleben, Plant Reproductive Biology, Seeland, Germany*

PIERRE DEVAUX • *Research & Innovation, Florimond Desprez, Cappelle en Pévèle, France*

KINGA DZIURKA • *The Franciszek Górski Institute of Plant Physiology, Polish Academy of Sciences, Kraków, Poland*

BEGOÑA ECHÁVARRI • *Departamento de Genética y Producción Vegetal, Estación Experimental de Aula Dei, Consejo Superior de Investigaciones Científicas, Zaragoza, Spain*

IRENE FERRERES • *Departament de Biologia Evolutiva, Ecologia i Ciencies Ambientals, Secció de Fisiologia Vegetal, Universitat de Barcelona, Barcelona, Spain*

Antoine Gaillard • *MAS Seeds, Haut-Mauco, France*

Marina Guillot Fernández • *Cell Biology Group – COMAV Institute, Universitat Politècnica de València, Valencia, Spain*

Hiroshi Hisano • *Institute of Plant Science and Resources, Okayama University, Kurashiki, Okayama, Japan*

Robert Eric Hoffie • *Plant Reproductive Biology, Leibniz Institute of Plant Genetics and Crop Plant Research (IPK), Gatersleben, Germany*

Nathanaël M. A. Jacquier • *Laboratoire Reproduction et Développement des Plantes, Univ Lyon, ENS de Lyon, UCB Lyon 1, CNRS, INRAE, Lyon, France; Limagrain, Limagrain Field Seeds, Research Center, Gerzat, France*

Katarzyna Juzoń • *The Franciszek Górski Institute of Plant Physiology, Polish Academy of Sciences, Kraków, Poland*

Jochen Kumlehn • *Plant Reproductive Biology, Leibniz Institute of Plant Genetics and Crop Plant Research (IPK), Gatersleben, Germany*

Csaba Lantos • *Department of Biotechnology, Cereal Research Non-profit Ltd., Szeged, Hungary*

John D. Laurie • *Agriculture and Agri-food Canada, Lethbridge, AB, Canada*

Priti Maheshwari • *Agriculture and Agri-food Canada, Lethbridge, AB, Canada*

Izabela Marcińska • *The Franciszek Górski Institute of Plant Physiology, Polish Academy of Sciences, Kraków, Poland*

Ricardo Mir • *Cell Biology Group – COMAV Institute, Universitat Politècnica de València, Valencia, Spain*

Andrea Müller • *Leibniz Institute of Plant Genetics and Crop Plant Research (IPK) Gatersleben, Plant Reproductive Biology, Seeland, Germany*

Mohsen Niazian • *Field and Horticultural Crops Research Department, Kurdistan Agricultural and Natural Resources Research and Education Center, Agricultural Research, Education and Extension Organization (AREEO), Sanandaj, Iran*

Salvador Nogués • *Departament de Biologia Evolutiva, Ecologia i Ciencies Ambientals, Secció de Fisiologia Vegetal, Universitat de Barcelona, Barcelona, Spain*

Sergio J. Ochatt • *UMR 1347 Agroécologie, AgroSup/INRAE/uB, Laboratoire de Physiologie Cellulaire, Morphogenèse et Validation (PCMV), Dijon, France*

Vivian Ott • *Leibniz Institute of Plant Genetics and Crop Plant Research (IPK) Gatersleben, Plant Reproductive Biology, Seeland, Germany*

Ingrid Otto • *Plant Reproductive Biology, Leibniz Institute of Plant Genetics and Crop Plant Research (IPK), Gatersleben, Germany*

János Pauk • *Department of Biotechnology, Cereal Research Non-profit Ltd., Szeged, Hungary*

Pooja Satpathy • *Leibniz Institute of Plant Genetics and Crop Plant Research (IPK) Gatersleben, Plant Reproductive Biology, Seeland, Germany*

Jose M. Seguí-Simarro • *Cell Biology Group, Ciudad Politécnica de la Innovación (CPI), COMAV Institute - Universitat Politècnica de València, Valencia, Spain*

Mehran E. Shariatpanahi • *Department of Tissue and Cell Culture, Agricultural Biotechnology Research Institute of Iran (ABRII), Agricultural Research, Education and Extension Organization (AREEO), Karaj, Iran*

Edyta Skrzypek • *The Franciszek Górski Institute of Plant Physiology, Polish Academy of Sciences, Kraków, Poland*

HAJER SLIM AMARA • *Laboratory of Genetics and Cereal Breeding LR14AGR01, National Agronomic Institute of Tunisia, Department of Agronomy and Plant Biotechnology, University of Carthage, Tunis, Tunisie*

SWAPAN K. TRIPATHY • *Department of Agricultural Biotechnology, College of Agriculture, OUAT, Bhubaneswar, India*

ISABEL VALERO-RUBIRA • *Department of Genetics and Plant Production. Aula Dei Experimental Station, Spanish National Research Council (EEAD-CSIC), Zaragoza, Spain*

MARÍA PILAR VALLÉS • *Department of Genetics and Plant Production. Aula Dei Experimental Station, Spanish National Research Council (EEAD-CSIC), Zaragoza, Spain*

PHILIPPE VERGNE • *Laboratoire Reproduction et Développement des Plantes, Université de Lyon, ENS de Lyon, UCB Lyon 1, CNRS, INRAE, Lyon, France*

MARZENA WARCHOŁ • *The Franciszek Górski Institute of Plant Physiology, Polish Academy of Sciences, Kraków, Poland*

JENS WEYEN • *HaploPlant™, Krefeld, Germany; Krefeld, Germany*

THOMAS WIDIEZ • *Laboratoire Reproduction et Développement des Plantes, Univ Lyon, ENS de Lyon, UCB Lyon 1, CNRS, INRAE, Lyon, France*

Part I

General Topics and Transversal Methods in DH Technology

Chapter 1

Overview of In Vitro and In Vivo Doubled Haploid Technologies

Jose M. Seguí-Simarro, Nathanaël M. A. Jacquier, and Thomas Widiez

Abstract

Doubled haploids (DH) have become a powerful tool to assist in different basic research studies, and also in applied research. The principal (but not the only) and routine use of DH by breeding companies is to produce pure lines for hybrid seed production in different crop species. Several decades after the discovery of haploid inducer lines in maize and of anther culture as a method to produce haploid plants from pollen precursors, the biotechnological revolution of the last decades allowed to the development of a variety of approaches to pursue the goal of doubled haploid production. Now, it is possible to produce haploids and DHs in many different species, because when a method does not work properly, there are several others to test. In this chapter, we overview the currently available approaches used to produce haploids and DHs by using methods based on in vitro culture, or involving the in vivo induction of haploid embryo development, or a combination of both.

Key words Androgenesis, Embryogenesis, Gynogenesis, Doubled haploid, Haploid, Haploid inducer, Tissue culture

1 Introduction

As opposed to animals, plants have a remarkable developmental plasticity. This is reflected in the totipotency of differentiated plant cells that can undergo a transition towards an undifferentiated, proliferative growth, forming callus masses, and/or switch towards other developmental programs, different from their original one. This way, new plants from individual cells can be regenerated through organogenesis, promoting the successive formation of all their organs. Alternatively, the expression of the embryogenic program (embryogenesis) can be promoted, thereby transforming plant cells into functional zygote-like structures which will become embryos and eventually plants. Virtually any plant cell type can be used, if optimal conditions (experimental treatments) are found to promote organogenesis and/or embryogenesis, which opens up a wealth of biotechnological possibilities. Among these cell types, the

Jose M. Seguí-Simarro (ed.), *Doubled Haploid Technology: Volume 1: General Topics, Alliaceae, Cereals*,
Methods in Molecular Biology, vol. 2287, https://doi.org/10.1007/978-1-0716-1315-3_1,
© Springer Science+Business Media, LLC, part of Springer Nature 2021

cells of the germ line (gametes or their precursors) are one of the most interesting from a biotechnological perspective, because they contain half of the chromosomes found in somatic cells, and are thus haploid.

Haploid generally refers to the product of meiosis, which, in a cell with a complete set of chromosomes ($2n$) leads to gametes with only half of this set (n). By convention "n" and "$2n$" refer to the gametic and sporophytic chromosome numbers, respectively, and "x" denotes the number of sets of chromosomes (Fig. 1a). Depending on the ploidy level of the organism, a haploid tissue/organism may have more than one copy of each homologous chromosome (Fig. 1a). However, haploidy is not restricted to gametes in the plant kingdom. The life cycle of bryophytes (Fig. 1b), for instance, relies mostly on the gametophyte which is the haploid phase with a short-lived $2n$ sporophyte dependent on the n gametophyte [1]. During evolution, plants progressively reduced their haploid phase. Haploid tissues (Fig. 1b) were restricted spatially and temporally and became dependent for development and nutrition on $2n$ sporophytic tissues [1]. The extended life cycle proportion of the sporophyte provided evolutionary advantages [2]. For instance, deleterious alleles, hence characters, might be hidden by the dominant ones held on homologous chromosomes.

From haploid cells, the artificial generation of new haploid individuals, which may become doubled haploid (DH), can be induced. To produce a DH, haploid cells (genetically unstable in essence) may undergo duplication of their genome at any time during their proliferation, becoming diploid with no need for additional treatments. Genome duplication is typically achieved by nuclear fusion after incomplete cytokinesis [3]. Since the second chromosome set is identical to the original one, they are true DHs, 100% homozygous. This is commonly known as *spontaneous* genome doubling, in the sense that nothing is done to specifically promote genome doubling, aside of the treatments applied to induce haploid embryogenesis itself, which may also have an indirect side effect on genome doubling. This is how DHs are produced in many instances, where the percentage of unaided genome doubling is high enough to make the direct disposal of haploid individuals cost-effective, keeping only those that effectively doubled their genome by themselves. However, different genetic backgrounds are differently prone to undergo such phenomenon without the application of additional treatments for genome doubling. In other cases, this percentage is very low, which makes mandatory the implementation in the DH protocol of such a treatment to stimulate doubling of haploid embryos. Among them, the most effective and widely used is by far the application of colchicine, an antimitotic drug. In this volume (Volume 1), Chapter 5 compiles the principal methods used for chromosome doubling applied to DH production.

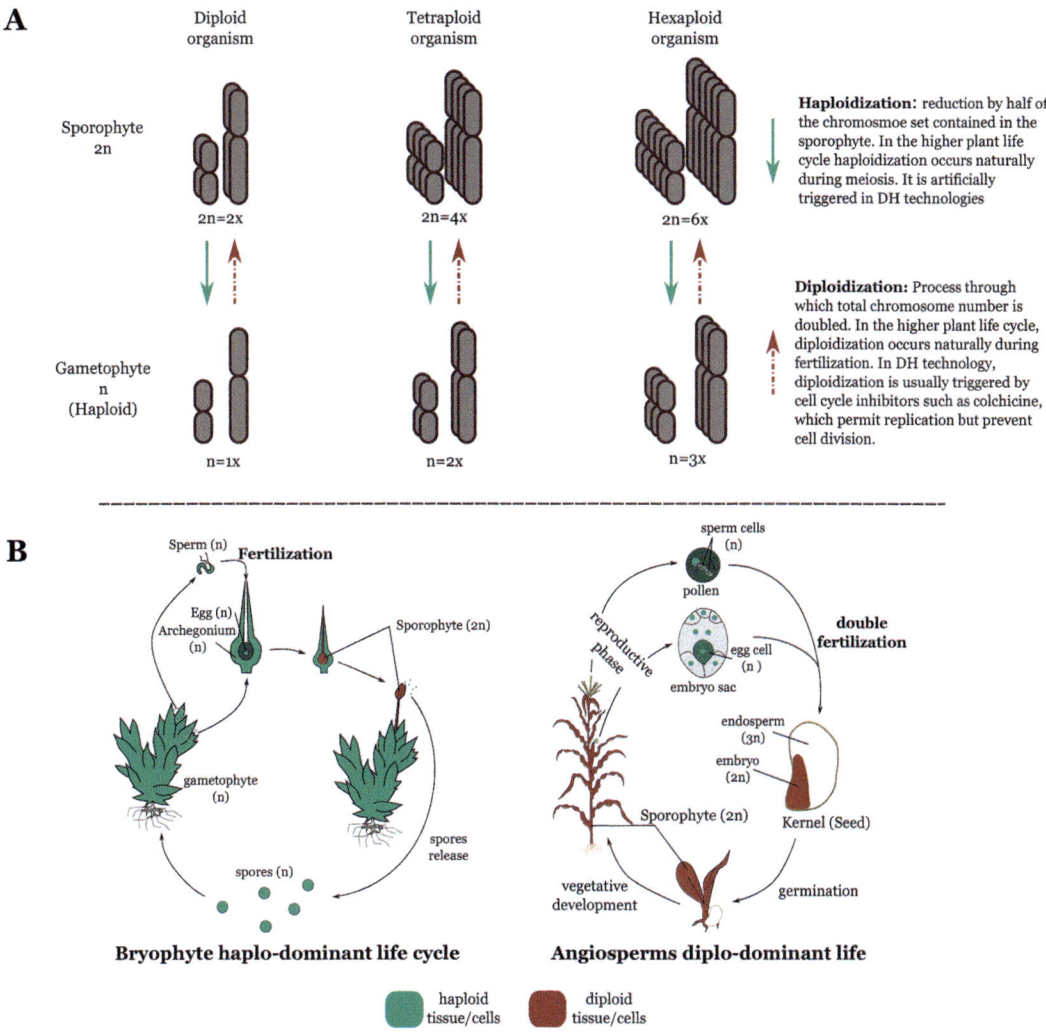

Fig. 1 Basic concepts concerning haploidy. (**a**) *Haploid* generally refers to a cell (or organism) containing half the chromosome number found in somatic cells. However, according to the ploidy level of the considered species, a haploid individual may harbor one or several versions of each chromosome. Examples are shown for a theoretical diploid, tetraploid, and hexaploid, each with a set of two different, nonhomologous chromosomes ($x = 2$). (**b**) Simple representation of the contrasted life cycles of bryophytes, left, and angiosperms, right, using maize as example. Turquoise color refers to the gametophytic (haploid) stage, and dark red to the sporophytic (diploid) stage. Note that some organisms' life cycle relies mostly on the haploid gametophytic stage. On the other hand, angiosperms drastically reduced their haploid phase to a few cells embedded in the diploid tissue

Haploids themselves are useful experimental systems to study, for example, the effects of recessive mutations, since their phenotypes are not masked by dominant alleles. However, their principal utility is to serve as the starting point to obtain DH lines. DHs are individuals whose diploid genome comes from a haploid set of chromosomes that has been duplicated, so that all their loci contain the same alleles. They are 100% homozygous individuals, and represent a valuable tool for basic and applied research, including breeding programs, where they are often used as pure lines for hybrid seed production. Chapter 2 of this same volume compiles the principal applications of DHs in both applied research and plant breeding.

Induced haploid embryos/plantlets have only one of the two copies of the genome of the male or female donor plant they come from. Due to this, it is common to refer to them as "male/paternal" or "female/maternal" haploids, respectively. Theoretically, haploid cells can only have two origins. They must come:

- Either from male haploid nuclei derived from meiosis of microspore mother cells. The origin of these haploid nuclei would potentially include meiotic products, still within the tetrad, not released as microspores, individual microspores once released from the tetrad, or any of the different cells of the pollen grain, the vegetative cell, the generative cell, or the sperm cells (the male gametes) produced from the latter.

- Or from female haploid nuclei derived from meiosis of megaspore mother cells. The origin of these haploid nuclei would potentially include the functional megaspore, either included or released from the meiocyte, and any of the haploid cells of the embryo sac: synergid cells, antipodal cells, and the egg cell, the female gamete.

In practice, only some of these haploid cells have been demonstrated to produce haploid/DH plants. They include microspores at different stages, young pollen grains and egg cells. However, we cannot rule out the possibility of other haploid cells being able to be induced [4]. These cells can be induced to produce haploid and/or DH plants by means of different techniques, which exploit a series of phenomena observed to occur in plants under natural or experimental conditions. These techniques can be grouped into two main categories, in vitro and in vivo methods, depending on whether they include exclusively in vitro procedures, or there are stages of in vivo development of haploid/DH individuals (Fig. 2). In this chapter, we will review the known in vitro and in vivo methods to produce DH individuals.

2   In Vitro Methods

The most widely extended approaches to obtain DHs have traditionally been based on the use of haploid cells of male and female origin to induce their development as haploid embryos by the application of different stresses in vitro and their subsequent in vitro culture. They are the so-called in vitro approaches

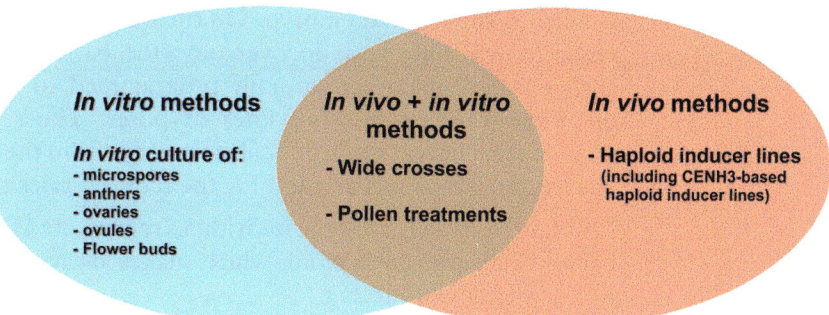

Fig. 2 Overview of in vitro and in vivo haploidization methods in plants. Intersection depicts in vivo methods that necessitate in vitro step(s) to prevent haploid embryo abortion

(Fig. 2). The production of haploid/DH plants from male haploid cells is commonly known as induction of in vitro *androgenesis*, whereas production of haploid/DH plants from female haploid cells is commonly known as induction of in vitro *gynogenesis*. The different strategies have in common the blockage of the normal development of these cells, whose natural fate is the production of functional gametes or accessory cells, and their in vitro reprogramming towards a different developmental fate, which is to become embryos without fertilization. This way, haploid and/or DH individuals can be produced in vitro.

In the last six decades, haploid cells of both male and female origins have been used to produce DHs in vitro, although with different success rates. In general, the haploid cells where in vitro haploid/DH induction has been most successful are male microspores and female egg cells. In particular, in vitro production of androgenic DHs has been more successful than production of gynogenic DHs due to several reasons:

- First, male haploid cells are by far more abundant. In a given hermaphrodite flower, thousands of microspores or pollen grains are present in each of the several anthers of a flower, whereas there is only one functional megaspore, which gives rise to six haploid cells per embryo sac, including the egg cell, per ovule. Different species may have different number of ovules per flower, and in some cases there may be up to hundreds of ovules, but this number will never be comparable to the enormous number of microspores/pollens produced by the same flower.

- Second, female haploid cells are confined in the interior of ovules, surrounded by layers of nucellar and tegument tissues, all this being included within the ovary. This confines the newly formed haploid embryos within a very small space that does not enlarge in parallel to the embryo, because the ovule and ovary do not receive the necessary developmental cues to develop in parallel to the haploid embryo. On the other hand, androgenic embryos will only have to overcome the barrier imposed by anther walls, which are naturally programmed to dehisce and open along the dehiscence lines or pores. Thus, gynogenic embryos will have it more difficult to emerge from the surrounding tissues, which will account for the reduced rates of success.

- Third, in nearly all cases, the individuals regenerated from female haploid cells remain haploid, which makes always mandatory a step of genome doubling.

Both androgenic and gynogenic embryos may be produced through different in vitro experimental approaches, as explained next.

2.1 Methods Based on In Vitro Androgenesis

Methods based on androgenesis exploit the possibility of switching the developmental fate of pollen precursors towards embryogenesis. The most used pollen precursors are, by far, microspores/young pollen grains. Microspore embryogenesis (also known as pollen embryogenesis) is by far the most used and efficient way to produce DHs in vitro. This experimental pathway was first discovered by Guha and Maheswari in 1964 [5], while working with in vitro-cultured anthers of *Datura innoxia*. Later on, many different research groups have reproduced their findings in many other species and genera, making this experimental phenomenon a powerful and widespread tool to produce DHs. However, not all the species respond equally to the induction of this process. Some species, considered models for the study of this phenomenon, respond fairly well. This is the case of certain lines of rapeseed (*Brassica napus*), tobacco (*Nicotiana tabacum*), or barley (*Hordeum vulgare*). Others, considered recalcitrant, present a low or very low response, and in other cases, a protocol to efficiently induce this process is still pending to be developed, as for scientifically or agronomically important species such as *Arabidopsis thaliana* or tomato (*Solanum lycopersicum*), respectively. Many other species are in between these two extreme situations, being possible to induce microspore embryogenesis, but with yet improvable protocols. Woody species are good examples of materials where some success has been achieved, but there is still a large room for improvement (reviewed in [6–8]). Chapter 3 of this same volume includes a list of species where production of haploids/DHs has been assessed by this approach.

Even within a species, there will be varieties, lines, and even individuals that respond differently. This strong influence of the genotype, together with the fact that this trait is transmitted across generations and segregates in the hybrids offspring [9, 10], indicates that it is under genetic control [11–17]. Furthermore, it was proposed that, at least for *Brassica napus*, the embryogenic competence of microspores is controlled by two loci with additive effects

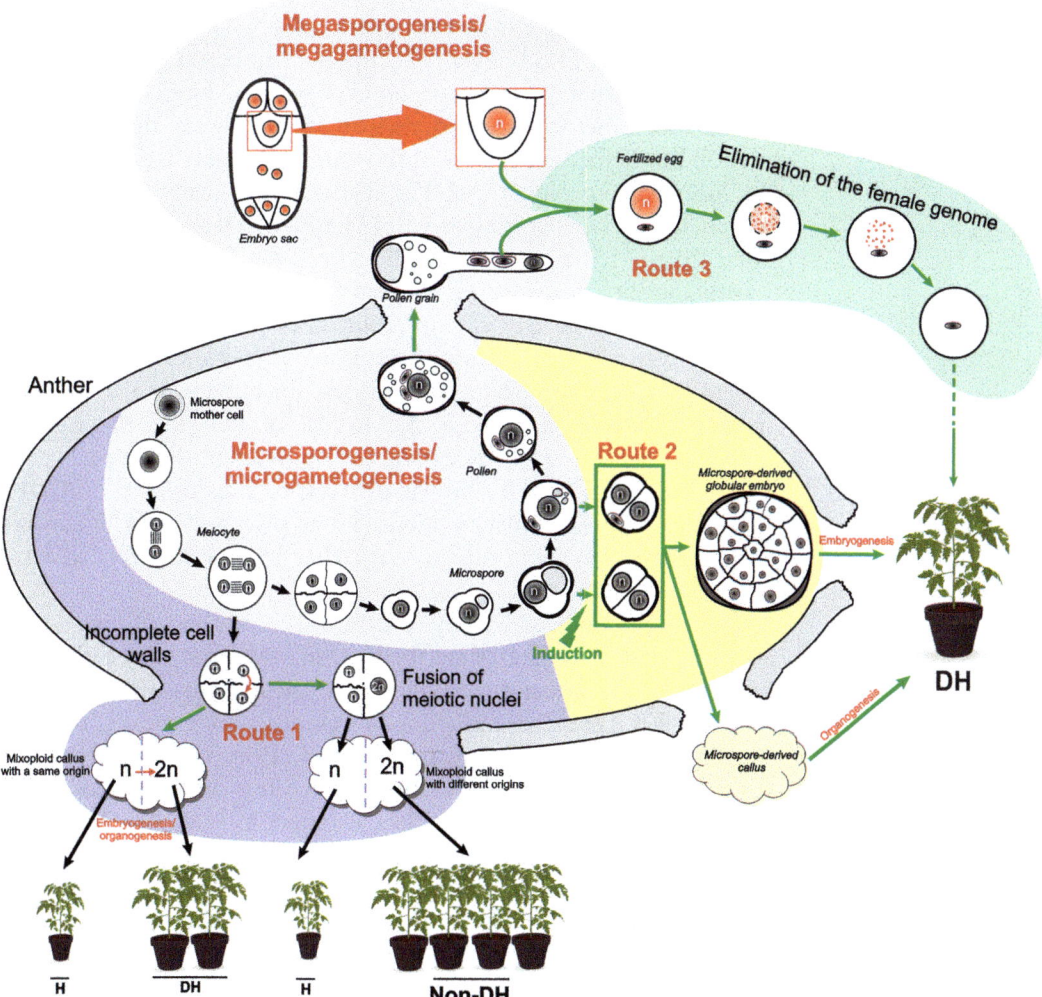

Fig. 3 Different alternatives to produce male-derived DHs. Microsporogenesis and microgametogenesis are natural pathways that normally take place to give rise to the male gametophyte (pollen grain) and male gametes (sperm cells). Male-derived haploids or DHs may arise from deviations of these pathways at three different levels: (1) diverting the meiocyte, (before microspore release) towards proliferation to induce in vitro the formation of callus, from which haploids and DHs, and also heterozygous diploids, can be produced by organogenesis (Route 1); (2) reprogramming the vacuolate microspore or young pollen grain towards embryogenesis (or alternatively callus formation + regeneration through organogenesis) by the in vitro application of a stress treatment and subsequent in vitro culture (Route 2); and (3) in vivo elimination of the female genome after egg fertilization by a sperm cell (Route 3). *See* text for further details

[10]. The gene or genes involved, however, remain to be elucidated.

In microspore embryogenesis, androgenic haploids/DHs are typically produced upon deviation of microspores towards embryogenesis or callus formation. However, not all species can be induced by isolation and culture of microspores at the same stage (Fig. 3).

Indeed, for a few species it has been shown that male-derived haploid and DH plants can be regenerated from meiocyte-derived callus [18–20]. This is a very exceptional route, much less frequent and studied than microspore embryogenesis. Under certain circumstances, meiocytes at late meiosis stages, always after recombination but before the release of the four individual microspores, can be induced to proliferation (Fig. 3, Route 1), forming undifferentiated haploid callus masses where haploid/DHs can be regenerated from. Meiocyte-derived callogenesis has been documented in *Arabidopsis thaliana, Vitis vinifera Digitalis purpurea, and Solanum lycopersicum* [21–32]. In addition to its rare occurrence and the scarcity of studies, the practical use of this alternative is hampered by a very low frequency of cases, since many of the regenerants produced come from the fusion of two meiotic haploid products separated by defective, incomplete, or absent cell walls (Fig. 3, Route 1) [25]. This will give rise to a majority of non-DH, useless plants that must be identified as such and then discarded. Altogether, these limitations make that in practice, this in vitro alternative is not used.

For some species, mostly cereals, the early stages of microspore development have been described as the only stage where embryogenesis can be induced [33], However, the majority of works point to the fact that the inducible developmental window revolves around the first pollen mitosis (Fig. 3, Route 2), which means that vacuolated microspores but also young, just divided pollen grains would also be inducible (reviewed in [19, 33, 34]). The identification and isolation of these stages is essential for the success of this process, since in the vast majority of species, they are the most sensitive to the induction treatments [35]. Beyond these stages, induction has only exceptionally been reported [36].

To be induced to embryogenesis, microspores/pollens must be stressed. The need for application of physicochemical stress treatments seems common to all inducible species. The variety of responses, depending principally on the genotype but also on the developmental stage of the microspore/pollen, makes that each species has its own specific inductive treatments to trigger the developmental switch. Some of these stresses (heat, cold or starvation) are common to many species, whereas others need more specific stressors or combinations of them [37]. As a rule of thumb, the more recalcitrant a species is, the more combined and more intense stresses are needed. Typically, induction of microspore embryogenesis produces microspore-derived embryos. However, there are some species where microspores proliferate in an undifferentiated manner, producing callus masses (Fig. 3, Route 2; reviewed in [19]). In other cases, embryos cannot progress as such and proliferate as callus masses [38–40]. As pointed out previously [19], this may be due the absence of developmental regulation in this type of in vitro embryogenesis, devoid of endosperm and other seed tissues that may interact with the embryo. When suboptimal culture conditions do not compensate for this, abnormal embryos or even calli may arise.

The process of microspore embryogenesis and the different factors, mentioned above, that influence its success, may be implemented in practice using two in vitro approaches: anther culture and isolated microspore culture.

2.1.1 Anther Culture

Anther culture is the most universal method to produce DHs. It is technically simple, consisting basically of the following steps: (1) flower bud collection, (2) isolation of anthers from flower buds, (3) inoculation and in vitro culture in agar-based culture medium, (4) isolation of embryos, (5) regeneration of plants, and (5) analysis of regenerants. Few weeks (months in many cases) later, microspore-derived embryos may be seen to emerge from anther walls, in parallel to the degradation and necrosis of these walls. In general, a given anther under optimal culture conditions may give rise to several tens of microspore-derived embryos during several months of culture. The presence of these walls (the tapetum principally) during the first stages of anther culture may protect and help microspores to undergo the first stages of haploid development, in a way similar to how they assist normal microspore development in vivo. Perhaps, this is the reason why anther culture works in many different species, including those where other DH methods do not work (*see* Chapter 3 of this volume).

However, anther cultures are not devoid of limitations. Perhaps, the main limitation comes from the fact that microspores are cultured together with anther walls. Anther walls (the tapetal layer mostly) may secrete molecules that may protect microspores or promote their growth, but it may also secrete inhibitory or even toxic compounds, as is the case of necrosing anther tissues. In any case, this secretory effect is uncontrollable in essence, and makes difficult a strict control of culture conditions. Moreover, when exposed to growth regulators, these walls are able to proliferate in vitro, producing calli. Indeed, some parts of the anther, such as the filament insertion, are especially prone to form calli when in vitro cultured. Therefore, we cannot rule out the possibility of occurrence of somatic embryos (very rare but possible) and calli (much more frequent) from anther walls. This implies that for every single plant confirmed as diploid (2C DNA content) by flow cytometry, we should check its origin. For this, the most reliable approach is the use of molecular markers previously confirmed as heterozygous in the donor plants used (*see* Chapter 3 of Volume 2). However, for very well-known cultivars, where repeated analyses have shown that no somatic embryos are produced, this step may be skipped simply by discarding all calli produced and using only embryos.

*2.1.2 Isolated
Microspore Culture*

Microspore embryogenesis can also be induced using microspore isolated from anthers. This alternative is more complex than anther culture because a step of microspore isolation and inoculation into liquid medium must be implemented to the protocol. In addition, the absence of anther tissues makes that proper microspore growth and development will exclusively depend on medium composition. Thus, in order to develop an efficient protocol for microspore culture, medium composition must include all the elements needed by microspores and must be adjusted to the particularities of the microspores of each species. The use of liquid culture media increase the risk of contamination compared with anther cultures, which use to be performed in agar-based, semisolid media.

Together with these limitations, isolated microspore cultures have also some advantages that, in some cases, largely surpass the limitations. Microspore cultures avoid the incontrollable contribution of anther walls, and the potential toxicity of anther wall degradation products. In addition, all nutrients and active compounds of the medium are easily available by microspores, suspended in the liquid medium instead of confined within the anther locule. These features are likely behind the fact that microspore cultures are notably faster than anther cultures. Indeed, in some model species where protocols are optimized, few weeks are needed produce hundreds of embryos from the microspores inoculated in a single dish [41], *see* Chapter 8 of Volume 2. In addition to this, microspore cultures avoid the routine, time-consuming procedure of checking the haploid origin of all the diploid plants obtained. Since microspores are isolated from all other anther tissues and only microspores are inoculated into the dishes, the only possible origin for embryos or calli must be the microspores. In other words, all diploid regenerants obtained will be DHs. All these advantages considered, isolated microspore culture is the method of choice in those materials where efficient protocols are well established.

*Methods Based on In Vitro
Gynogenesis*

Gynogenesis would exploit the ability of egg cells to develop in the embryo sac as a haploid zygote without fertilization (Fig. 4, Route 4). This alternative to the normal development of the megagametophyte was first described in vitro in 1976 by San et al. [42]. It would therefore be a form of female haploid parthenogenesis (from the Greek words *parthenos*, meaning "virgin" and *genesis*, meaning "origin." In some species, the gynogenic embryo is believed to originate from antipodal or synergid cells, but in the vast majority of cases the gynogenic embryo is derived from the egg cell (reviewed in [4]). Gynogenic embryos are mostly haploid, which implies that in order to obtain the desired double haploid, the application of additional treatments for chromosome duplication should be considered in nearly all cases. As in the case of

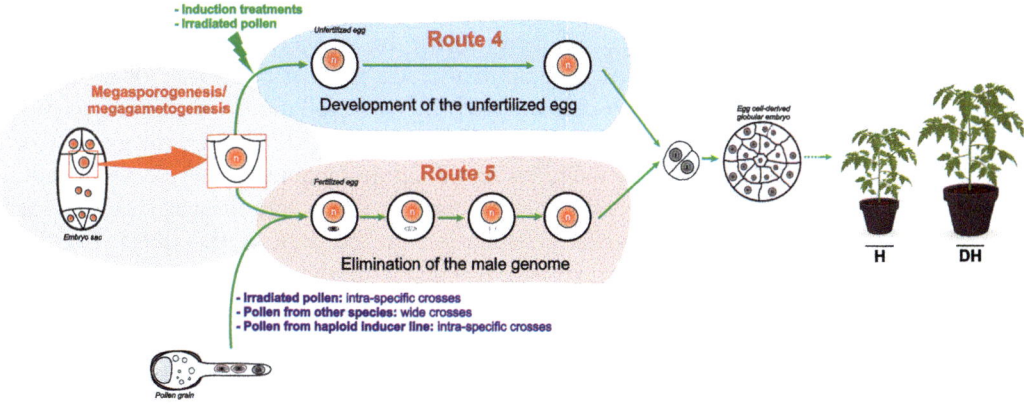

Fig. 4 Different alternatives to produce female-derived haploids and DHs. Megasporogenesis and mega-gametogenesis are pathways that normally take place to give rise to the female gametophyte (embryo sac) and female gametes (egg cell and central cell). Female-derived haploids or DHs may arise by two main means: (1) Reprogramming of egg cell development into haploid embryogenesis, either by in vitro induction or in vivo pollination with haploid inducer lines (Route 4); and (2) the sperm cell fertilizes the egg cell, but the male genome is progressively eliminated during early embryogenesis. Some methods lead to viable seeds with haploid embryos and are thus fully in vivo, whereas other methods need additional steps of in vitro culture

androgenesis, colchicine is the most effective and therefore the most widely used antimitotic (Chapter 5 of this same volume).

The success of gynogenesis induction is influenced by many different factors, including the developmental stage of the embryo sac and the in vitro culture conditions. However, the genotype is the most important, even more than for microspore embryogenesis. In fact, this is the most limiting factor for the practical application of this technique, since there are very few responsive genotypes, much less than those that respond to microspore embryogenesis. Other limitations include a low efficiency, much lower than microspore embryogenesis (there are much less egg cells than microspores in a flower), a very low rate of spontaneous duplication of the genome, and low levels of embryo regeneration, perhaps due to the instability of haploid genomes, prone to chromosomal alterations.

All these limitations make in vitro gynogenesis-based approaches a secondary alternative, used in a reduced range of species where other in vitro approaches (microspore embryogenesis) have proven ineffective. Chapter 3 of this same volume includes a list of species where production of haploids/DHs has been assessed by this approach. Among them, the model species for the study of this process is onion (*Allium cepa*) [43, 44], *see* Chapter 6 of this same volume.

From a methodological point of view, this technique is implemented by in vitro culture of ovules [45, 46], ovaries [47], or even full immature flowers [43], not yet open and therefore unpollinated, until the embryo sac matures and the gynogenic embryo develops. In some species, mere in vitro culture seems not enough, and an "extra" factor must be applied to trigger the process. Examples of this factor include pollination before ovary excision and in vitro culture [48], in vitro pollination with mentor pollen from other species [49], with pollen irradiated with gamma or X rays to inactivate its fertilization capability [50] or with triploid pollen, still able to germinate and stimulate the egg cell, but not to fertilize. In vitro pollination can be done at the apical part of the stigma of entire pistils isolated from the flower and cultured in vitro [51]. After several months, gynogenic embryos will be visible. Alternatively, pollen may be applied by placental pollination, which implies isolating the ovules from the ovary, but maintaining a fragment of the placenta to help the viability of the egg cell. In many cases of gynogenesis in cucurbits and fruit trees, treated pollen is applied directly in situ, on the emasculated flower in the plant [52–56]. Then, seeds or haploid embryos are rescued and in vitro cultured (*see* also Subheading 3.2). These approaches would be half way between the in vitro approaches and the in vivo approaches described next (Fig. 2).

3   In Vivo Approaches

Alternatives to the in vitro approaches to generate haploid plantlets are in vivo approaches (Fig. 2). Overall, in vivo methods are less numerous as compared to in vitro methods and thus less plants/crops are concerned. In vivo methods look attractive because they appear to be "simpler," since the plant is mainly doing the job instead of labor-intensive in vitro work. Nevertheless, improvement and optimization are usually needed between the discovery of an in vivo induction system and its application in breeding scale. Indeed, long-standing in vivo haploid induction methods exist [33, 57, 58] and some have been improved and are currently used on a routine basis in breeding programs. Examples of these methods include the use of maize haploid inducer lines (*see* Chapter 2 of Volume 2) or the so-called wide crosses. More recently (10 years ago), new in vivo methods have been discovered [59], and are currently tested for translation to crops (e.g. centromere engineering, *see* Chapter 1 of Volume 3). In vivo haploid induction methods could be divided into two main broad categories: (1) the "wide-crosses" and (2) the intra-specific crosses (Fig. 2).

3.1   Haploidization by Wide-Crosses

Wide crosses consist to force crosses between species spanning wide taxonomic boundaries. It could thus involve inter-generic or inter-specific pollinations, and often concern crosses between a cultivated crop and a wild-relative species. Wide crosses are also named *wide hybridization*, and imply to overcome pre-fertilization and post-fertilization barriers [60]. They have been reported more frequently in monocotyledonous species as compared to dicotyledonous [57].

Two different outcomes, both useful for breeders, need to be distinguished from these wide crosses. Firstly, in case of successful hybridization, which could be helped by embryo rescue or other techniques [60], the production of hybrid embryos is a starting material in order to introduce agronomic traits of interest (disease resistance, stress tolerance, etc.) across species. Secondly, it could be used for the production of haploid embryos. Due to the unstable nature of hybrid embryos generated by joining two different genetic materials, chromosome elimination from one parent occurs during early embryogenesis in some "wide-crosses" [19, 57, 58, 61]. Although the paternal chromosomes are eliminated in most of the cases to give rise to maternal haploid embryos (Fig. 4, route 5), some rare cases were reported in which the paternal genome remains, being the maternal genome eliminated (Fig. 3, route 3) (reviewed in [19]). A pioneering discovery in haploidization by wide-crosses was the *Bulbosum method* (Chapter 10 of this same volume), which has been well studied and is now widely used in barley breeding [57, 62]. Cross of cultivated barley (*Hordeum vulgare*) using pollen from its wild relative *Hordeum bulbosum* leads to the production of *H. vulgare* haploid embryos [63, 64]. The success of barley haploid embryo production thanks to wide crosses was then extended to other species, especially using maize pollen which appears to display low-intensity fertilization barriers [57, 65]. For example, wheat and triticale haploid embryos are currently obtained in some plant breeding programs by the following crosses: wheat × maize and triticale × maize respectively [57, 66, 67]. Overall, wide crosses are limited to some crops at breeding scale, but once they are well-established methods, they have the advantage to be effective across a wide range of genotypes, as opposed to the in vitro approaches, which are highly genotype-dependent within a given species. Nevertheless, a limitation of the wide crosses method is that it is not fully in vivo (Fig. 2), since it necessitates to include a stage of in vitro tissue culture to prevent embryo abortion. Indeed, successful seed development relies on the correct development of the two fertilization products: the embryo and the endosperm. The endosperm tightly interacts, both physically and chemically, with the embryo and it is thus of vital importance to sustain embryo development [68–70]. In haploid embryo-producing wide crosses, the endosperm fails to develop properly, probably due to selective parental chromosome elimination occurring in this tissue as well, and consequently haploid embryos must be rescued by in vitro culture.

3.2 Haploidization by Intra-specific Crosses

Two main methods could be differentiated in order to induce haploid embryos via intra-specific crosses: (1) pollination with treated pollen and (2) the use of haploid inducer lines. While methods based on treated pollen usually necessitate haploid embryo rescue due to early seed abortion [33, 71], the haploid inducer lines present the advantage to be fully in planta because the output is the production of viable seeds containing haploid embryos (Fig. 2) [72, 73]. Depending on the methods and species considered, haploidization by intra-specific crosses could produce two different kinds of haploid embryos: maternal haploid embryos with the cytoplasm and nuclear genome from the female parent (Fig. 4, route 4 and/or route 5), and paternal haploid embryos, having the cytoplasm of the egg cell (maternal) but the nuclear genome from the male parent (Fig. 3, route 3). In the latter case, male haploid embryos produced by intra-specific crosses might be additionally useful for other biotechnological purposes beyond conventional production of DH pure lines. Indeed, since mitochondrial defects are behind cytoplasmic male sterility (CMS), CMS is maternally inherited through the cytoplasm [74, 75]. CMS is a valuable tool in hybrid seed production, since it avoids the time-consuming process of emasculation to prevent self-pollination [76]. This trait is traditionally transferred from one germplasm to another through multiple rounds of backcrossing. Obtaining a nuclear male genome within a *"maternal"* cytoplasm in just a single cross reduces CMS conversion to just one step, thus accelerating hybrid seed production.

Pollinations with treated pollen induce maternal haploid embryos (Fig. 4, routes 4, 5). This method consists in the treatment of pollen, prior to pollination, with physical or chemical agents, irradiation being the most used treatment [33, 71]. Although haploidization via pollen treatments has been reported in more than 15 species [33, 71], it works ineffectively (low haploid induction rate). Thus, this method is used in plant breeding only when no alternative efficient methods are available, for example in melon and cucumber [77]. Haploid inducer lines could be seen as an exception since they exist in few species only [58, 73, 78]. Moreover, haploid inducer lines are routinely used in plant breeding in maize (Chapter 2 of Volume 2) and potato [79]. Intensive researches are currently being done to extend the CENH3-based inducer line to others crops (Chapter 1 of Volume 3). In addition, the identification of the causal genes leading to embryo haploid induction in maize allowed for the translation of this trait to two other crops: wheat and rice [80–82]. Lastly, recent patents and publications reported on the identification of sorghum haploid inducer lines [83, 84].

In maize, two different types of haploid inducer lines have been reported, which are able to produce either maternal or paternal haploid embryos (Figs. 3, route 3 and 4, route 5, respectively) [85, 86]. Chapter 2 of Volume 2 details the properties and uses of these two maize haploid inducer lines. These two lines gained their haploid induction phenotypes due to mutations in genes involved in male and female gametophyte development, and in double fertilization.

In potato (*Solanum tuberosum* L.), the haploid induction system relies on a cross between a diploid male haploid inducer line (*S. tuberosum* Andigenum group, previously referred to as cultivar *S. Phureja*) with a tetraploid cultivated potato of interest used as female parent (Fig. 4, route 5) [79, 87–89]. It thus refers to an interploidy cross ($4x$ potato $\times$ $2x$ haploid inducer line), and the haploid embryos found in some of the viable seeds of this cross are commonly called dihaploids to indicate that they contain two sets of chromosomes (from maternal origin). Such dihaploid plants are not homozygous, but allow breeders to work at the diploid level for simpler genetic analysis/mapping, or for introgression of valuable traits from the wild species. Since wild species are mostly diploid, they could be then crossed with the dihaploid by inter-specific hybridization [87, 88].

The CENH3-based haploid inducer lines originate from the manipulation of the centromeric histone protein CENH3 in *Arabidopsis thaliana* [59]: it was reported that the genome of the parental having the engineered CENH3 is eliminated after the cross with wild-type plants with intact CENH3, creating haploid inducer lines (Chapter 1 Volume 3 and for recent reviews: [58, 61, 73, 90]). These CENH3-based haploid inducer lines are thus able to induce either maternal or paternal haploid embryos (Figs. 3, route 3 and 4, route 5), although they seem more efficient in producing paternal haploid embryos, in *Arabidopsis* at least [59]. Although CENH3 is conserved across plant species, efficient translation of this haploid induction method to crops remain to be achieved, since very low haploid induction rates have been observed so far in crops [58, 73, 78].

To sum-up, haploidization by intra-specific crosses is attractive since haploid embryos are formed within a viable seed, but remain limited to few crops. Once haploid inducer lines have been reported or created, the main limitations/constraints for the use haploid inducer lines in breeding programs are the need for a relatively high haploid induction rate, and the existence of a system to identify the seeds having haploid embryos among the seeds having diploid embryos. The history of the maize male haploid inducer line exemplifies the important need of research and development to achieve a "good" haploidization method by intra-specific crosses: indeed more than 20 years separate the discovery of the first maize haploid inducer line by Ed Coe [86] from its use in maize breeding programs [91, 92]. The improvements of both haploid induction rate (from ~2–3% to ~10%) and color markers to accurately identify haploid embryos were some of the key steps in the successful use of maize haploid inducer lines in breeding programs (Chapter 2 of Volume 2). Recently, the knowledge gained in the mode of action behind the maize in vivo haploid induction system [72, 73], allowed for the successful translation of this feature to two new crops: rice and wheat [80–82].

4   Concluding Remarks

In this chapter we have revised the principal approaches currently available to produce haploids and DHs (Fig. 2) for different purposes, principally focused on the rapid generation of pure lines to accelerate hybrid seed production or CMS conversion and to simplify breeding programs by producing dihaploids, as for potato. These approaches imply the use of methods exclusively based on in vitro culture, in vivo induction of haploid development, or a combination of them to induce haploid embryos in vivo and then rescue them in vitro. Practical examples of the application of these methods to particular species or varieties are presented throughout the book. The choice of the best performing approach will depend on the species used, and to what extent these methods have been developed and adapted to this species. Together, these approaches illustrate how a given goal can be accomplished by different biotechnological means, and are a good example of the power of combining different biotechnological for solving specific applied problems of industry and in general, of society.

Acknowledgments

We thank Luca Comai for the helpful discussions on potato haploid inducer lines. This work was supported by grants AGL2017-88135-R and PID2020-115763RB-I00 to JMSS from Spanish MICINN jointly funded by FEDER. T.W. was supported by the ANR grant (ANR-19-CE20-0012), and by "pack ambition recherche" from the Région Auvergne-Rhone-Alpes ("HD-INNOV"). N.M.A.J. was supported by CIFRE PhD fellow-ship from ANRT funding agency (grant no. 2019/0771).

References

1. Li W, Ma H (2002) Gametophyte development. Curr Biol 12(21):R718–R721. https://doi.org/10.1016/s0960-9822(02)01245-9

2. Bennici A (2008) Origin and early evolution of land plants. Commun Integr Biol 1 (2):212–218. https://doi.org/10.4161/cib.1.2.6987

3. Seguí-Simarro JM, Nuez F (2008) Pathways to doubled haploidy: chromosome doubling during androgenesis. Cytogenet Genome Res 120 (3-4):358–369. https://doi.org/10.1159/000121085

4. Bohanec B (2009) Doubled haploids via gynogenesis. In: Touraev AF, Forster BP, Jain, SM (ed) Advances in haploid production in higher plants. Springer, New York pp 35–46

5. Guha S, Maheshwari SC (1964) In vitro production of embryos from anthers of *Datura*. Nature 204:497

6. Srivastava P, Chaturvedi R (2008) *In vitro* androgenesis in tree species: an update and

prospect for further research. Biotechnol Adv 26(5):482–491. https://doi.org/10.1016/j.biotechadv.2008.05.006

7. Andersen SB (2005) Haploids in the improvement of woody species. In: Palmer CE, Keller WA, Kasha KJ (eds) Haploids in crop improvement II, Biotechnology in agriculture and forestry, vol 56. Springer-Verlag, Berlin, pp 243–257

8. Germanà MA (2009) Haploids and doubled haploids in fruit trees. In: Touraev A, Forster BP, Jain SM (eds) Advances in haploid production in higher plants. Springer, Dordrecht, pp 241–263

9. Rivas-Sendra A, Campos-Vega M, Calabuig-Serna A, Seguí-Simarro JM (2017) Development and characterization of an eggplant (*Solanum melongena*) doubled haploid population and a doubled haploid line with high androgenic response. Euphytica 213(4):89. https://doi.org/10.1007/s10681-017-1879-3

10. Zhang FL, Takahata Y (2001) Inheritance of microspore embryogenic ability in Brassica crops. Theor Appl Genet 103(2-3):254–258

11. Barret P, Brinkman M, Dufour P, Murigneux A, Beckert M (2004) Identification of candidate genes for in vitro androgenesis induction in maize. Theor Appl Genet 109(8):1660–1668

12. Kitashiba H, Taguchi K, Kaneko I, Inaba K, Yokoi S, Takahata Y, Nishio T (2016) Identification of loci associated with embryo yield in microspore culture of Brassica rapa by segregation distortion analysis. Plant Cell Rep:1–8. https://doi.org/10.1007/s00299-016-2029-4

13. Malik MR, Wang F, Dirpaul J, Zhou N, Hammerlindl J, Keller W, Abrams SR, Ferrie AMR, Krochko JE (2008) Isolation of an embryogenic line from non-embryogenic Brassica napus cv. Westar through microspore embryogenesis. J Exp Bot 59(10):2857–2873. https://doi.org/10.1093/jxb/ern149

14. Beckert M (1998) Genetic analysis of *in vitro* androgenetic response in maize. In: Chupeau Y, Caboche M, Henry Y (eds) Androgenesis and haploid plants. Springer-Verlag, Berlin, pp 24–37

15. Rudolf K, Bohanec B, Hansen M (1999) Microspore culture of white cabbage, *Brassica oleracea* var. capitata L.: genetic improvement of non-responsive cultivars and effect of genome doubling agents. Plant Breed 118(3):237–241

16. Yamagishi M, Otani M, Higashi M, Fukuta Y, Fukui K, Shimada T (1998) Chromosomal

regions controlling anther culturability in rice (*Oryza sativa* L.). Euphytica 103(2):227–234. https://doi.org/10.1023/a:1018328708322

17. Zhang FL, Aoki S, Takahata Y (2003) RAPD markers linked to microspore embryogenic ability in *Brassica crops*. Euphytica 131(2):207–213

18. Seguí-Simarro JM, Corral-Martínez P, Parra-Vega V, González-García B (2011) Androgenesis in recalcitrant solanaceous crops. Plant Cell Rep 30(5):765–778. https://doi.org/10.1007/s00299-010-0984-8

19. Seguí-Simarro JM (2010) Androgenesis revisited. Bot Rev 76(3):377–404. https://doi.org/10.1007/s12229-010-9056-6

20. Seguí-Simarro JM (2016) Androgenesis in solanaceae. In: Germanà MA, Lambardi M (eds) In vitro embryogenesis, Methods in molecular biology, vol 1359. Springer Science + Business Media, New York, pp 209–244. https://doi.org/10.1007/978-1-4939-3061-6_9

21. Gresshoff PM, Doy CH (1974) Derivation of a haploid cell line from *Vitis vinifera* and importance of stage of meiotic development of anthers for haploid culture of this and other genera. Z Pflanzenphysiol 73(2):132–141

22. Gresshoff PM, Doy CH (1972) Haploid *Arabidopsis thaliana* callus and plants from anther culture. Aust J Biol Sci 25(2):259

23. Corduan G, Spix C (1975) Haploid callus and regeneration of plants from anthers of *Digitalis purpurea* L. Planta 124(1):1–11

24. Corral-Martínez P, Nuez F, Seguí-Simarro JM (2011) Genetic, quantitative and microscopic evidence for fusion of haploid nuclei and growth of somatic calli in cultured *ms10³⁵* tomato anthers. Euphytica 178(2):215–228. https://doi.org/10.1007/s10681-010-0303-z

25. Seguí-Simarro JM, Nuez F (2007) Embryogenesis induction, callogenesis, and plant regeneration by *in vitro* culture of tomato isolated microspores and whole anthers. J Exp Bot 58(5):1119–1132

26. Bal U, Abak K (2007) Haploidy in tomato (*Lycopersicon esculentum* Mill.): a critical review. Euphytica 158(1-2):1–9

27. Seguí-Simarro JM, Nuez F (2005) Meiotic metaphase I to telophase II is the most responsive stage of microspore development for induction of androgenesis in tomato (*Solanum lycopersicum*). Acta Physiol Plant 27(4B):675–685

28. Zagorska NA, Shtereva LA, Kruleva MM, Sotirova VG, Baralieva DL, Dimitrov BD (2004) Induced androgenesis in tomato

(*Lycopersicon esculentum* Mill.). III. Characterization of the regenerants. Plant Cell Rep 22 (7):449–456

29. Zagorska NA, Shtereva A, Dimitrov BD, Kruleva MM (1998) Induced androgenesis in tomato (*Lycopersicon esculentum* Mill.)—I. Influence of genotype on androgenetic ability. Plant Cell Rep 17(12):968–973

30. Shtereva LA, Zagorska NA, Dimitrov BD, Kruleva MM, Oanh HK (1998) Induced androgenesis in tomato (*Lycopersicon esculentum* Mill). II. Factors affecting induction of androgenesis. Plant Cell Rep 18(3-4):312–317

31. Sharp WR, Raskin RS, Sommer HW (1972) The use of nurse culture in the development of haploid clones in tomato. Planta 104 (4):357–361

32. Gresshoff PM, Doy CH (1972) Development and differentiation of haploid *Lycopersicon esculentum* (tomato). Planta 107(2):161–170

33. Dunwell JM (2010) Haploids in flowering plants: origins and exploitation. Plant Biotechnol J 8(4):377–424. https://doi.org/10.1111/j.1467-7652.2009.00498.x

34. Germanà MA (2011) Anther culture for haploid and doubled haploid production. Plant Cell Tissue Organ Cult 104(3):283–300. https://doi.org/10.1007/s11240-010-9852-z

35. Touraev A, Pfosser M, Heberle-Bors E (2001) The microspore: a haploid multipurpose cell. Adv Bot Res 35:53–109

36. Binarova P, Hause G, Cenklova V, Cordewener JHG, van Lookeren-Campagne MM (1997) A short severe heat shock is required to induce embryogenesis in late bicellular pollen of *Brassica napus* L. Sex Plant Reprod 10(4):200–208

37. Shariatpanahi ME, Bal U, Heberle-Bors E, Touraev A (2006) Stresses applied for the re-programming of plant microspores towards *in vitro* embryogenesis. Physiol Plant 127 (4):519–534

38. Rivas-Sendra A, Corral-Martínez P, Camacho-Fernández C, Seguí-Simarro JM (2015) Improved regeneration of eggplant doubled haploids from microspore-derived calli through organogenesis. Plant Cell Tissue Organ Cult 122(3):759–765. https://doi.org/10.1007/s11240-015-0791-6

39. Corral-Martínez P, Seguí-Simarro JM (2014) Refining the method for eggplant microspore culture: effect of abscisic acid, epibrassinolide, polyethylene glycol, naphthaleneacetic acid, 6-benzylaminopurine and arabinogalactan proteins. Euphytica 195(3):369–382. https://doi.org/10.1007/s10681-013-1001-4

40. Corral-Martínez P, Seguí-Simarro JM (2012) Efficient production of callus-derived doubled haploids through isolated microspore culture in eggplant (*Solanum melongena* L.). Euphytica 187(1):47–61. https://doi.org/10.1007/s10681-012-0715-z

41. Custers JBM, Cordewener JHG, Fiers MA, Maassen BTH, van Lookeren-Campagne MM, Liu CM (2001) Androgenesis in *Brassica*: a model system to study the initiation of plant embryogenesis. In: Bhojwani SS, Soh WY (eds) Current trends in the embryology of angiosperm. Kluwer Academic Publishers, Dordrecht, pp 451–470

42. San Noeum LH (1976) Haploides d'*Hordeum vulgare* L. par culture *in vitro* d'ovaries non fécondés. Annu Amélior Plantes 26:751–754

43. Fayos O, Vallés MP, Garcés-Claver A, Mallor C, Castillo AM (2015) Doubled haploid production from Spanish onion (*Allium cepa* L.) germplasm: embryogenesis induction, plant regeneration and chromosome doubling. Front Plant Sci 6:384. https://doi.org/10.3389/fpls.2015.00384

44. Bohanec B (2002) Doubled-haploid onions. In: Rabinowitch HD, Currah L (eds) Allium crop science: recent advances. CABI Publishing, Wallingford, pp 145–157

45. Li JW, Si SW, Cheng JY, Li JX, Liu JQ (2013) Thidiazuron and silver nitrate enhanced gynogenesis of unfertilized ovule cultures of *Cucumis sativus*. Biol Plant 57(1):164–168. https://doi.org/10.1007/s10535-012-0269-x

46. Doi H, Hoshi N, Yamada E, Yokoi S, Nishihara M, Hikage T, Takahata Y (2013) Efficient haploid and doubled haploid production from unfertilized ovule culture of gentians (*Gentiana* spp.). Breed Sci 63(4):400–406. https://doi.org/10.1270/jsbbs.63.400

47. Raquin C (1985) Induction of haploid plants by in vitro culture of *Petunia ovaries* pollinated with irradiated pollen. Z Pflanzenzüchtg 94 (2):166–169

48. Tang F, Tao Y, Zhao T, Wang G (2006) In vitro production of haploid and doubled haploid plants from pollinated ovaries of maize (*Zea mays*). Plant Cell Tissue Organ Cult 84 (2):233–237

49. Kantartzi S, Roupakias D (2009) In vitro gynogenesis in cotton (Gossypium sp.). Plant Cell Tissue Organ Cult 96(1):53–57. https://doi.org/10.1007/s11240-008-9459-9

50. Troung-Andre I (1998) In vitro haploid plants derived from pollination by irradiated pollen on cucumber. Paper presented at the Cucurbitaceae 88: proceedings of the IXth Eucarpia

meeting on genetics and breeding of Cucurbitaceae, France

51. Germanà MA, Chiancone B (2001) Gynogenetic haploids of *Citrus* after in vitro pollination with triploid pollen grains. Plant Cell Tissue Organ Cult 66(1):59–66

52. Bouvier L, Zhang Y-X, Lespinasse Y (1993) Two methods of haploidization in pear, *Pyrus communis* L.: greenhouse seedling selection and in situ parthenogenesis induced by irradiated pollen. Theor Appl Genet 87 (1-2):229–232

53. Zhang YX, Lespinasse Y (1991) Pollination with gamma-irradiated pollen and development of fruits, seeds and parthenogenetic plants in apple. Euphytica 54(1):101–109. https://doi.org/10.1007/bf00145636

54. Hooghvorst I, Torrico O, Hooghvorst S, Nogués S (2020) In situ parthenogenetic doubled haploid production in Melon "Piel de Sapo" for breeding purposes. Front Plant Sci 11:378. https://doi.org/10.3389/fpls.2020.00378

55. Kurtar ES, Balkaya A (2010) Production of in vitro haploid plants from in situ induced haploid embryos in winter squash (*Cucurbita maxima* Duchesne ex Lam.) via irradiated pollen. Plant Cell Tissue Organ Cult 102 (3):267–277. https://doi.org/10.1007/s11240-010-9729-1

56. Germanà MA (2006) Doubled haploid production in fruit crops. Plant Cell Tissue Organ Cult 86(2):131–146

57. Ishii T, Karimi-Ashtiyani R, Houben A (2016) Haploidization via chromosome elimination: means and mechanisms. Annu Rev Plant Biol 67(1):421–438. https://doi.org/10.1146/annurev-arplant-043014-114714

58. Kalinowska K, Chamas S, Unkel K, Demidov D, Lermontova I, Dresselhaus T, Kumlehn J, Dunemann F, Houben A (2019) State-of-the-art and novel developments of in vivo haploid technologies. TAG Theor Appl Genet 132(3):593–605. https://doi.org/10.1007/s00122-018-3261-9

59. Ravi M, Chan SWL (2010) Haploid plants produced by centromere-mediated genome elimination. Nature 464(7288):615–618. https://doi.org/10.1038/nature08842

60. Baum M, Lagudah ES, Appels R (1992) Wide crosses in cereals. Annu Rev Plant Physiol Plant Mol Biol 43:117–143

61. Comai L, Tan EH (2019) Haploid induction and genome instability. Trends Genet 35 (11):791–803. https://doi.org/10.1016/j.tig.2019.07.005

62. Devaux P (2003) The *Hordeum bulbosum* (L.) method. In: Maluszynski M, Kasha KJ, Forster BP, Szarejko I (eds) Doubled haploid production in crop plants. A manual. Kluwer Academic Publishers, Dordretch, pp 15–19

63. Lange W (1971) Crosses between Hordeum vulgare L. and H. bulbosum L. I. Production, morphology and meiosis of hybrids, haploids and dihaploids. Euphytica 20(1):14–29. https://doi.org/10.1007/bf00146769

64. Kasha KJ, Kao KN (1970) High frequency haploid production in barley (*Hordeum vulgare* L.). Nature 225:874–876

65. Wedzony M, Forster BP, Zur I, Golemiec E, Szechynska-Hebda M, Dubas E, Gotebiowska G (2009) Progress in doubled haploid technology in higher plants. In: Touraev A, Forster BP, Jain SM (eds) Advances in haploid production in higher plants. Springer, Dordrecht, pp 1–33

66. Inagaki MN (2003) Doubled haploid production in wheat through wide hybridization. In: Maluszynski M, Kasha KJ, Forster BP, Szarejko I (eds) Doubled haploid production in crop plants: a manual. Springer, Dordrecht, pp 53–58

67. Wędzony M (2003) Protocol for doubled haploid production in hexaploid triticale (x Triticosecale Wittm.) by crosses with maize. In: Maluszynski M, Kasha KJ, Forster BP, Szarejko I (eds) Doubled haploid production in crop plants: a manual. Springer, Dordrecht, pp 135–140

68. Berger F, Grini PE, Schnittger A (2006) Endosperm: an integrator of seed growth and development. Curr Opin Plant Biol 9(6):664–670. https://doi.org/10.1016/j.pbi.2006.09.015

69. Marsollier A-C, Ingram G (2018) Getting physical: invasive growth events during plant development. Curr Opin Plant Biol 46:8–17. https://doi.org/10.1016/j.pbi.2018.06.002

70. Widiez T, Ingram GC, Gutierrez-Marcos J (2017) Embryo-endosperm-sporophyte interaction in maize seeds. In: Larkins BA (ed) Maize kernel development. CABI, Boston, MA, pp 95–107

71. Murovec J, Bohanec B (2012) Haploids and doubled haploids in plant breeding. In: Abdurakhmonov I (ed) Plant breeding. InTech, Rijeka

72. Gilles LM, Khaled A, Laffaire J-B, Chaignon S, Gendrot G, Laplaige J, Bergès H, Beydon G, Bayle V, Barret P, Comadran J, Martinant J-P, Rogowsky PM, Widiez T (2017) Loss of pollen-specific phospholipase NOT LIKE DAD triggers gynogenesis in maize. EMBO J 36(6):707–717. https://doi.org/10.15252/embj.201796603

73. Jacquier NMA, Gilles LM, Pyott DE, Martinant J-P, Rogowsky PM, Widiez T (2020) Puzzling out plant reproduction by haploid induction for innovations in plant breeding. Nat Plant 6(6):610–619. https://doi.org/10.1038/s41477-020-0664-9

74. Schnable PS, Wise RP (1998) The molecular basis of cytoplasmic male sterility and fertility restoration. Trends Plant Sci 3(5):175–180. https://doi.org/10.1016/S1360-1385(98)01235-7

75. Chase CD, Gabay-Laughnan S (2004) Cytoplasmic male sterility and fertility restoration by nuclear genes. In: Daniell H, Chase C (eds) Molecular biology and biotechnology of plant organelles: chloroplasts and mitochondria. Springer, Dordrecht, pp 593–621. https://doi.org/10.1007/978-1-4020-3166-3_22

76. Bohra A, Jha UC, Adhimoolam P, Bisht D, Singh NP (2016) Cytoplasmic male sterility (CMS) in hybrid breeding in field crops. Plant Cell Rep 35(5):967–993. https://doi.org/10.1007/s00299-016-1949-3

77. Dong Y-Q, Zhao W-X, Li X-H, Liu X-C, Gao N-N, Huang J-H, Wang W-Y, Xu X-L, Tang Z-H (2016) Androgenesis, gynogenesis, and parthenogenesis haploids in cucurbit species. Plant Cell Rep 35:1991–2019. https://doi.org/10.1007/s00299-016-2018-7

78. Gilles LM, Martinant J-P, Rogowsky PM, Widiez T (2017) Haploid induction in plants. Curr Biol 27(20):R1095–R1097

79. Maine MJD (2003) Potato haploid technologies. In: Maluszynski M, Kasha KJ, Forster BP, Szarejko I (eds) Doubled haploid production in crop plants: a manual. Springer, Dordrecht, pp 241–247

80. Yao L, Zhang Y, Liu C, Liu Y, Wang Y, Liang D, Liu J, Sahoo G, Kelliher T (2018) OsMATL mutation induces haploid seed formation in indica rice. Nat Plant 4:530–533. https://doi.org/10.1038/s41477-018-0193-y

81. Liu H, Wang K, Jia Z, Gong Q, Lin Z, Du L, Pei X, Ye X (2020) Efficient induction of haploid plants in wheat by editing of TaMTL using an optimized Agrobacterium-mediated CRISPR system. J Exp Bot 71(4):1337–1349. https://doi.org/10.1093/jxb/erz529

82. Liu C, Zhong Y, Qi X, Chen M, Liu Z, Chen C, Tian X, Li J, Jiao Y, Wang D, Wang Y, Li M, Xin M, Liu W, Jin W, Chen S (2020) Extension of the in vivo haploid induction system from diploid maize to hexaploid wheat. Plant Biotechnol J 18(2):316–318. https://doi.org/10.1111/pbi.13218

83. Hussain T, Franks C (2019) Discovery of sorghum haploid induction system. In: Zhao Z-Y, Dahlberg J (eds) Sorghum: methods and protocols. Springer, New York, NY, pp 49–59. https://doi.org/10.1007/978-1-4939-9039-9_4

84. Kloiber-Maitz M, Wieckhorst S, Bolduan C, Ouzunova M (2018) Haploidisierung in sorghum. EP3366778A1

85. Kermicle JL (1969) Androgenesis conditioned by a mutation in maize. Science 166 (3911):1422–1424. https://doi.org/10.1126/science.166.3911.1422

86. Coe EH (1959) A line of maize with high haploid frequency. Am Nat 93(873):381–382. https://doi.org/10.1086/282098

87. Rokka VM (2009) Potato haploids and breeding. In: Touraev A, Forster BP, Jain SM (eds) Advances in haploid production in higher plants. Springer, Dordrecht, pp 199–208

88. Spooner DM, Ghislain M, Simon R, Jansky SH, Gavrilenko T (2014) Systematics, diversity, genetics, and evolution of wild and cultivated potatoes. Bot Rev 80(4):283–383. https://doi.org/10.1007/s12229-014-9146-y

89. Amundson KR, Ordoñez B, Santayana M, Tan EH, Henry IM, Mihovilovich E, Bonierbale M, Comai L (2020) Genomic outcomes of haploid induction crosses in potato (Solanum tuberosum L.). Genetics 214(2):369–380. https://doi.org/10.1534/genetics.119.302843

90. Britt AB, Kuppu S (2016) Cenh3: an emerging player in haploid induction technology. Front Plant Sci 7:357. https://doi.org/10.3389/fpls.2016.00357

91. Jackson D (2017) No sex please, we're (in)breeding. EMBO J 36(6):703–704. https://doi.org/10.15252/embj.201796735

92. Geiger HH, Gordillo GA (2009) Doubled haploids in hybrid maize breeding. Maydica 54(4):485–499

Chapter 2

Applications of Doubled Haploids in Plant Breeding and Applied Research

Jens Weyen

Abstract

Manifold and diverse applications of doubled haploid (DH) plants have emerged in academy and in the plant breeding industry since the first discovery of a haploid mutant in the Jimson Weed (*Datura stramonium*), followed by the first reports about anther culture in the same species, maternal haploids by wide crosses in tobacco (*Nicotiana tabacum* L.) and barley (*Hordeum vulgare* L.), interspecific hybridization, ovary culture (gynogenesis), isolated microspore culture, and more recently the CENH3 approach in thale cress (*Arabidopsis thaliana* L.) and other species. Research and development efforts were and are still significant in both user groups. Luckily, often academic and industrial partners cooperate in challenging and sometimes voluminous projects worldwide. Not only to develop innovative DH protocols and technologies per se, but also to exploit the advantages of DH plants in a huge variety of research and development experiments. This review concentrates not on the DH technologies per se, but on the application of DHs in plant-related research and development projects.

Key words DH, Doubled haploids, Homozygosity, Molecular markers, Selection, Genetic variability, Recombination, Biostatistics, Genome editing, Genomic selection, Breeding strategy

1 Introduction

The first report about a haploid mutant was in *Datura stramonium* [1], and the first report of haploid production through anther culture was in a related species, *Datura innoxia* [2]. Maternal haploids by wide crosses were then described in tobacco [3] and barley [4]. Interspecific hybridization [5], ovary culture [6], isolated microspore culture [7, 8], and the use of haploid inducer lines, recently including the CENH3 approach [9] are the most prominent additional doubled haploid (DH) technologies. While the primary application of DHs was to fix genetic variability as fast as possible by immediately reaching full homozygosity (one "step" versus several selfing generations), this advantage of DHs was later and is still widely used in marker-assisted selection in academy. Later, with the development of cheaper and easier molecular

Jose M. Seguí-Simarro (ed.), *Doubled Haploid Technology: Volume 1: General Topics, Alliaceae, Cereals*, Methods in Molecular Biology, vol. 2287, https://doi.org/10.1007/978-1-0716-1315-3_2,

biological and genomic tools and technologies, the link between DHs and marker applications in commercial breeding programs was accelerated as well. Today, DHs in many plant species are routinely used, often combined with diverse tool kits from the fields of genomics, transgenics (e.g. reverse breeding), bioinformatics, tissue culture, genome editing, epigenetics, but phenotyping and sophisticated field nursery technologies as well. This is often possible in species in which the efficiency of DH production was improved by optimization of several steps, being it either in vitro or in vivo. Those improvements (described in other chapters of this book) led mainly to accelerated in vitro haploid cell induction, better quality and quantity of organogenesis and regeneration, in vivo haploid induction, and genome doubling. Earlier reviews were published [10, 11].

Due to their genetic characteristics, DH plant lines are currently used in manifold applications in academic and commercial breeding programs, as well as in research and development (R&D) in plant biology, genetics, physiology, phenotyping, biostatistics, etc. Big data applications as genome sequencing, genomics, high throughput, or deep phenotyping are benefiting from DHs as well. Protocols for breeding of DH lines are available for almost 400 species (*see* Chapter 3 of Volume 1), and over 300 DH-derived cultivars have been developed in 12 species worldwide [12]. In maize, for example, methods for inducing, selecting, and doubling haploid plants are advanced and are in widespread use [13]. Some authors published studies about the epigenetic effects in DH lines. Especially, the developments in genomics (genome sequencing), the increasing speed and volume of data handling procedures, the new methods for hybrid mechanisms, and the improvement of DH protocols in formerly nonfunctional or minimally efficient DH protocols in "recalcitrant" crop species, has led to immense progress in selection and hybrid breeding.

This review article here starts with the use of DHs to analyze their genetics and agronomic characters and the comparison of DH populations with conventionally generated (selfed) populations under field conditions. In the following sections, the use of DHs in diverse genetic mapping studies and gene cloning approaches (and other genomic applications) is described, as well as their use in breeding and research of transgenic plants. The most recent reports from large biostatistical projects, genome editing, and phenotyping applications are mentioned too. In particular, the following applications of DHs in plant breeding and applied research will be discussed:

- Recombination and fixation of genetic variance.
- Use of haploid and DH technologies to develop wide crosses and use of hybridization.

- DHs for mapping and diverse range of MAS procedures and strategies.
- DHs for genomic selection and genomic prediction.
- Haploid tissues used for genetic transformation and development of stable transgenic lines and DHs for breeding with transgenic lines.
- Use of haploid cells and tissues for increasing genetic variation by mutation.
- Genome editing by the use of DH technologies.
- Epigenetics and DHs.
- Reverse Breeding.

2 Recombination and Fixation of Genetic Variance

The agronomic and molecular comparison of DH and conventional lines started as soon as the first relatively efficient DH production methods were at hand. For example, this was the case in the late 1970s and beginning of the 1980s in barley. It was found that the traits of barley DH lines distributed as in single-seed descent (SSD) lines. These traits included grain yield, heading date, and plant height [14, 15]. The results indicated that although the SSD method has more opportunities for genetic recombination than the DH method, it did not produce a sample of recombinants significantly different from the DH sample; thus, both methods were equally efficient to derive homozygous lines from F_1 hybrids in a relatively short time. One example is that barley SSD and DH progenies were used to develop lines with high kernel weight, but low β-glucan content [16]. Since then, hundreds if not thousands of registered varieties were developed worldwide, not only in barley. Despite the fact that the influence and share of DH inbred lines in hybrid breeding is significant, detailed data and lists of varieties developed by the use of DHs are not available due to trade secrets or nonpublic data at variety registration authorities and breeding companies. The COST Action 851 of the European Commission started to collect those data [17] but there was no continuation of this effort, unfortunately.

Theoretically, in a recurrent selection program, the use of DHs can increase the genetic advance per unit of time. To evaluate the efficiency expected from the use of DHs for grain yield improvement in a maize population, two recurrent selection programs for testcross performance were initiated, using testcross progenies from DH lines and S_1 families. Several selection cycles using DH and S_1 families were carried out and testcross genetic variance was twice as high among DH lines as among S_1 families. A year advantage of 29% for the S_1 family method over the DH method with a cycle of four years was calculated while in a 3-year cycle for the DH method, both methods were expected to be equivalent. With a 3-year cycle for the DH method, the advantage would have been in favor of DH method. Furthermore, the DH method has the advantage of simultaneously producing lines that are directly usable as hybrid parents. Thus, if the genetic advance per unit of time is evaluated at the level of developed varieties even with the same or lower genetic advance in population improvement, the DH method appears to be the most efficient [18].

Although there is still a lot of space for improvements and cost reduction, a range of different routine and highly efficient DH production systems are currently at hand in barley, wheat, triticale, several *brassica* species and in maize. In maize, for example, a comparison between the frequency of recombination events and genetic variance in DH F_1 and F_2 populations generated by haploid inducer lines revealed that DHF_2 showed a higher mean recombination [19]. It was shown that for several traits, the means between DH F_1 and DH F_2 lines did not differ, while the genetic variance was higher among DH F_2 lines than among DH F_1 lines for one trait. The ratio of repulsion to coupling linkages was higher among DH F_1 lines than among DH F_2 lines for one of the analyzed traits. These results indicated that the decision of inducing DH lines from F_1 or F_2 plants should be made from considerations other than the performance of the resulting DH F_1 or DH F_2 lines [19].

In maize, the F_1 generation has been the most frequently used for haploid induction, due to the facility in the process. However, using F_2 generations would be a good alternative to increase genetic variability owing to the additional recombination in meiosis. The effect of F_1 and F_2 generations on DH production in tropical germplasm was explored, evaluating the expression of the R1-navajo anthocyan marker in seeds, the working steps of the methodology, and the genetic variability of the DH lines obtained by assessing haploid induction rate, inhibition seed rate, and diploid seed rate [20]. Estimates of population parameters in DH lines from F_1 were higher than from F_2. Furthermore, it was shown that one additional generation was not enough to create new genotype subgroups. Additionally, the relative efficiency of the response to selection in the F_1 was higher than in the F_2 due to the number of cycles that are used to obtain the DH. The results showed that in tropical maize, the use of the F_1 generation is recommended due to a superior balance between time and genetic variability. It can be concluded that joint R&D projects between academic and commercial partners can lead to useful results for both parties. It is without doubt that DHs in the breeding of self- and cross-pollinating plant species have many advantages in terms of both time and data quality.

3 Use of Haploid and DH Technologies to Develop Wide Crosses and Use of Hybridization

As a DH technology, isolated microspore culture is often used on material with meiotic instability, such as interspecific hybrids. As a result of chromosome missegregation and homologous exchanges, DH progenies might lose their homozygous status. For example, the fertility, meiosis, and genetic variability were assessed in a self-pollinated progeny set (the MDL2 population) resulting from first-generation plants (the MDL1 population) derived from microspores of a near-allohexaploid interspecific hybrid from the cross (*Brassica napus* × *B. carinata*) × *B. juncea*. Seed fertility and viability decreased substantially from the MDL1 to the MDL2 generation. In the MDL2 population, 87% of individuals differed genetically from their MDL1 parent. These genetic differences resulted from novel homologous exchanges between chromosomes, chromosome loss and gain, and segregation and instability of pre-existing karyotype abnormalities. Novel karyotype change was extremely common, with 2.2 new variants observed per MDL2 individual. Significant differences between progeny sets in the number of novel genetic variants were also observed. Thus, meiotic instability clearly has the potential to dramatically change karyotypes (often without detectable effects on the presence or absence of alleles) in putatively homozygous, microspore-derived lines, resulting in loss of fertility and viability [21].

4 DHs for Mapping and Marker-Assisted Selection

Since the beginning of DH technology, DHs have been used for mapping and marker-assisted selection. Full homozygosity and fixation of recombination events offer extreme advantages in mapping and many marker-assisted selection projects, at both the academic and commercial levels. The number of reports about the use of DH lines in mapping studies is therefore enormous, covering a diverse range of traits, phenotyping, and genotyping technologies (*see* Chapter 3 of Volume 2), which gives an idea of how useful DHs are for mapping. Besides the exceedingly early mapping studies (*see* above) some newer reports are mentioned next. Marker-aided breeding and DH technology have been used to improve host plant resistance in barley, rice, and wheat [22]. RILs (recombinant inbred lines) and DHs induced by an engineered haploid inducer in *Arabidopsis thaliana* were developed for QTL mapping [23]. DHs in triticale were used for exemplary QTL mapping of days to heading, revealing loci on chromosomes 2BL and 2R responsible for extended vernalization requirement, and identifying candidate genes [24]. DHs have also been described as advantageous in a pyramiding strategy of several BaMMV/BaYMV resistance genes in winter barley. For pyramiding of resistance genes *rym4*, *rym5*, *rym9*, and *rym11*, two different crossing strategies were applied and compared. Besides DH plants carrying all possible two-gene combinations, 20 DH plants out of 107 analyzed were found to carry *rym4*, *rym9*, and *rym11* in homozygosity, and 27 out of 187 tested were found to carry *rym5*, *rym9*, and rym11 in homozygosity [25]. There are many other reports in a significant range of species and for a huge diversity of traits.

Most frequently, the reasons to invest in the production of DHs include the easier phenotyping in DHs or the number of samples to be analyzed for certain genetic compositions of genes or haplotypes. For example, mapping of breeding traits using DH lines in oat was described in another study [26]. The root system architecture in seedlings of a population out of 300 maize DHs was phenotyped and significant single nucleotide polymorphism associations could be identified [27]. The application of DH melon lines for the development of multiple virus resistances was also described [28]. DH lines obtained from F_1 hybrids of reciprocal crosses between yellow- and black-seeded lines in *Brassica napus* were used to analyze tocopherols, plastochomanol-8, and phytosterols [29, 30], observing significantly positive correlations between the seed color and PC-8 content. According to the range of genetic variation among DHs of two populations, selected DH lines may be good parents for further breeding programs focused on increasing the amount and improving the quality of oilseed rapeseed oil. Furthermore, DHs in tobacco were reported to combined resistance to black root rot fungus and tomato spotted wilt virus (TSWV) [31].

DHs have also been proved useful to gain knowledge on the genetics of male fertility. A genome wide association study in two DH maize populations derived from a diversity panel (481 inbred lines) crossed with two parental lines was performed to analyze the genetics of haploid male fertility [32]. It was found out that this trait has a complex quantitative genetic structure (only one association larger than 11%), concluding that recurrent phenotypic selection coupled with marker-assisted selection for individual QTL might be the best strategy to improve haploid male fertility. On the other hand, a large-scale genome-wide association study to analyze spontaneous chromosome doubling in haploids derived from tropical maize inbred lines [33] concluded that the presence of large variation for both haploid male fertility and haploid fertility can be potentially exploited to improve the efficiency of DH derivation in tropical maize germplasm.

5   DHs for Genomic Selection and Genomic Prediction

Fast and affordable genome sequencing technologies and powerful software tools, in combination with high-speed processing units, made it possible to develop new biostatistical approaches in academia and plant breeding industry. The integration of genomic resources with DH technology provides new opportunities for the improvement of selection methods, maximizing selection gains and accelerating the development of new cultivars. Just as an example, it is now possible to estimate the additive variance for line value and the variance of additive by additive epistasis for line value from an experimental design with several lines per DH plant randomly taken from a population. Variances of higher-order epistasis can be estimated with a two-factor mating design in which a cross is replaced by the population of lines that can be derived from it. With a diallelic or a factorial design, a direct test for the presence of homozygous by homozygous epistasis is possible. A brief consideration of these expressions leads to the conclusion that recurrent selection of single DH descents will be one of the most efficient methods for low heritability, together with a rapid development of DH lines [34]. Another example of a more complex combination of genomics and DHs is the use of microsatellite marker analysis of DH progenies to predict heterosis in snowball cauliflower [35]. It was evident that the development of DH lines could broaden the genetic base of any crop through creating more diversity in the existing population. In addition to this, the study further suggested that genetic distances based on genomic and EST-SSRs can be used as a predictor of heterosis for commercial traits in CMS and DH-based F_1 cauliflower.

DHs have also been used in genomic selection strategies using both phenotypes and genotypes [36]. It was also shown that there exist additional opportunities to combine genomic prediction methods with the creation of DHs. The authors proposed an extension to genomic selection, optimal haploid value (OHV) selection, which predicts the best DH that can be produced from a segregating plant. This method focuses selection on the haplotype and optimizes the breeding program toward its end goal of generating an elite fixed line. The authors rigorously tested OHV selection breeding programs, using computer simulation, and showed that this approach results in more genetic gain than genomic selection. OHV selection preserved a substantially greater amount of genetic diversity in the population than genomic selection, which is important to achieve long-term genetic gain in breeding populations.

It was argued that due to the large estimates of genotypic variance among DH lines derived from maize landraces, individual lines with superior performance for agronomic and morphological traits can be selected and introgressed into the elite material [37]. Further, the improvement of seed set and other traits related to fitness in synthetic populations suggest that the DH technique might help in purging detrimental alleles present in landraces, apparently without strongly affecting the phenotypic diversity. Creation of DH lines from landraces shows great promise to broaden and improve the genetic basis of the Elite Flint breeding material without necessarily introducing negative agronomic features present in the landraces. Furthermore, the rapid decay of linkage disequilibrium (LD) together with the high genotypic variances and absence of population structure within the populations of DH lines derived from landraces make these lines an ideal tool for high-resolution association mapping. Genomic prediction within and among doubled haploid libraries from maize landraces was reported as well [38]. Altogether, the DH technology combined with genomic prediction offers a powerful approach to exploit the idle genetic diversity within landraces, but substantial investments are needed to mine this "gold reserve" for future breeding. It was concluded, that selected DH lines averaged similar testcross performance as their original landraces, and the best of them approached the yields of elite inbreds, demonstrating their potential to broaden the narrow genetic diversity of the flint germplasm pool. As to trait correlations of DH lines, correlation of test cross performance (TP) with line per se (LP) performance was zero for grain yield, underpinning the need to evaluate TP in addition to LP. For all traits, the authors observed substantial variation for TP among the DH lines and the best showed TP yields similar to the elite inbreds. Their results demonstrate the high potential of landraces for broadening the narrow genetic base of the flint heterotic pool and the usefulness of the DH technology for exploiting genetic resources from gene banks.

The use of DH maize lines for hybrid maize breeding was explored in a series of papers. In hybrid maize breeding, DHs are increasingly replacing inbreds developed by recurrent selfing. The authors analyzed the optimum allocation of the number of lines, test locations, as well as the number and type of testers in hybrid maize breeding using DHs. The production costs of DHs had only a minor effect on the optimum number of locations and on values of the optimization criteria [39]. In a second paper the authors stated that the optimum allocation of test resources is of crucial importance for the efficiency of breeding programs [40].

Early testing prior to DH production is a promising approach in hybrid maize breeding. In this third report of the series the authors determined the optimum allocation of the number of S (1) families, DH lines, and test locations for two different breeding schemes, compared the maximum selection gain achievable under both breeding schemes, and investigated limitations in the current method of DH production. Different assumptions were made regarding the budget, variance components, and time of DH production within S(1) families. The large potential of early testing prior to DH production was indicated. Substantial increases in haploid induction and chromosome doubling rates as well as reduction in costs of DH production would allow for early testing of S(1) lines and subsequent production and testing of DH lines in a breeding scheme that combines high selection gain with a short cycle length [41].

Parental selection influences the gain from selection and the optimum allocation of test resources in breeding programs. Two hybrid maize breeding schemes with evaluation of testcross progenies were analyzed. One with DH lines in both stages (DHTC) and the second with S(1) families in the first stage and DH lines within S(1) families in the second stage (S(1)TC-DHTC). The breeding Scheme S(1)TC-DHTC led to a larger selection gain but had a longer cycle length than DHTC. However, with further improvements in the DH technique and the realization of more than two generations per year, early testing of S(1) families prior to production of DH lines would become very attractive in hybrid maize breeding [42]. DHs may be developed directly from S(0) plants in the parental cross or via S(1) families. The superiority of S(1)TC-DHTC was increased when the selection was done among all DH lines ignoring their cross and family structure, and using variable sizes of crosses and S(1) families. In DHTC, the best selection strategy was to ignore cross structures and use uniform size of crosses [43].

DH lines were used in a comparison of five different genomic selection strategies for grain yield. Different variables were considered, including the available budget, the costs for DH (DH) line and hybrid seed production as well as variance components for grain yield in a wide range. A nursery selection for disease resistance just before genomic selection (GS) on grain yield was included. Owing to the extremely high number of test candidates entering breeding strategies with GS, the costs for DH line production had a larger impact on the annual selection gain than the hybrid seed production costs. A specific genomic selection procedure for cereals was concluded [44].

Assuming a finite number of unlinked loci and a given total number of individuals to be genotyped, three methods of marker-assisted selection (MAS) for gene stacking in DH lines derived from biparental crosses were compared by theory and simulations [45]. With best linear unbiased prediction (BLUP), information from genetically related candidates is combined to obtain more precise estimates of genotypic values of test candidates and thereby increase progress from selection. The breeding schemes involved selection for testcross performance either of DH lines at both stages or of S(1) families at the first stage and DH lines at the second stage [46].

6   DHs and Genetic Transformation

Easier inheritance and faster fixation of transgenes in homozygosis are the reasons to combine the use of genetic transformation and DH technologies. Physiological and genetic parameters of DNA synthesis in barley microspores were analyzed to optimize the particle bombardment technology. By maintaining the temperature low during the 4 h osmotic adjustment period following the cold plus mannitol pre-treatment, it was expected to find a frequency of homozygous DHs higher than following a 21-day cold pre-treatment. The best procedure for obtaining transgenic barley plants from this study was to pre-treat the cultures at either 4 °C or 25 °C during 4 h, using the actin promoter and adding arabinogalactan proteins in the microspore culture medium [47, 48]. Beyond this, anther culture-derived haploid embryos were used as explants for *Agrobacterium*-mediated genetic transformation of bread wheat to develop stable drought-tolerant transgenics. Stable transgenic DH plants showed faster seed germination and seedling establishment, and better drought tolerance in comparison with nontransgenic DH plants [49]. Embryogenic pollen cultures of barley were also used for *Agrobacterium*-mediated genetic transformation to achieve transgene homozygosity immediately. The routine application of the method based on cultivar 'Igri' over a period of over 10 years has achieved an average yield of about two transgenic plants per donor spike. The whole procedure from pollen isolation to nonsegregating transgenic, mature grain takes less than 12 months [50]. Genetic transformation with isolated wheat microspores and microspore-derived embryos has been described as well [51]. Thus, DH technologies have the potential to facilitate and simplify breeding of transgenic lines and registered varieties as in nontransgenic breeding. Back crossing programs and pyramiding and stacking several transgenes by crossing may benefit from DH technology as well.

7   Use of Haploid Cells and Tissues for Increasing Genetic Variation by Mutation

DH lines can also be used to accelerate and simplify the handling of mutations, as these are easier to handle in homozygous, DH individuals. DH lines can be generated from mutated parental lines, but haploid cells are optimal targets for direct mutagenesis as well. For example, anther culture in a japonica rice variety was used with donor plants derived from EMS-mutagenized fertilized egg cells. The authors were able to generate stable mutants for a diverse range of traits and many of those lines showed the same yield as the original variety "Mankeumbeo" [52]. Flower buds of Chinese cabbage were soaked in a range of EMS solutions at different concentrations, isolated microspore culture was applied, and plants regenerated therefrom. Embryo production rate and seedling rate were evaluated in five of the genotypes. Mutations in four color-related genes were identified. In total, 142 mutants with distinct variations in leaf shape, leaf color, corolla size, flower color, bolting time, and downy mildew resistance were identified from 475 DH lines [53].

Many more examples are published and definitely DHs are a very useful tool to generate variability by mutation technologies and to accelerate the development of advanced breeding lines with new, useful traits. Indeed, routine DH breeding with mutated genetic material is the daily operation of many plant breeders.

8 Genome Editing Using DH Technologies

The benefits of DH lines or haploid and later doubled cell and plant material in genome editing experiments is identical as the benefits of DH material in the development of mutations by more traditional technologies. Interestingly, genome editing in haploid cells or cells in vitro can generally be applied to further explore new basic processes in in vitro development of plant cells. Optimized methods for genome editing of haploid microspores and production of DH plants by microspore culture has been reported for wheat [54]. Many plant species and/or genotypes are still recalcitrant to conventional transformation methods. This, together with the long generation time of crop plants, poses a significant obstacle to effective application of gene editing technology, as it takes a long time to produce modified homozygous genotypes. As an alternative, the haploid, single-celled microspores are an attractive target for gene editing experiments, as they enable generation of homozygous DH mutants in one generation.

A different strategy to deliver edited genomic information into haploid cells was used in field and sweet corn and wheat [55]. The aberrant reproductive process of haploid induction was co-opted to induce edits in nascent seeds of diverse monocot and dicot species, which enables direct genomic modification of commercial crop varieties. The technology was tested in field and sweet corn using a native haploid-inducer line and extended to dicots using an engineered CENH3 system. They also recovered edited wheat embryos using Cas9 delivered by maize pollen and their data indicated that a transient hybrid state precedes uniparental chromosome elimination in a maize haploid inducer system. Edited haploid plants lack both the haploid-inducer parental DNA and the editing machinery. Therefore, edited plants could be used in trait testing and directly integrated into commercial variety development.

When combined, DHs and genome editing are two powerful game-changing technologies to generate pure inbred lines with multiple desired traits: DHs for acceleration and genome editing for new genetic variability. This is the case of the Haploid-Inducer Mediated Genome Editing (IMGE) approach [56], which utilizes a maize haploid inducer line carrying a CRISPR/Cas9 cassette targeting for a desired agronomic trait to pollinate an elite maize inbred line and to generate genome-edited haploids in the elite maize background. Homozygous pure DH lines with the desired trait improved could be generated within two generations, thus bypassing the lengthy procedure of repeated crossing and back-crossing used in conventional breeding [56].

9   Epigenetics and DHs

Epigenetic factors play an important role in gene regulation. It is interesting to find out the impact of DH technologies on epigenetics during haploid embryogenesis and vice versa. Indeed, tissue culture-induced genetic and epigenetic variation in triticale has been described [57]. Regenerated plants out of androgenesis and somatic embryogenesis were analyzed by metAFLP and RP-HPLC in four distinct genotypes and it was shown that regeneration via in vitro culture was error-prone and affected DNA sequence and methylation patterns, irrespective of the culture method used. Similar results have been reported in chrysanthemum, an important ornamental species. Its highly heterozygous state complicates molecular analysis, so it is interesting to derive haploid forms. A total of 2579 nonfertilized chrysanthemum ovules pollinated by *Argyranthemum frutescens* were cultured in vitro to isolate a haploid progeny [58]. 105 calli were produced, and in three of them, one single regenerant emerged. Only one of them was a true haploid. Nine DH derivatives were subsequently generated by colchicine treatment of 80 in vitro-cultured haploid nodal segments. Morphological screening showed that the haploid plant was shorter than the DHs, developed smaller leaves, flowers, and stomata, and only few of its pollen grains were able to germinate, although they were abnormal. Both the haploid and the DHs produced yellow flowers, whereas those of the donor cultivar were mauve. Methylation-sensitive amplification polymorphism (MSAP) profiling showed 52.2% of cytosine-methylated amplified fragments in the donor genome, whereas in haploid and DH genomes these percentages were 47.0 and 51.7%, respectively. In other words, there was a reduction in global cytosine methylation caused by haploidization, and a partial recovery following chromosome doubling.

Inhibition of Histone Deacytelase activity is sufficient to induce embryogenic growth in cultured pollen of *B. napus* and *Arabidopsis* [59]. Proteins in the range of 10–25 kD were differentially acetylated after TSA treatment compared with a control. Thus, in this respect, the deregulation of HDACs or HDAC-mediated pathways by stress and the accompanying changes in histone acetylation status could provide a single, common regulation point for the induction of haploid embryogenesis.

In conclusion, more research will be necessary to identify and to understand the role of epigenetic factors and to be able to influence them to improve DH technologies even more.

10 Reverse Breeding

Reverse breeding [60] is defined as the combination of a technology to control genetic recombination in hybrids through engineered meiosis, followed by DH production from nonrecombinant gamete precursors, in order to generate perfectly complementing homozygous parental lines, useful to reconstruct the hybrid background they come from. This is why it is called "reverse." The method is based on reducing genetic recombination in the selected heterozygote by eliminating meiotic crossing over. Male or female spores obtained from such plants contain combinations of nonrecombinant parental chromosomes which can be cultured in vitro to generate homozygous DH plants (DHs). From these DHs, complementary parents can be selected and used to reconstitute the heterozygote in perpetuity. The advantages and possibilities of such a method for on-demand reconstruction of hybrid backgrounds in the context of private breeding companies are evident.

11 Future Perspectives

Although in many plant species the progress in DH technology to produce DHs and to link DHs with other research and breeding tools was immense in the last decades, there is still a lot of space for further R&D, in both basic academic studies and more applied projects in commercial breeding companies. The use of DH populations to study genetic effects and inheritance of breeding traits will increase with new improved DH protocols and new DH technologies. Mapping and marker-assisted development using DHs are routine today in breeding companies and will increase, if phenotyping and data handling and processing technologies develop in parallel. New software tools are also necessary for this, and new algorithms co-developed by using artificial intelligence and machine learning will surely help. DHs would then in parallel serve as very suitable genetic material for complex biostatistical R&D projects in academia, industry, or both in joint consortia. For example, the prediction of yield in self and cross- pollinating species would be simplified. By having new hybrid breeding technologies, the genetic analysis of heterosis, genetic diversity, pool development, etc., will be easier even in self-pollinating crops. Sugar beet, sunflower, and rice are only a few of the important crops in which DH routines are highly demanded. Many vegetables, herbs, fruits, nuts, and ornamental species are so far difficult to handle in androgenic or gynogenic protocols. Haploid inducer lines, either natural or engineered, will support DH technology development in those species mainly for academic purposes, but in part for industrial applications as well. Tree breeding will benefit as well.

It is without doubt that DHs will accelerate the breeding of new registered varieties in many plant species. Even in species where DH technologies work "routinely" today as in barley, wheat, triticale, and rapeseed, there is much to improve, as there is still a strong genotype dependency in some of the key steps in the procedures. In so far recalcitrant crops or crops in which several genotype-dependent tissue culture steps as microspore or egg cell induction, regeneration, rooting, and even more importantly ploidy doubling new DH technologies would of course lead to a significant step forward in the breeding process. The development of new DH protocols and improvements in already functional protocols depend very much on the analysis of the genetics and physiology of gametophytic cell development in vivo and in vitro. Not only the sequencing and bioinformatic analysis of genomes but also other technologies will support those. Microscopic technologies, cell phenotyping and image analysis, new software algorithms, and automation are under steady development also in the plant field. Both, academic and industrial R&D projects, separate or in joint efforts, public or protected by trade secrets or patents will lead to a continuous growth of knowledge, and this will lead to more registered improved plant varieties, which are and will be desperately needed under the future climate conditions.

References

1. Blakeslee AF, Belling J, Farnham ME, Bergner AD (1922) A haploid mutant in the Jimson Weed, "*Datura stramonium*". Science 55 (1433):646–647. https://doi.org/10.1126/science.55.1433.646

2. Guha S, Maheshwari SC (1964) In vitro production of embryos from anthers of datura. Nature 204(4957):497–497. https://doi.org/10.1038/204497a0

3. Burk LG, Gerstel DU, Wernsman EA (1979) Maternal haploids of Nicotiana tabacum L. from seed. Science 206(4418):585. https://doi.org/10.1126/science.206.4418.585

4. Kasha KJ, Kao KN (1970) High frequency haploid production in barley (Hordeum vulgare L.). Nature 225(5235):874–876. https://doi.org/10.1038/225874a0

5. Kalinowska K, Chamas S, Unkel K, Demidov D, Lermontova I, Dresselhaus T, Kumlehn J, Dunemann F, Houben A (2019) State-of-the-art and novel developments of in vivo haploid technologies. Theor Appl Genet 132(3):593–605. https://doi.org/10.1007/s00122-018-3261-9

6. Van Geyt J, Speckmann GJ Jr, D'Halluin K, Jacobs M (1987) In vitro induction of haploid plants from unpollinated ovules and ovaries of the sugarbeet (Beta vulgaris L.). Theor Appl Genet 73(6):920–925. https://doi.org/10.1007/bf00289399

7. Pescitelli SM, Johnson CD, Petolino JF (1990) Isolated microspore culture of maize: effects of isolation technique, reduced temperature, and sucrose level. Plant Cell Rep 8(10):628–631. https://doi.org/10.1007/bf00270070

8. Lichter R (1982) Induction of haploid plants from isolated pollen of Brassica napus. Z Pflanzenphysiol 105(5):427–434. https://doi.org/10.1016/S0044-328X(82)80040-8

9. Ravi M, Chan SW (2010) Haploid plants produced by centromere-mediated genome elimination. Nature 464(7288):615–618. https://doi.org/10.1038/nature08842

10. Germanà MA (2011) Gametic embryogenesis and haploid technology as valuable support to plant breeding. Plant Cell Rep 30(5):839–857. https://doi.org/10.1007/s00299-011-1061-7

11. L-q Y, Fu S-h, Yang J, Li Y, J-s W, M-l W (2016) Generation, identification, formation mechanism and application of plant haploids. Yi Chuan 38(11):979–991. https://doi.org/10.16288/j.yczz.16-121

12. Forster BP, WTB T (2005) Doubled haploids in genetics and plant breeding. Plant Breed Rev 25:57–88. https://doi.org/10.1002/9780470650301.ch3

13. Chang M-T, Coe EH (2009) Doubled haploids. In: Kriz AL, Larkins BA (eds) Molecular genetic approaches to maize improvement. Springer, Berlin, pp 127–142. https://doi.org/10.1007/978-3-540-68922-5_10

14. Choo TM (1981) Doubled haploids for studying the inheritance of quantitative characters. Genetics 99(3–4):525–540

15. Choo TM, Reinbergs E, Park SJ (1982) Comparison of frequency distributions of doubled haploid and single seed descent lines in barley. Theor Appl Genet 61(3):215–218. https://doi.org/10.1007/BF00273777

16. Powell W, Caligari PD, Swanston JS, Jinks JL (1985) Genetical investigations into β-glucan content in barley. Theor Appl Genet 71(3):461–466. https://doi.org/10.1007/BF00251188

17. COST

18. Bordes J, Charmet G, de Vaulx RD, Pollacsek M, Beckert M, Gallais A (2006) Doubled haploid versus S1 family recurrent selection for testcross performance in a maize population. Theor Appl Genet 112(6):1063–1072. https://doi.org/10.1007/s00122-006-0208-3

19. Sleper JA, Bernardo R (2016) Recombination and genetic variance among maize doubled haploids induced from F(1) and F(2) plants. Theor Appl Genet 129(12):2429–2436. https://doi.org/10.1007/s00122-016-2781-4

20. Couto EGO, Cury MN, Bandeira e Souza M, Granato ÍSC, Vidotti MS, Domingos Garbuglio D, Crossa J, Burgueño J, Fritsche-Neto R (2019) Effect of F1 and F2 generations on genetic variability and working steps of doubled haploid production in maize. PLoS One 14(11):e0224631. https://doi.org/10.1371/journal.pone.0224631

21. Mwathi MW, Schiessl SV, Batley J, Mason AS (2019) "Doubled-haploid" allohexaploid Brassica lines lose fertility and viability and accumulate genetic variation due to genomic instability. Chromosoma 128(4):521–532. https://doi.org/10.1007/s00412-019-00720-w

22. Dwivedi SL, Britt AB, Tripathi L, Sharma S, Upadhyaya HD, Ortiz R (2015) Haploids: constraints and opportunities in plant breeding. Biotechnol Adv 33(6 Pt 1):812–829. https://doi.org/10.1016/j.biotechadv.2015.07.001

23. Filiault DL, Seymour DK, Maruthachalam R, Maloof JN (2017) The generation of doubled haploid lines for QTL mapping. Methods Mol Biol 1610:39–57. https://doi.org/10.1007/978-1-4939-7003-2_4

24. Tyrka M, Oleszczuk S, Rabiza-Swider J, Wos H, Wedzony M, Zimny J, Ponitka A, Ślusarkiewicz-Jarzina A, Metzger RJ, Baenziger PS, Lukaszewski AJ (2018) Populations of doubled haploids for genetic mapping in hexaploid winter triticale. Mol Breed 38(4):46–46. https://doi.org/10.1007/s11032-018-0804-3

25. Werner K, Friedt W, Ordon F (2005) Strategies for pyramiding resistance genes against the barley yellow mosaic virus complex (BaMMV, BaYMV, BaYMV-2). Mol Breed 16(1):45–55. https://doi.org/10.1007/s11032-005-3445-2

26. Kiviharju E, Moisander S, Tanhuanpää P (2017) Oat anther culture and use of DH-lines for genetic mapping. Methods Mol Biol 1536:71–93. https://doi.org/10.1007/978-1-4939-6682-0_6

27. Sanchez DL, Liu S, Ibrahim R, Blanco M, Lübberstedt T (2018) Genome-wide association studies of doubled haploid exotic introgression lines for root system architecture traits in maize (Zea mays L.). Plant Sci 268:30–38. https://doi.org/10.1016/j.plantsci.2017.12.004

28. Lotfi M, Alan AR, Henning MJ, Jahn MM, Earle ED (2003) Production of haploid and doubled haploid plants of melon (Cucumis melo L) for use in breeding for multiple virus resistance. Plant Cell Rep 21(11):1121–1128. https://doi.org/10.1007/s00299-003-0636-3

29. Cegielska-Taras T, Nogala-Kałucka M, Szala L, Siger A (2016) Study of variation of tocochromanol and phytosterol contents in black and yellow seeds of Brassica napus L. doubled haploid populations. Acta Sci Pol Technol Aliment 15(3):321–332. https://doi.org/10.17306/J.AFS.2016.3.31

30. Siger A, Michalak M, Lembicz J, Nogala-Kałucka M, Cegielska-Taras T, Szała L (2018) Genotype × environment interaction on tocochromanol and plastochromanol-8 content in seeds of doubled haploids obtained from F1 hybrid black × yellow seeds of winter oilseed rape (Brassica napus L.). J Sci Food Agric 98 (9):3263–3270. https://doi.org/10.1002/jsfa.8829

31. Trojak-Goluch A, Laskowska D, Kursa K (2016) Morphological and chemical characteristics of doubled haploids of flue-cured tobacco combining resistance to Thielaviopsis basicola and TSWV. Breed Sci 66(2):293–299. https://doi.org/10.1270/jsbbs.66.293

32. Ma H, Li G, Würschum T, Zhang Y, Zheng D, Yang X, Li J, Liu W, Yan J, Chen S (2018) Genome-wide association study of haploid male fertility in maize (Zea Mays L.). Front Plant Sci 9:974–974. https://doi.org/10.3389/fpls.2018.00974

33. Chaikam V, Gowda M, Nair SK, Melchinger AE, Boddupalli PM (2019) Genome-wide association study to identify genomic regions influencing spontaneous fertility in maize haploids. Euphytica 215(8):138–138. https://doi.org/10.1007/s10681-019-2459-5

34. Gallais A (1990) Quantitative genetics of doubled haploid populations and application to the theory of line development. Genetics 124 (1):199–206

35. Singh S, Dey SS, Bhatia R, Kumar R, Sharma K, Behera TK (2019) Heterosis and combining ability in cytoplasmic male sterile and doubled haploid based Brassica oleracea progenies and prediction of heterosis using microsatellites. PLoS One 14(8):e0210772. https://doi.org/10.1371/journal.pone.0210772

36. Daetwyler HD, Hayden MJ, Spangenberg GC, Hayes BJ (2015) Selection on optimal haploid value increases genetic gain and preserves more genetic diversity relative to genomic selection. Genetics 200(4):1341. https://doi.org/10.1534/genetics.115.178038

37. Strigens A, Schipprack W, Reif JC, Melchinger AE (2013) Unlocking the genetic diversity of maize landraces with doubled haploids opens new avenues for breeding. PLoS One 8(2):e57234–e57234. https://doi.org/10.1371/journal.pone.0057234

38. Brauner PC, Müller D, Schopp P, Böhm J, Bauer E, Schön C-C, Melchinger AE (2018) Genomic prediction within and among doubled-haploid libraries from maize landraces. Genetics 210(4):1185. https://doi.org/10.1534/genetics.118.301286

39. Longin CFH, Utz HF, Reif JC, Schipprack W, Melchinger AE (2006) Hybrid maize breeding with doubled haploids: I. One-stage versus two-stage selection for testcross performance. Theor Appl Genet 112(5):903–912. https://doi.org/10.1007/s00122-005-0192-z

40. Longin CFH, Utz HF, Melchinger AE, Reif JC (2007) Hybrid maize breeding with doubled haploids: II. Optimum type and number of testers in two-stage selection for general combining ability. Theor Appl Genet 114 (3):393–402. https://doi.org/10.1007/s00122-006-0422-z

41. Longin CFH, Utz HF, Reif JC, Wegenast T, Schipprack W, Melchinger AE (2007) Hybrid maize breeding with doubled haploids: III. Efficiency of early testing prior to doubled haploid production in two-stage selection for testcross performance. Theor Appl Genet 115 (4):519–527. https://doi.org/10.1007/s00122-007-0585-2

42. Wegenast T, Longin CFH, Utz HF, Melchinger AE, Maurer HP, Reif JC (2008) Hybrid maize breeding with doubled haploids. IV. Number versus size of crosses and importance of parental selection in two-stage selection for testcross performance. Theor Appl Genet 117(2):251–260. https://doi.org/10.1007/s00122-008-0770-y

43. Wegenast T, Utz HF, Longin CFH, Maurer HP, Dhillon BS, Melchinger AE (2010) Hybrid maize breeding with doubled haploids: V. Selection strategies for testcross performance with variable sizes of crosses and S(1) families. Theor Appl Genet 120 (4):699–708. https://doi.org/10.1007/s00122-009-1187-y

44. Marulanda JJ, Mi X, Melchinger AE, Xu J-L, Würschum T, Longin CFH (2016) Optimum breeding strategies using genomic selection for hybrid breeding in wheat, maize, rye, barley, rice and triticale. Theor Appl Genet 129 (10):1901–1913. https://doi.org/10.1007/s00122-016-2748-5

45. Melchinger AE, Technow F, Dhillon BS (2011) Gene stacking strategies with doubled haploids derived from biparental crosses: theory and simulations assuming a finite number of loci. Theor Appl Genet 123(8):1269–1279. https://doi.org/10.1007/s00122-011-1665-x

46. Mi X, Wegenast T, Utz HF, Dhillon BS, Melchinger AE (2011) Best linear unbiased prediction and optimum allocation of test resources in maize breeding with doubled haploids. Theor Appl Genet 123(1):1–10. https://doi.org/10.1007/s00122-011-1561-4

47. Shim Y-S, Pauls KP, Kasha KJ (2009) Transformation of isolated barley (Hordeum vulgare L.) microspores: I. the influence of pretreatments and osmotic treatment on the time of DNA synthesis. Genome 52(2):166–174. https://doi.org/10.1139/g08-112

48. Shim Y-S, Pauls KP, Kasha KJ (2009) Transformation of isolated barley (Hordeum vulgare L.) microspores: II. Timing of pretreatment and temperatures relative to results of bombardment. Genome 52(2):175–190. https://doi.org/10.1139/g08-113

49. Chauhan H, Khurana P (2011) Use of doubled haploid technology for development of stable drought tolerant bread wheat (Triticum aestivum L.) transgenics. Plant Biotechnol J 9 (3):408–417. https://doi.org/10.1111/j.1467-7652.2010.00561.x

50. Otto I, Müller A, Kumlehn J (2015) Barley (Hordeum vulgare L.) transformation using embryogenic pollen cultures. Methods Mol Biol 1223:85–99. https://doi.org/10.1007/978-1-4939-1695-5_7

51. Rustgi S, Ankrah NO, Brew-Appiah RAT, Sun Y, Liu W, von Wettstein D (2017) Doubled haploid transgenic wheat lines by microspore transformation. Methods Mol Biol 1679:213–234. https://doi.org/10.1007/978-1-4939-7337-8_13

52. Lee SY, Cheong JI, Kim TS (2003) Production of doubled haploids through anther culture of M1 rice plants derived from mutagenized fertilized egg cells. Plant Cell Rep 22(3):218–223. https://doi.org/10.1007/s00299-003-0663-0

53. Lu Y, Dai S, Gu A, Liu M, Wang Y, Luo S, Zhao Y, Wang S, Xuan S, Chen X, Li X, Bonnema G, Zhao J, Shen S (2016) Microspore induced doubled haploids production from ethyl methanesulfonate (EMS) soaked flower buds is an efficient strategy for mutagenesis in Chinese cabbage. Front Plant Sci 7:1780–1780. https://doi.org/10.3389/fpls.2016.01780

54. Ferrie AMR, Bhowmik P, Rajagopalan N, Kagale S (2020) CRISPR/Cas9-mediated targeted mutagenesis in wheat doubled haploids. Methods Mol Biol 2072:183–198. https://doi.org/10.1007/978-1-4939-9865-4_15

55. Kelliher T, Starr D, Su X, Tang G, Chen Z, Carter J, Wittich PE, Dong S, Green J, Burch E, McCuiston J, Gu W, Sun Y, Strebe T, Roberts J, Bate NJ, Que Q (2019) One-step genome editing of elite crop germplasm during haploid induction. Nat Biotechnol 37(3):287–292. https://doi.org/10.1038/s41587-019-0038-x

56. Wang B, Zhu L, Zhao B, Zhao Y, Xie Y, Zheng Z, Li Y, Sun J, Wang H (2019) Development of a haploid-inducer mediated genome editing system for accelerating maize breeding. Mol Plant 12(4):597–602. https://doi.org/10.1016/j.molp.2019.03.006

57. Machczyńska J, Zimny J, Bednarek PT (2015) Tissue culture-induced genetic and epigenetic variation in triticale (× Triticosecale spp. Wittmack ex A. Camus 1927) regenerants. Plant Mol Biol 89(3):279–292. https://doi.org/10.1007/s11103-015-0368-0

58. Wang H, Dong B, Jiang J, Fang W, Guan Z, Liao Y, Chen S, Chen F (2014) Characterization of in vitro haploid and doubled haploid Chrysanthemum morifolium plants via unfertilized ovule culture for phenotypical traits and DNA methylation pattern. Front Plant Sci 5:738–738. https://doi.org/10.3389/fpls.2014.00738

59. Li H, Soriano M, Cordewener J et al (2014) The histone deacetylase inhibitor trichostatin a promotes totipotency in the male gametophyte. Plant Cell 26(1):195–209. https://doi.org/10.1105/tpc.113.116491

60. Dirks R, van Dun K, de Snoo CB, van den Berg M, Lelivelt CLC, Voermans W, Woudenberg L, de Wit JPC, Reinink K, Schut JW, van der Zeeuw E, Vogelaar A, Freymark G, Gutteling EW, Keppel MN, van Drongelen P, Kieny M, Ellul P, Touraev A, Ma H, de Jong H, Wijnker E (2009) Reverse breeding: a novel breeding approach based on engineered meiosis. Plant Biotechnol J 7(9):837–845. https://doi.org/10.1111/j.1467-7652.2009.00450.x

Chapter 3

Species with Haploid or Doubled Haploid Protocols

Jose M. Seguí-Simarro, Javier Belinchón Moreno, Marina Guillot Fernández, and Ricardo Mir

Abstract

In this chapter, we present a list of species (and few interspecific hybrids) where haploids and/or doubled haploids have been published, including the method by which they were obtained and the corresponding references. This list is an update of the compilation work of Maluszynski et al. published in 2003, including new species for which protocols were not available at that time, and also novel methodologies developed during these years. The list includes 383 different backgrounds. In this book, we present full protocols to produce DHs in 43 of the species included in this list. In addition, this book includes a chapter for one species not included in the list. This makes a total of 384 species where haploids and/or DHs have been reported up to date.

Key words Androgenesis, Anther culture, Gynogenesis, Haploid inducers, Intraspecific hybridization, Irradiated pollen, Microspore culture, Ovary culture, Ovule culture, Spontaneous haploids, Uniparental genome elimination, Wide hybridization

1 Introduction

We hereby present a list of species (and few interspecific hybrids) where haploids and/or doubled haploids have been published, including the corresponding references. This list (Table 1) includes major crops such as potato, rapeseed, tomato, pepper, wheat, maize, and barley, and also many different minor crops and species with lower agricultural impact. This list was born as an attempt to update, 17 years later, the enormous work that Maluszynski et al. published in 2003 [1], including new species for which protocols were not available at that time, and also novel methodologies developed thanks to the revolutionary advance of plant biotechnology tools in the last two decades.

For each of the listed species, the method by which the haploids/DHs have been obtained is mentioned. Methods include anther and microspore culture, induction of gynogenesis in vitro by ovary or ovule culture, or with irradiated pollen, the use of

Jose M. Seguí-Simarro (ed.), *Doubled Haploid Technology: Volume 1: General Topics, Alliaceae, Cereals*,
Methods in Molecular Biology, vol. 2287, https://doi.org/10.1007/978-1-0716-1315-3_3,
© Springer Science+Business Media, LLC, part of Springer Nature 2021

haploid inducer lines, intraspecific hybridization, wide (either inter-specific or intergeneric) hybridization, uniparental genome elimination via genetic manipulation, and also the spontaneous occurrence of haploids in natural or cultivated populations. It is important to note that in some cases, one method may be included in more than one of these groups. For example, uniparental genome elimination via genetic manipulation may imply the use of genetically engineered haploid inducer lines. In turn, haploid inducer lines are typically used to produce haploids via uniparental genome elimination. However, we considered important to differentiate the use of spontaneously produced haploid inducer lines, as in maize, from the use of artificially generated lines through genetic manipulation of *CENH3* or *NLD/MATL/ZmPLA1* genes, for example.

After more than 50 years of research on haploid and DH technology, the compilation of all the published works in a single chapter is impossible. In this chapter, we put together the published works we had in our own reference collection, together with an updated version of those previously published by Maluszynski et al. in 2003 [1], and those additional we were able to find in public databases. For the most important species, where protocols are more developed and therefore used, to the best of our knowledge, we restricted the list to the most recent or relevant in terms of protocol development, excluding many references using DHs (or earlier stages in the process of DH production) for the study of basic aspects of plant cell physiology, biochemistry, cell or developmental biology, or genetics. In this chapter, we present a list of 383 species (including few interspecific hybrids) where publications reporting the production of haploids and/or DHs are available. For 43 of the species included in this list, full protocols to produce DHs are also included in this book. Despite this, we considered that it is also important to include references for these 43 species, because it will give an idea of how deep a protocol in these species has been studied, and will also provide different perspectives and additional resources for the adaption of a protocol to a particular variety in these species. In addition to these 43 species, there is one chapter on a species not included in the list, leek (*Allium ampeloprasum*). This makes a total of 384 species where haploids and/or DHs have been reported.

Some of the references included in this list do not describe a full protocol to produce DHs. A few of them report the occurrence of spontaneous (not induced) haploids, or describe a protocol to produce haploid plants or embryos, but not its conversion to DHs. However, we think that for researchers interested in working in a particular species, this list may be useful to check whether the species has been studied in advance or not, and if so, to what extent protocols have been optimized. For those trying to optimize a protocol in these species, the knowledge of earlier attempts can be of great help in order to design their own starting points. Hopefully, this compilation may pave the way for those trying to enter into the fascinating challenge of developing a full DH protocol in a new or recalcitrant species.

Table 1
List of plant species where protocols for production of haploids and/or doubled haploids have been published

Species	Method	Reference
Aconitum carmichaeli	a	[2]
Actinidia arguta	a	[3]
Actinidia deliciosa	g, irr	[4, 5]
Aesculus carnea	a	[6]
Aesculus hippocastanum	a, m	[7–12]
Agropyron cristatum	a	[13]
Agropyron desertorum	a	[13]
Agropyron fragile	a	[13]
Agropyron glaucum	a	[14]
Albizia lebbeck	a	[15]
Allium cepa	g, o, spont, wh, irr	[16–37]
Allium fistulosum L.	g, o	[38]
Allium giganteum	a	[39]
Allium sativum	a	[40]
Allium schoenoprasum	o	[41]
Althaea officinalis	a	[42]
Ammi majus	m	[43, 44]
Ammi visnaga	m	[43]
Anacardium occidentale	a	[45]
Ananas comosus	o	[46]
Anemone canadensis	a	[47]
Anemone coronaria	a	[48]
Anethum graveolens	m	[44]
Angelica archangelica	m	[43, 44]
Annona squamosa	a	[49]

(continued)

Table 1
(continued)

Species	Method	Reference
Anethum graveolens	m	[43, 44, 50]
Anthriscus cerefolium	m	[43, 44]
Antirrhinum majus	a	[51]
Apium graveolens	a, m	[43, 52]
Arabidopsis griffithiana	a	[53]
Arabidopsis korshinskyi	a	[53]
Arabidopsis pumila	a	[53]
Arabidopsis thaliana	a, u	[53–63]
Arachis hypogaea	a	[64–67]
Arachis villosa	a	[64]
Asparagus officinalis	a	[68, 69]
Atriplex glauca	a	[70]
Atropa belladonna	a	[71–73]
Avena sativa	a, m, wh	[74–85]
Avena sterilis	a	[86, 87]
Azadirachta indica	a	[88]
Begonia × *hiemalis*	a	[89]
Beta vulgaris	g, o	[90–95]
Betula pendula	a	[96]
Bletilla ochracea	g	[97]
Bletilla striata	g	[97]
Borago officinalis	a	[44, 98, 99]
Boswellia serrata	a	[100]
Brassica campestris	a, m	[101–105]
Brassica carinata	a, m	[106–110]
Brassica carinata × *Brassica rapa*	m	[111]
Brassica juncea	m	[106, 112–117]
Brassica napus	a, m, hi, ih	[106, 118–136]
Brassica napus × *Brassica carinata*	m	[137, 138]
Brassica napus × *Brassica carinata* × *Brassica juncea*	m	[111, 138, 139]
Brassica napus × *Brassica nigra*	m	[111, 138]

(continued)

Table 1
(continued)

Species	Method	Reference
Brassica nigra	a	[106, 140]
Brassica oleracea	a, m	[106, 141–162]
Brassica rapa	m	[146, 163–175]
Brassicoraphanus	a	[176]
Bupleurum falcatum	a	[177, 178]
Cajanus cajan	a, m	[179–182]
Camellia assamica	a	[183]
Camellia japonica	a, m	[184–189]
Camellia sinensis	a, o	[190–193]
Capsicum annuum	a, m, spont	[194–223]
Capsicum baccatum	a	[198]
Capsicum chinense	a	[198]
Capsicum frutescens	a	[224–226]
Carica papaya	a	[227–229]
Carthamus tinctorius	a	[230]
Carum carvi L.	a, m	[43, 44, 50, 231, 232]
Cassia siamea	a	[233]
Cassia fistula	a	[234]
Catharanthus roseus	a	[235–237]
Ceratonia siliqua L.	a	[238]
Chrysanthemum morifolium	o	[239]
Chrysanthemum spp.	a, g	[240]
Cicer arietinum	a, ih	[241–248]
Cichorium intybus	a, m, wh	[249–251]
Citrullus lanatus	g	[252, 253]
Citrus aurantifolia	a	[254]
Citrus aurantium	a	[255]
Citrus clementina	a, g, o, m, irr	[256–265]
Citrus deliciosa × *C. paradisi*	a	[266]
Citrus lemon	a	[267]
Citrus madurensis	a	[268]

(continued)

Table 1
(continued)

Species	Method	Reference
Citrus maxima	wh	[269]
Citrus natsudaidai	irr	[270]
Citrus reticulata	a, g, irr	[193, 258, 271]
Citrus sinensis	a, g	[193, 256, 272–275]
Clausena excavata	a	[276]
Cocos nucifera	a, m	[277–281]
Coffea arabica	m	[282]
Coffea canephora	spont, g	[283–285]
Coffea arabica	a, spont	[286]
Corchorus olitorius	m	[287]
Corylus avellana	m	[288]
Crepis capillaris	a	[289, 290]
Crepis tectorum	irr	[291]
Crotalaria pallida	a	[292]
Cryptotaenia japonica	a, m	[293]
Cucumis melo	a, g, wh, irr	[294–306]
Cucumis sativus	a, g, irr, o	[307–332]
Cucurbita maxima	irr	[333–336]
Cucurbita moschata	g	[333–335, 337]
Cucurbita pepo	g, irr, o	[16, 325, 338–346]
Cupressus dupreziana	ih	[347]
Cyclamen persicum × *Cyclamen purpurascens*	a	[348]
Cynara scolymus	a, g	[349]
Dactylis glomerata	a	[350]
Datura ferox	a	[351]
Datura innoxia	a, m	[352–366]
Datura metel	a, m	[367–374]
Datura meteloides	a	[375]
Datura stramonium	spont	[376]
Daucus carota	a, m, o, wh, hi	[43, 293, 377–387]
Dendranthema grandiflorum	a	[388]

(continued)

Table 1
(continued)

Species	Method	Reference
Dianthus caryophyllus	a, o	[389]
Digitalis lanata	a	[390, 391]
Digitalis obscura	a	[392]
Digitalis purpurea	a	[393]
Dolichos biflorus	a	[394]
Echinacea purpurea	a	[395]
Elaeis guineensis	s	[396]
Ephedra foliata	g	[397]
Eragrostis tef	a, g	[398, 399]
Eriobotrya japonica	a	[400, 401]
Eruca sativa	m	[402]
Eucalyptus globulus	a	[403]
Euphoria longana	a	[404]
Fagopyrum esculentum	a, g	[405–407]
Fagus sylvatica	a	[7]
Feijoa sellowiana	a	[408–410]
Festuca arundinacea	*a*	[411–413]
Festuca arundinacea × *Lolium multiflorum*	a	[413, 414]
Festuca pratensis	a	[415, 416]
Festuca pratensis × *Lolium multiflorum*	a	[417]
Foeniculum vulgare	a, m	[43, 50, 293]
Fragaria × *ananassa*	a, wh	[418–422]
Fragaria orientalis	a	[423]
Fragaria × *Potentilla*	wh	[424]
Gentiana scabra	o	[425, 426]
Gentiana triflora	o, a	[425–427]
Gerbera jamesonii	g	[428–436]
Gingko biloba	m	[437]
Glycine max	a	[438–445]
Gossypium arboreum	a	[446, 447]
Gossypium barbadense	spont	[446, 448, 449]

(continued)

Table 1
(continued)

Species	Method	Reference
Gossypium hirsutum	a, o, m	[448–457]
Guizotia abyssinica	a, o	[458–463]
Haemanthus katherinae	a	[464]
Helianthus annuus	a, wh, m, o, irr	[465–477]
Helianthus annuus × H. smithii	a	[478]
Helianthus annuus × H. eggerttii	a	[478]
Helianthus mollis	a	[479]
Helianthus salicifolius	a	[479]
Helianthus smithii	a	[479]
Hemerocallis fulva	a	[480]
Hepatica nobilis	a	[481]
Hevea brasiliensis	a, o	[482–484]
Hibiscus cannabinus	a, m, o	[485–487]
Hibiscus sabdariffa	m	[488]
Hieracium pilosella	a	[489]
Hordeum bulbosum	a	[490, 491]
Hordeum marinum	a, wh	[491, 492]
Hordeum murinum	a, wh	[491, 493, 494]
Hordeum secalinum	wh	[493]
Hordeum spontaneum	a, wh	[495–497]
Hordeum vulgare	a, hi, m, o, u, wh	[498–530]
Hyoscyamus muticus	a, g	[531–534]
Hyoscyamus niger	a	[531, 535–541]
Hylocereus polyrhizus	o	[542]
Hylocereus undatus	o	[542]
Hypericum perforatum	a	[543]
Iochroma warscewiczii	a	[544]
Ipomoea batatas	a	[545, 546]
Jacaranda acutifolia	a	[234]
Jatropha curcas	a	[547]
Juglans regia	irr	[548]

(continued)

Table 1
(continued)

Species	Method	Reference
Lactuca sativa	wh	[549]
Larix decidua	o	[550–552]
Larix decidua × *Larix leptolepis*	o	[552]
Larix leptolepis	o	[552]
Lathyrus sp.	m	[553]
Lens culinaris	m	[554]
Lesquerella fendleri	a	[555]
Levisticum officinale	m	[43, 44]
Lilium auratum	a	[556, 557]
Lilium davidii	o	[558]
Lilium formosanum	a	[559]
Lilium longiflorum	a	[560, 561]
Lilium speciosum	irr	[562]
Lilium spp.	a, o	[563, 564]
Linum usitatissimum	a, o	[565–568]
Litchi chinensis	a	[569]
Lolium multiflorum	a	[570]
Lolium multiflorum × *Festuca arundinacea*	a	[412, 415, 571, 572]
Lolium perenne	a	[570]
Lolium perenne × *Festuca pratensis*	a	[573]
Lolium temulentum	a	[415, 416]
Lotus corniculatus	a, wh	[574–576]
Lupinus albus	a, m	[577–579]
Lupinus angustifolius	a, m	[577, 580]
Lupinus luteus	m	[577]
Lupinus polyphyllus	a	[581]
Lycium barbarum	a	[582]
Lycium chinense	a	[582]
Malus domestica	m, a	[583–591]
Malus prunifolia	a	[592]
Mammillaria elongata	a	[593]

(continued)

Table 1
(continued)

Species	Method	Reference
Manihot esculenta	a	[594–597]
Medicago sativa	a, ih, m, g	[598–608]
Medicago truncatula	m	[553]
Melandrium album	a, g	[609–611]
Mentha piperita	a	[612]
Mentha spicata	a	[612]
Mimulus aurantiacus	g, irr	[613]
Momordica charantia	a	[614]
Morus alba	g	[615]
Morus indica	a, o	[616–618]
Musa acuminata	a	[619]
Musa balbisiana	a	[619, 620]
Nephelium lappaceum	a	[621]
Nicotiana attenuata	a	[622]
Nicotiana glutinosa	wh	[623]
Nicotiana knightiana	a	[622]
Nicotiana langsdorfii	spont.	[624]
Nicotiana raimondii	a	[622]
Nicotiana repanda	wh	[623]
Nicotiana sylvestris	ih, wh	[625, 626]
Nicotiana tabacum	a, m, spont, wh	[625, 627–644]
Oenothera hookeri	a	[645]
Olea europaea	a	[646, 647]
Opuntia ficus-indica	a	[648]
Oryza aha	a	[649]
Oryza glaberrima	a	[650]
Oryza perennis	a	[649, 651]
Oryza punctata	a	[649]
Oryza ridleyi	a	[649]
Oryza rufipogon	a	[649]
Oryza sativa	a, m, u	[650, 652–673]

(continued)

Table 1
(continued)

Species	Method	Reference
Oryza sativa × *O. glaberrima*	a	[674]
Paeonia albiflora	a	[675]
Panax ginseng	a	[676]
Panax quinquifolius	a	[677]
Papaver somniferum	a	[678]
Parthenium argentatum	ih	[679]
Passiflora edulis	a, g, o	[680, 681]
Pastinaca sativa	m	[43, 44]
Pelargonium hortorum	a	[682]
Pelargonium roseum	a	[683]
Pelargonium zonale	a	[684]
Peltophorum pterocarpum	a	[685]
Pennisetum americanum	a	[686]
Pennisetum glaucum	a, m	[687–690]
Pennisetum purpureum	a	[691]
Pennisetum typhoides	m	[692, 693]
Petunia axillaris	a, g	[694–696]
Petunia hybrida	a, m, g, wh	[373, 695–701]
Petunia parodii	g, wh	[699]
Petunia spp.	a	[696, 702, 703]
Petunia violacea	a	[704]
Phaseolus vulgaris	a	[705, 706]
Phleum pratense	a, m	[707, 708]
Phlox drummondii	a	[709]
Physalis ixocarpa	a	[710, 711]
Physalis peruviana	a	[712]
Pimpinella anisum	m	[43, 44]
Picea sitchensis	g	[713]
Pisum sativum	a, m	[553, 714, 715]
Poinciana regia	a	[234]
Poncirus trifoliata	a	[716]

(continued)

Table 1
(continued)

Species	Method	Reference
Populus alba	a	[717]
Populus balsamifera	a	[718]
Populus beijingensis	a	[719]
Populus deltoides	a	[720–722]
Populus glandulosa	a	[723]
Populus maximowiczii	a	[718, 724, 725]
Populus nigra	a	[721, 722, 726, 727]
Populus simonii × *P. nigra*	a	[728]
Populus spp.	a, m	[717, 729–731]
Populus tremula	a	[732]
Populus trichocarpa	a	[718, 733]
Portulacea grandiflora	ih	[734]
Primula forbesii	a	[735]
Prunus armeniaca	a	[736]
Prunus avium	a	[587, 737]
Prunus dulcis	a	[738]
Prunus persica	a	[739–742]
Pseudotsuga menziesii	g, spont	[743, 744]
Psidium guajava	a	[745]
Psophocarpus tetragonolobus	a, m	[746, 747]
Psoralea corylifolia	g	[748]
Potentilla argentea	ih	[749]
Punica granatum	a	[750]
Pyropia haitanensis	spont	[751]
Pyrus communis	g, irr	[586, 752–754]
Pyrus pyrifolia	a	[586]
Quercus ilex	a	[755]
Quercus petraea	a	[7]
Quercus suber	a	[756–758]
Raphanus sativus	m	[759–764]
Ribes nigrum × *Ribes Holland*	a	[765]

(continued)

Table 1
(continued)

Species	Method	Reference
Ricinus communis	s	[766]
Rosa damascena	a	[767]
Rosa elliptica	m	[768]
Rosa hybrida	irr	[767, 769]
Rosa micrantha	m	[768]
Saccharum officinarum	a	[770, 771]
Saccharum spontaneum	a	[771–773]
Salvia sclarea	a	[774]
Saintpaulia ionantha	a	[775–779]
Saponaria vaccaria	m	[44, 780]
Scilla indica	a	[781]
Secale cereale	a, m, wh	[782–784]
Selenicereus megalanthus	o	[542]
Sesamum indicum	a	[785–787]
Setaria italica	a	[788]
Silene latifolia	a	[789]
Sinapis alba	a	[790]
Sinocalamus latiflora	a	[791]
Sisymbrium irio	ih	[792]
Solanum acaule ssp. *Acaule*	a	[793]
Solanum bulbocastanum	a	[794]
Solanum carolinense	a	[795–800]
Solanum chacoense	a	[801–806]
Solanum dulcamara	a	[807, 808]
Solanum goniocalyx	a	[809]
Solanum iopetalum	a	[810]
Solanum lycopersicum	a, m	[811–829]
Solanum lycopersicum × *Solanum etuberosum*	a	[830]
Solanum melongena	a, m	[16, 831–857]
Solanum peruvianum	*a*	[858]

(continued)

Table 1
(continued)

Species	Method	Reference
Solanum pimpinellifolium	*g, irr*	[859]
Solanum phureja	*a*	[805, 806, 860–865]
Solanum surattense	a	[866]
Solanum torvum	a	[867]
Solanum tuberosum	a, hi	[505, 793, 803–806, 860, 861, 865, 868–881]
Solanum viarum	a	[882]
Sorbus domesticus	a	[883]
Sorghum bicolor	a, hi, m, spont	[884–893]
Spathiphyllum wallisii	g, o	[894]
Spinacia oleracea	a, wh	[895]
Stevia rebaudiana	a	[896]
Streptocarpus hybridus	a	[897]
Taeniatherum caput-medusae	wh	[898]
Tagetes erecta	a, o	[899–903]
Tagetes patula	a	[899, 902]
Theobroma cacao	g, o, irr	[904–907]
Thuja gigantea	a	[684]
Trifolium alexandrinum	a	[908]
Trifolium hybridum	s	[909]
Trifolium rubens	g	[910]
Thinopyrum bessarabicum	a	[911]
Thinopyrum intermedium	a	[13]
Thinopyrum ponticum	a	[13]
Tradescantia paludosa	irr	[912, 913]
Triticosecale spp.	a, m, wh	[221, 529, 783, 914–937]
Triticosecale × *Triticum aestivum*	wh	[783]
Triticum aestivum	a, m, wh	[529, 530, 918, 924, 926, 929, 930, 938–961]
Triticum compactum	wh	[962]
Triticum monococcum	a, ih	[963, 964]

(continued)

Table 1
(continued)

Species	Method	Reference
Triticum spelta	a, m	[965–967]
Triticum turgidum	m, g, o, wh	[968–971]
Triticum ventricosum	wh	[972]
Tritordeum spp.	a	[973]
Tulipa gesneriana	m	[974]
Tulipa spp.	a, o, m	[564, 975]
Ulmus americana	a	[976]
Vaccinium spp.	a	[977]
Vasconcellea pubescens	a	[978]
Vicia faba	a	[979]
Vigna radiata	a	[980]
Vigna unguiculata	a	[981]
Viola odorata	o	[982]
Vitis latifolia	a	[983]
Vitis riparia	a	[984]
Vitis rupestris	a	[985]
Vitis vinifera	a, o	[986–990]
Zantedeschia aethiopica	a	[991]
Zea mays	a, g, hi, m, o, spont, u, wh	[992–1020]
Zingiber officinale	a	[1021]

a anther culture, *g* gynogenesis, *hi* haploid inducers, *ih* intraspecific hybridization, *irr* induction of gynogenesis with irradiated pollen, *m* microspore culture, *o* ovary or ovule culture (in vitro gynogenesis), *spont* spontaneous haploids, *u* uniparental genome elimination via genetic manipulation, *wh* wide hybridization, either interspecific or intergeneric

Acknowledgments

This work was supported by grants AGL2017-88135-R and PID2020-115763RB-I00 to JMSS from Spanish MICINN jointly funded by FEDER. R.M. is a recipient of a contract of the CDIGENT program of the Valencian Government.

References

1. Maluszynski M, Kasha KJ, Szarejko I (2003) Published doubled haploid protocols in plant species. In: Maluszynski M, Kasha KJ, Forster BP, Szarejko I (eds) Doubled haploid production in crop plants. A manual. Kluwer Academic, Dordrecht, pp 309–335

2. Hatano K, Shoyama Y, Nishioka I (1987) Somatic embryogenesis and plant regeneration from the anther of *Aconitum carmichaeli* Debx. Plant Cell Rep 6(6):446–448

3. Wang GF, Qin HY, Sun D, Fan ST, Yang YM, Wang ZX, Xu PL, Zhao Y, Liu YX, Ai J (2018) Haploid plant regeneration from hardy kiwifruit (*Actinidia arguta* Planch.) anther culture. Plant Cell Tissue Organ Cult 134(1):15–28. https://doi.org/10.1007/s11240-018-1396-7

4. Chat J, Decroocq S, Petit RJ (2003) A one-step organelle capture: gynogenetic kiwifruits with paternal chloroplasts. Proc R Soc Lond Ser B Biol Sci 270 (1517):783–789

5. Pandey KK, Przywara L, Sanders PM (1990) Induced parthenogenesis in kiwifruit (*Actinidia deliciosa*) through the use of lethally irradiated pollen. Euphytica 51(1):1–9. https://doi.org/10.1007/bf00022886

6. Marinkovié N, Radojevié L (1992) The influence of bud length, age of the tree and culture media on androgenesis induction in *Aesculus carnea* Hayne anther culture. Plant Cell Tissue Organ Cult 31(1):51–59

7. Jörgensen J (1991) Androgenesis in *Quercus petraea*, *Fagus sylvatica* and *Aesculus hippocastanum*. In: Ahuja MR (ed) Woody plant biotechnology. NATO ASI series (series A: life sciences). Springer, New York, NY, pp 353–354

8. Radojevic L, Marinkovic N, Jervremovic S (2000) Influence of the sex of flowers on androgenesis in *Aesculus hippocastanum* L. anther culture. In Vitro Cell Dev Biol Plant 36(6):464–469

9. Ćalić D, Zdravković-Korać S, Jevremović S, Guć-Šćekić M, Radojević L (2003) Variability and bimodal distribution of size in microspores of *Aesculus hippocastanum*. Biol Plant 47(2):289

10. Calic D, Zdravkovic-Korac S, Pemac D, Radojevic L (2005) The effect of low temperature on germination of androgenic embryos of *Aesculus hippocastanum* L. Biol Plant 49(3):431–433

11. Radojevic L, Marinkovic N, Jevremovic S, Calic D (1999) Plant regeneration from uninuclear microspore suspension cultures of *Aesculus hippocastanum* L. In: Plant biotechnology and in vitro biology in the 21st century. Springer, New York, NY, pp 201–204

12. Calic D, Zdravkovic-Korac S, Radojevic L (2005) Secondary embryogenesis in androgenic embryo cultures of *Aesculus hippocastanum* L. Biol Plant 49(3):435–438

13. Marburger J, Wang R-C (1988) Anther culture of some perennial triticeae. Plant Cell Rep 7(5):313–317

14. Chekurov V, Razmakhnin E (1999) Effect of inbreeding and growth regulators on the in vitro androgenesis of wheatgrass, *Agropyron glaucum*. Plant Breed 118 (6):571–573

15. Gharyal P, Rashid A, Maheshwari S (1983) Production of haploid plantlets in anther cultures of *Albizzia lebbeck* L. Plant Cell Rep 2(6):308–309

16. Gémes-Juhasz A, Venczel G, Sagi ZS, Gajdos L, Kristof Z, Vagi P, Zatyko L (2006) Production of doubled haploid breeding lines in case of paprika, spice paprika, eggplant, cucumber, zucchini and onion. Acta Hortic 725:845–854

17. Jakše M, Hirschegger P, Bohanec B, Havey MJ (2010) Evaluation of gynogenic responsiveness and pollen viability of selfed doubled haploid onion lines and chromosome doubling via somatic regeneration. J Am Soc Hortic Sci 135(1):67–73

18. Bohanec B, Jakse M, Havey MJ (2003) Genetic analyses of gynogenetic haploid production in onion. J Am Soc Hortic Sci 128 (4):571–574

19. Bohanec B (2002) Doubled-haploid onions. In: Rabinowitch HD, Currah L (eds) Allium crop science: recent advances. CABI Publishing, Wallingford, pp 145–157

20. Grzebelus E, Adamus A (2004) Effect of anti-mitotic agents on development and genome doubling of gynogenic onion (*Allium cepa* L.) embryos. Plant Sci 167 (3):569–574. https://doi.org/10.1016/j.plantsci.2004.05.001

21. Havey MJ (2007) Onion inbred line 'b8667 a&b' and synthetic populations 'Sapporo-Ki-1 A&B' and 'onion haploid-1'. HortScience 42(7):1731–1732

22. Campion B, Azzimonti MT, Vicini E, Schiavi M, Falavigna A (1992) Advances in haploid plant induction in onion (*Allium cepa* L.) through in vitro gynogenesis. Plant Sci 86(1):97–104. https://doi.org/10.1016/0168-9452(92)90183-M

23. Dunstan DI, Short KC (1977) Improved growth of tissue cultures of the onion, *Allium cepa*. Physiol Plant 41(1):70–72. https://doi.org/10.1111/j.1399-3054.1977.tb01525.x

24. Ponce MT (2007) Ginogénesis en cebolla. Adv Hortic 5:1–12

25. Foschi ML, Martínez L, Ponce MT, Galmarini CR (2009) Doblehaploides, una estrategia biotecnológica para el mejoramiento genético en cebolla (*Allium cepa*). Hortic Argentina 28(66):40–47

26. Alan AR, Lim W, Mutschler MA, Earle ED (2007) Complementary strategies for ploidy manipulations in gynogenic onion (*Allium cepa L.*). Plant Sci 173(1):25–31. https://doi.org/10.1016/j.plantsci.2007.03.010

27. Geoffriau E, Kahane R, Bellamy C, Rancillac M (1997) Ploidy stability and in vitro chromosome doubling in gynogenic clones of onion (*Allium cepa* L.). Plant Sci 122 (2):201–208. https://doi.org/10.1016/S0168-9452(96)04556-6

28. Michalik B, Adamus A, Nowak E (2000) Gynogenesis in Polish onion cultivars. J Plant Physiol 156(2):211–216. https://doi.org/10.1016/s0176-1617(00)80308-9

29. Martínez L (2003) In vitro gynogenesis induction and doubled haploid production in onion (*Allium cepa L.*). In: Maluszynski M, Kasha KJ, Forster BP, Szarejko I (eds) Doubled haploid production in crop plants: a manual. Springer, Dordrecht, pp 275–279. https://doi.org/10.1007/978-94-017-1293-4_40

30. Fayos O, Vallés MP, Garcés-Claver A, Mallor C, Castillo AM (2015) Doubled haploid production from Spanish onion (*Allium cepa* L.) germplasm: embryogenesis induction, plant regeneration and chromosome doubling. Front Plant Sci 6. https://doi.org/10.3389/fpls.2015.00384

31. Jakše M, Havey MJ, Bohanec B (2003) Chromosome doubling procedures of onion (*Allium cepa* L.) gynogenic embryos. Plant Cell Rep 21(9):905–910

32. Alan AR, Mutschler MA, Brants A, Cobb E, Earle ED (2003) Production of gynogenic plants from hybrids of *Allium cepa L.* and *A. roylei* Stearn. Plant Sci 165 (6):1201–1211

33. Alan AR, Brants A, Cobb E, Goldschmied PA, Mutschler MA, Earle ED (2004) Fecund gynogenic lines from onion (*Allium cepa L.*) breeding materials. Plant Sci 167 (5):1055–1066

34. Dore C, Marie F (1993) Production of gynogenetic plants of onion (*Allium cepa* L.) after crossing with irradiated pollen. Plant Breed 111(2):142–147

35. Jakse M, Bohanec B (2003) Haploid induction in onion via gynogenesis. In: MMe (ed) Doubled haploid producion in crop plants. Springer, Dordrecht

36. Sulistyaningsih E, Aoyagi Y, Tashiro Y (2006) Flower bud culture of shallot (*Allium cepa L.* Aggregatum group) with cytogenetic analysis of resulting gynogenic plants and somaclones. Plant Cell Tissue Organ Cult 86(2):249–255

37. Cho KS, Hong SY, Yun BK, Kwon YS, Huh EJ (2006) Production and analysis of doubled haploid lines in long-day onion (*Allium cepa*) through in vitro gynogenesis. Hortic Environ Biotechnol 47(3):110–116

38. Ibrahim AM, Kayat F, Susanto D, Kashiani P, Arifullah M (2016) Haploid induction in spring onion (*Allium fistulosum* L.) via gynogenesis. Biotechnology 15(1):10–16

39. Inagaki N, Matsunaga H, Kanechi M, Maekawa S (1994) In vitro micropropagation of *Allium giganteum* R. 2: Embryoid and plantlet regeneration through the anther culture of *Allium giganteum* R. Science Reports of Faculty of Agriculture Kobe University

40. Suh S, Park H (1986) Studies on the anther culture of garlic (*Allium sativum* L.). 1. Callus formation and plant regeneration. J Korean Soc Hortic Sci 27:89–95

41. Kim CK, Oh JY, Chung JD (1998) Plant regeneration of Korean native Chinese chive by unpollinated ovule culture. J Korean Soc Hortic Sci 39(6):693–696

42. Ewais EA, Ismail MA, Amin MA, Abd-El-moety ES (2019) Phytochemical contents of white and pink flowers of marshmallow (*Althaea officinalis* L) plants and their androgenesis potential on anther culture in response to chemical elicitors. Biosci Res 16 (2):1276–1289

43. Ferrie AMR, Bethune TD, Mykytyshyn M (2011) Microspore embryogenesis in Apiaceae. Plant Cell Tissue Organ Cult 104 (3):399–406. https://doi.org/10.1007/s11240-010-9770-0

44. Ferrie AMR, Bethune T, Kernan Z (2005) An overview of preliminary studies on the development of doubled haploid protocols for nutraceutical species. Acta Physiol Plant 27(4B):735–741

45. Fialho JS, Bueno DM, Júnior AT, daSilveira Carvalho P (2005) Methodology

development to obtain cashwe tree haploids (*Anacardium occidentale* L.) through rising of anthers. Rev Ciênc Agron 36(2):195

46. Benega R, Isidrón M, Arias E, Cisneros A, Martínez J, Companioni L, Borroto CG (1997) Plant regeneration from pineapple ovules (*Ananas comosus* L. Merr.). International Society for Horticultural Science (ISHS), Leuven, pp 247–250. https://doi.org/10.17660/ActaHortic.1997.425.27

47. Johansson LB, Calleberg E, Gedin A (1990) Correlations between activated-charcoal, Fe-EDTA and other organic media ingredients in cultured anthers of *Anemone canadensis*. Physiol Plant 80(2):243–249

48. Ari E, Buyukalaca S, Abak K, Cetiner S (2007) Callus initiation for indirect pollen embryogenesis in *Anemone coronaria*. Proceedings of the 22nd international eucarpia symposium section ornamentals: breeding for beauty Pt II 743:87–90

49. Nair S, Gupta PK, Mascarenhas AF (1983) Haploid plants from *in vitro* anther culture of *Annona squamosa* Linn. Plant Cell Rep 2 (4):198–200. https://doi.org/10.1007/bf00270103

50. Ferrie AMR, Bethune TD, Arganosa GC, Waterer D (2011) Field evaluation of doubled haploid plants in the Apiaceae: dill (*Anethum graveolens* L.), caraway (*Carum carvi* L.), and fennel (*Foeniculum vulgare* Mill.). Plant Cell Tissue Organ Cult 104 (3):407–413. https://doi.org/10.1007/s11240-010-9821-6

51. Sharma R, Babber S (1990) In vitro studies of anther culture of *Antirrhinum majus*. Ann Biol (Ludhiana) 6(2):175–178

52. Dohya N, Matsubara S, Murakami K (1997) Callus formation and regeneration of adventitious embryos from celery microspores by anther and isolated microspore cultures. J Jpn Soc Hortic Sci 65(4):747–752

53. Amos JA, Scholl RL (1978) Induction of haploid callus from anthers of four species of *Arabidopsis*. Z Pflanzenphysiol 90 (1):33–43

54. Ravi M, Chan SWL (2010) Haploid plants produced by centromere-mediated genome elimination. Nature 464(7288):615–618. https://doi.org/10.1038/nature08842

55. Ravi M, Marimuthu MP, Tan EH, Maheshwari S, Henry IM, Marin-Rodriguez B, Urtecho G, Tan J, Thornhill K, Zhu F, Panoli A, Sundaresan V, Britt AB, Comai L, Chan SW (2014) A haploid genetics toolbox for *Arabidopsis thaliana*. Nat Commun 5:5334. https://doi.org/10.1038/ncomms6334

56. Ravi M, Bondada R (2016) Genome elimination by tailswap CenH3: in vivo haploid production in *Arabidopsis thaliana*. In: Murata M (ed) Chromosome and genomic engineering in plants: methods and protocols. Springer, New York, NY, pp 77–99. https://doi.org/10.1007/978-1-4939-4931-1_6

57. Avetisov V (1976) Production of haploids during in vitro culturing of *Arabidopsis thaliana* (L.) Heynh. anthers and isolated protoplasts. Genetika (USSR) 12:17–25

58. Avetisov V (1976) Production of haploids during in vitro culturing of Arabidopsis thaliana (L.) Heynh. anthers and isolated protoplasts. Genetika (USSR) 12:17–25

59. Karimi-Ashtiyani R, Ishii T, Niessen M, Stein N, Heckmann S, Gurushidze M, Banaei-Moghaddam AM, Fuchs J, Schubert V, Koch K, Weiss O, Demidov D, Schmidt K, Kumlehn J, Houben A (2015) Point mutation impairs centromeric CENH3 loading and induces haploid plants. Proc Natl Acad Sci 112(36):11211–11216. https://doi.org/10.1073/pnas.1504333112

60. Gresshoff PM, Doy CH (1972) Haploid *Arabidopsis thaliana* callus and plants from anther culture. Aust J Biol Sci 25(2):259

61. Kelliher T, Starr D, Su X, Tang G, Chen Z, Carter J, Wittich PE, Dong S, Green J, Burch E, McCuiston J, Gu W, Sun Y, Strebe T, Roberts J, Bate NJ, Que Q (2019) One-step genome editing of elite crop germplasm during haploid induction. Nat Biotechnol 37(3):287–292. https://doi.org/10.1038/s41587-019-0038-x

62. Kuppu S, Ron M, Marimuthu MPA, Li G, Huddleson A, Siddeek MH, Terry J, Buchner R, Shabek N, Comai L, Britt AB (2020) A variety of changes, including CRISPR/Cas9-mediated deletions, in CENH3 lead to haploid induction on outcrossing. Plant Biotechnol J. https://doi.org/10.1111/pbi.13365

63. Seymour DK, Filiault DL, Henry IM, Monson-Miller J, Ravi M, Pang A, Comai L, Chan SW, Maloof JN (2012) Rapid creation of *Arabidopsis* doubled haploid lines for quantitative trait locus mapping. Proc Natl Acad Sci 109 (11):4227–4232

64. Bajaj Y, Ram A, Labana K, Singh H (1981) Regeneration of genetically variable plants from the anther-derived callus of *Arachis*

hypogaea and *Arachis villosa*. Plant Sci Lett 23(1):35–39

65. Bansal U, Bassi G, Gosal S, Satija D (1991) Induction of pollen embryogenesis and cytological variability in *Arachis hypogaea* L. through anther culture. Ind J Genet 51:125–129

66. Willcox MC, Reed SM, Burns JA, Wynne JC (1990) Microsporogenesis in peanut (*Arachis hypogaea*). Am J Bot 77 (10):1257–1259. https://doi.org/10. 1002/j.1537-2197.1990.tb11377.x

67. Lee J-K, Yeh M-S (2001) Studies on the anther culture of peanut IV. Pollen development, Somatic embryogenesis and shoot regeneration from anther culture in *Arachis hypogaea* L. J Agric Forest Taichung 50 (2):65–79

68. Falavigna A, Casali PE, Valente MT (2012) Recent progress of asparagus breeding in Italy. International Society for Horticultural Science (ISHS), Leuven, pp 133–142. https://doi.org/10.17660/ActaHortic. 2012.950.14

69. Delaitre C, Ochatt S, Deleury E (2001) Electroporation modulates the embryogenic responses of asparagus (*Asparagus officinalis* L.) microspores. Protoplasma 216 (1):39–46. https://doi.org/10.1007/ bf02680129

70. Kenny L, Caligari P (1996) Androgenesis of the salt tolerant shrub *Atriplex glauca*. Plant Cell Rep 15(11):829–832

71. Bajaj YPS (1978) Effect of super-low temperature on excised anthers and pollen-embryos of *Atropa*. Phytomorphology 28 (2):171–176

72. Mazzolani G, Pasqua G, Monacelli B (1981) Condizioni per la formazione di piante aploidi da pollini coltivati *in vitro* [*Nicotiana tabacum* e *Atropa belladonna*]. Ann Bot 38 (2):107–117

73. Zenkteler M (1971) In vitro production of haploid plants from pollen grains of *Atropa belladonna* L. Experientia 27 (9):1087–1087. https://doi.org/10.1007/ bf02138897

74. Kiviharju E, Moisander S, Tanhuanpää P (2017) Oat anther culture and use of DH-lines for genetic mapping. In: Gasparis S (ed) Oat Methods Protoc. Springer New York, New York, NY, pp 71–93. https://doi.org/10.1007/978-1-4939-6682-0_6

75. Sidhu PK, Davies PA (2009) Regeneration of fertile green plants from oat isolated microspore culture. Plant Cell Rep 28

(4):571–577. https://doi.org/10.1007/ s00299-009-0684-4

76. Sidhu PK, Howes NK, Aung T, Zwer PK, Davies PA (2006) Factors affecting oat haploid production following oat × maize hybridization. Plant Breed 125:243–247

77. Warchoł M, Czyczyło-Mysza I, Marcińska I, Dziurka K, Noga A, Kapłoniak K, Pilipowicz M, Skrzypek E (2019) Factors inducing regeneration response in oat (Avena sativa L.) anther culture. In Vitro Cell Dev Biol Plant 55(5):595–604. https://doi.org/10.1007/s11627-019-09987-1

78. Warchoł M, Czyczyło-Mysza I, Marcińska I, Dziurka K, Noga A, Skrzypek E (2018) The effect of genotype, media composition, pH and sugar concentrations on oat (*Avena sativa* L.) doubled haploid production through oat × maize crosses. Acta Physiol Plant 40(5). https://doi.org/10.1007/ s11738-018-2669-9

79. Ferrie A, Irmen K, Beattie A, Rossnagel B (2014) Isolated microspore culture of oat (*Avena sativa* L.) for the production of doubled haploids: effect of pre-culture and post-culture conditions. Plant Cell Tissue Organ Cult 116(1):89–96

80. Marcińska I, Nowakowska A, Skrzypek E, Czyczyło-Mysza I (2013) Production of double haploids in oat (*Avena sativa* L.) by pollination with maize (*Zea mays* L.). Centr Eur J Biol 8(3):306–313

81. Warchol M, Czyczylo-Mysza I, Marcinska I, Dziurka K, Noga A, Kaploniak K, Pilipowicz M, Skrzypek E (2019) Factors inducing regeneration response in oat (*Avena sativa* L.) anther culture. In Vitro Cell Dev Biol Plant 55(5):595–604. https://doi.org/10.1007/s11627-019-09987-1

82. Dziurka K, Dziurka M, Warchol M, Czyczylo-Mysza I, Marcinska I, Noga A, Kaploniak K, Skrzypek E (2019) Endogenous phytohormone profile during oat (*Avena sativa* L.) haploid embryo development. In Vitro Cell Dev Biol Plant 55 (2):221–229. https://doi.org/10.1007/ s11627-019-09967-5

83. Noga A, Skrzypek E, Warchol M, Czyczylo-Mysza I, Dziurka K, Marcinska I, Juzon K, Warzecha T, Sutkowska A, Nita Z, Werwinska K (2016) Conversion of oat (*Avena sativa* L.) haploid embryos into plants in relation to embryo developmental stage and regeneration media. In Vitro Cell Dev Biol Plant 52(6):590–597. https://doi.org/10. 1007/s11627-016-9788-z

84. Skrzypek E, Warchol M, Czyczylo-Mysza I, Marcinska I, Nowakowska A, Dziurka K, Juzon K, Noga A (2016) The effect of light intensity on the production of oat (*Avena sativa* L.) doubled haploids through oat × maize crosses. Cereal Res Commun 44 (3):490–500. https://doi.org/10.1556/0806.44.2016.007

85. Nowakowska A, Skrzypek E, Marcinska I, Czyczylo-Mysza I, Dziurka K, Juzon K, Cyganek K, Warchol M (2015) Application of chosen factors in the wide crossing method for the production of oat doubled haploids. Open Life Sci 10(1):112–118. https://doi.org/10.1515/biol-2015-0014

86. Kiviharju E, Puolimatka M, Pehu E (1997) Regeneration of anther-derived plants of *Avena sterilis*. Plant Cell Tissue Organ Cult 48(2):147–152

87. Kiviharju E, Pehu E (1998) The effect of cold and heat pretreatments on anther culture response of *Avena sativa* and *A. sterilis*. Plant Cell Tissue Organ Cult 54(2):97–104

88. Chaturvedi R, Razdan M, Bhojwani S (2003) Production of haploids of neem (*Azadirachta indica* A. Juss.) by anther culture. Plant Cell Rep 21(6):531–537

89. Khoder M, Villemur P, Jonard R (1984) Obtainment of monoploid and triploid plants by in vitro androgenesis in *Begonia* × *hiemalis* Fotsch cv. (A). Bull Soc Bot Fr Lett Bot 131:43–48

90. Eujayl I, Strausbaugh C, Lu C (2016) Registration of Sugarbeet doubled haploid line KDH13 with resistance to beet curly top. J Plant Registr 10(1):93–96. https://doi.org/10.3198/jpr2015.09.0055crgs

91. Pazuki A, Aflaki F, Gürel S, Ergül A, Gürel E (2018) Production of doubled haploids in sugar beet (*Beta vulgaris*): an efficient method by a multivariate experiment. Plant Cell Tissue Organ Cult 132(1):85–97. https://doi.org/10.1007/s11240-017-1313-5

92. Pazuki A, Aflaki F, Gürel E, Ergül A, Gürel S (2018) Gynogenesis induction in sugar beet (*Beta vulgaris*) improved by 6-benzylaminopurine (BAP) and synergized with cold pretreatment. Sugar Tech 20 (1):69–77

93. Levites E, Svirshchevskaya A, Kirikovich S, Mil'ko L (2005) Variation at isozyme loci in culturedin vitro sugar beet regenerants of gynogenetic origin. Sugar Tech 7(1):71–75

94. Pazuki A, Aflaki F, GÜREL S, ERGÜL A, GÜREL E (2018) The effects of proline on in vitro proliferation and propagation of doubled haploid sugar beet (*Beta vulgaris*). Turk J Bot 42(3):280–288

95. Aflaki F, Pazuki A, Gurel S, Stevanato P, Biancardi E, Gurel E (2017) Doubled haploid sugar beet: an integrated view of factors influencing the processes of gynogenesis and chromosome doubling. Int Sugar J 119 (1427):884–895

96. Huhtinen O (1978) Callus and plantlet regeneration from anther culture of *Betula pendula* Roth. 4th Int Cong Plant Tissue Cell Culture 1978:20–25

97. Kato J, Ichihashi S (2018) Haploid seed formation via parthenogenesis in Bletilla. In: Lee Y-I, Yeung EC-T (eds) Orchid propagation: from laboratories to greenhouses—methods and protocols. Springer New York, New York, NY, pp 303–315. https://doi.org/10.1007/978-1-4939-7771-0_16

98. Hoveida ZS, Abdollahi MR, Mirzaie-Asl A, Moosavi SS, Seguí-Simarro JM (2017) Production of doubled haploid plants from anther cultures of borage (*Borago officinalis* L.) by the application of chemical and physical stress. Plant Cell Tissue Organ Cult 130 (2):369–378. https://doi.org/10.1007/s11240-017-1233-4

99. Chardoli Eshaghi Z, Abdollahi MR, Moosavi SS, Deljou A, Seguí-Simarro JM (2015) Induction of androgenesis and production of haploid embryos in anther cultures of borage (*Borago officinalis* L.). Plant Cell Tissue Organ Cult 2015:1–9. https://doi.org/10.1007/s11240-015-0768-5

100. Prakash DVSSR, Chand S, Kishor PBK (1999) In vitro response from cultured anthers of *Boswellia serrata* Roxb. In: Plant tissue culture and biotechnology: emerging trends. Universities Press Ltd, Hyderabad, pp 226–231

101. Niu L, Shi F, Feng H, Zhang Y (2019) Efficient doubled haploid production in microspore culture of Zengcheng flowering Chinese cabbage (*Brassica campestris* L. ssp. *chinensis* [L.] Makino var. utilis Tsen et Lee). Sci Hortic 245:57–64. https://doi.org/10.1016/j.scienta.2018.09.076

102. Guo YD, Pulli S (1995) *In vitro* pollen culture and the regeneration of *Brassica campestris* L plants. Agric Sci Finl 4 (5–6):513–518

103. Aslam FN, Macdonald MV, Loudon P, Ingram DS (1990) Rapid-cycling Brassica species – inbreeding and selection of *Brassica campestris* for anther culture ability. Ann Bot 65(5):557–566

104. Aslam FN, Macdonald MV, Ingram DS (1990) Rapid-cycling Brassica species – anther culture potential of *Brassica campestris* L and *Brassica napus* L. New Phytol 115 (1):1–9

105. Gao Y, Jia J, Cong J, Ma Y, Feng H, Zhang Y (2019) Non-ionic surfactants improved microspore embryogenesis and plant regeneration of recalcitrant purple flowering stalk (*Brassica campestri*s ssp. *chinensis* var. *purpurea* Bailey). In Vitro Cell Dev Biol Plant. https://doi.org/10.1007/s11627-019-10033-3

106. Ferrie AMR, Keller WA (2007) Optimization of methods for using polyethylene glycol as a non-permeating osmoticum for the induction of microspore embryogenesis in the Brassicaceae. In Vitro Cell Dev Biol Plant 43(4):348–355. https://doi.org/10.1007/s11627-007-9053-6

107. Ferrie AMR, Dirpaul J, Krishna P, Krochko J, Keller WA (2005) Effects of brassinosteroids on microspore embryogenesis in *Brassica* species. In Vitro Cell Dev Biol Plant 41(6):742–745

108. Yadav R, Sareen P, Chowdhury J (1988) High frequency induction of androgenesis in Ethiopian mustard (*Brassica carinata* A. Br.). Cruciferae Newsl 13:77

109. Chuong PV, Beversdorf WD (1985) High frecuency embryogenesis through isolated microspore culture in *Brassica napus* L. and *B. carinata* Braun. Plant Sci 39:219–226

110. Yazdi EJ, Falk KC, Séguin-Swartz G (2013) Improvement in efficiency of microspore culture to produce doubled haploid lines of Ethiopian mustard. In Vitro Cell Dev Biol Plant 49(6):682–689

111. Geng X, Chen S, Astarini I, Yan G, Tian E, Meng J, Li Z, Ge X, Nelson M, Mason A (2013) Doubled haploids of novel trigenomic *Brassica* derived from various interspecific crosses. Plant Cell Tissue Organ Cult 113(3):501–511

112. Agarwal PK, Agarwal P, Custers JBM, Liu CM, Bhojwani SS (2006) PCIB an antiauxin enhances microspore embryogenesis in microspore culture of *Brassica juncea*. Plant Cell Tissue Organ Cult 86(2):201–210

113. Prem D, Gupta K, Agnihotri A (2005) Effect of various exogenous and endogenous factors on microspore embryogenesis in Indian mustard (*Brassica juncea* (L.) Czern and Coss). In Vitro Cell Dev Biol Plant 41 (3):266–273

114. Chanana NP, Dhawan V, Bhojwani SS (2005) Morphogenesis in isolated microspore cultures of *Brassica juncea*. Plant Cell Tissue Organ Cult 83 (2):169–177

115. Lionneton E, Beuret W, Delaitre C, Ochatt S, Rancillac M (2001) Improved microspore culture and doubled-haploid plant regeneration in the brown condiment mustard (*Brassica juncea*). Plant Cell Rep 20 (2):126–130

116. Liu C, Xu ZH, Chua NH (1993) Proembryo culture: in vitro development of early globular-stage zygotic embryos from *Brassica juncea*. Plant J 3(2):291–300

117. Prem D, Gupta K, Sarkar G, Agnihotri A (2008) Activated charcoal induced high frequency microspore embryogenesis and efficient doubled haploid production in *Brassica juncea*. Plant Cell Tissue Organ Cult 93 (3):269–282. https://doi.org/10.1007/s11240-008-9373-1

118. Ahmadi B, Shariatpanahi M, Ojaghkandi M, Heydari A (2014) Improved microspore embryogenesis induction and plantlet regeneration using putrescine, cefotaxime and vancomycin in *Brassica napus* L. Plant Cell Tissue Organ Cult 118(3):497–505. https://doi.org/10.1007/s11240-014-0501-9

119. Prem D, Solís MT, Bárány I, Rodríguez-Sanz H, Risueño MC, Testillano PS (2012) A new microspore embryogenesis system under low temperature which mimics zygotic embryogenesis initials, expresses auxin and efficiently regenerates doubled-haploid plants in *Brassica napus*. BMC Plant Biol 12:127

120. Custers JBM (2004) Efficient in vitro production of embryos with suspensors in a refined *Brassica napus* microspore culture procedure. In: Proceedings of the COST action 851. Technology advancement in gametic embryogenesis of recalcitrant genotypes, vol 1. Workshop of the Working Group, Palermo

121. Custers J (2003) Microspore culture in rapeseed (*Brassica napus* L.). In: Maluszynski M, Kasha KJ, Forster BP, Szarejko I (eds) Doubled haploid production in crop plants. Kluwer Academic, Dordrecht, pp 185–193

122. Custers JBM, Cordewener JHG, Fiers MA, Maassen BTH, van Lookeren-Campagne MM, Liu CM (2001) Androgenesis in *Brassica*: a model system to study the initiation of plant embryogenesis. In: Bhojwani SS, Soh WY (eds) Current trends in the embryology of angiosperm. Kluwer Academic, Dordrecht, pp 451–470

123. Binarova P, Hause G, Cenklova V, Cordew-ener JHG, van Lookeren-Campagne MM (1997) A short severe heat shock is required to induce embryogenesis in late bicellular pollen of *Brassica napus* L. Sex Plant Reprod 10(4):200–208

124. Zhao JP, Simmonds DH, Newcomb W (1996) Induction of embryogenesis with colchicine instead of heat in microspores of *Brassica napus* L cv Topas. Planta 198 (3):433–439

125. Hansen NJP, Andersen SB (1996) *In vitro* chromosome doubling potential of colchi-cine, oryzalin, trifluralin, and APM in *Brassica napus* microspore culture. Euphytica 88 (2):159–164

126. Zhao JP, Simmonds DH (1995) Application of trifluralin to embryogenic microspore cul-tures to generate doubled haploid plants in *Brassica napus*. Physiol Plant 95 (2):304–309

127. Simmonds DH, Long NE, Keller WA (1991) High plating efficiency and plant regeneration frequency in low density proto-plast cultures derived from an embryogenic *Brassica napus* cell suspension. Plant Cell Tissue Organ Cult 27(3):231–241. https://doi.org/10.1007/bf00157586

128. Lichter R (1982) Induction of haploid plants from isolated pollen of *Brassica napus*. Z Pflanzenphysiol 105(5):427–434

129. Lichter R (1981) Anther culture of *Brassica napus* in a liquid culture medium. Z Pflan-zenphysiol 103(3):229–237

130. Keller WA, Armstrong KC (1977) Embryo-genesis and plant regeneration in *Brassica napus* anther cultures. Can J Bot 55 (10):1383–1388. https://doi.org/10. 1139/b77-160

131. Fu SH, Yin LQ, Xu MC, Li Y, Wang ML, Yang J, Fu TD, Wang JS, Shen JX, Ali A, Zou Q, Yi B, Wen J, Tao LR, Kang ZM, Tang R (2018) Maternal doubled haploid production in interploidy hybridization between *Brassica napus* and *Brassica* allooc-taploids. Planta 247(1):113–125. https:// doi.org/10.1007/s00425-017-2772-y

132. Liu S, Wang H, Zhang J, Fitt BDL, Xu Z, Evans N, Liu Y, Yang W, Guo X (2005) In vitro mutation and selection of doubled-haploid *Brassica napus* lines with improved resistance to *Sclerotinia sclerotiorum*. Plant Cell Rep 24(3):133–144. https://doi.org/ 10.1007/s00299-005-0925-0

133. Mohammadi P, Moieni A, Ebrahimi A, Javidfar F (2012) Doubled haploid plants following colchicine treatment of microspore-derived embryos of oilseed rape (*Brassica napus* L.). Plant Cell Tissue Organ Cult 108(2):251–256. https://doi.org/10. 1007/s11240-011-0036-2

134. Ahuja I, Borgen BH, Hansen M, Honne BI, Muller C, Rohloff J, Rossiter JT, Bones AM (2011) Oilseed rape seeds with ablated defence cells of the glucosinolate-myrosinase system. Production and characteristics of double haploid *MINELESS* plants of *Brassica napus* L. J Exp Bot. https://doi.org/10. 1093/jxb/err195

135. Takahira J, Cousin A, Nelson MN, Cowling WA (2011) Improvement in efficiency of microspore culture to produce doubled hap-loid canola (*Brassica napus* L.) by flow cyto-metry. Plant Cell Tissue Organ Cult 104 (1):51–59. https://doi.org/10.1007/ s11240-010-9803-8

136. Weber S, ÜNker F, Friedt W (2005) Improved doubled haploid production pro-tocol for *Brassica napus* using microspore colchicine treatment in vitro and ploidy determination by flow cytometry. Plant Breed 124:511–513. https://doi.org/10. 1111/j.1439-0523.2005.01114.x

137. Nelson M, Mason A, Castello M-C, Thomson L, Yan G, Cowling W (2009) Microspore culture preferentially selects unreduced (2n) gametes from an interspe-cific hybrid of *Brassica napus* L. × *Brassica carinata* Braun. Theor Appl Genet 119 (3):497–505. https://doi.org/10.1007/ s00122-009-1056-8

138. Yang S, Chen S, Zhang KN, Li L, Yin YL, Gill RA, Yan GJ, Meng JL, Cowling WA, Zhou WJ (2018) A high-density genetic map of an allohexaploid *Brassica* doubled haploid population reveals quantitative trait loci for pollen viability and fertility. Front Plant Sci 9. https://doi.org/10.3389/fpls. 2018.01161

139. Mwathi MW, Schiessl SV, Batley J, Mason AS (2019) "Doubled-haploid" allohexa-ploid *Brassica* lines lose fertility and viability and accumulate genetic variation due to genomic instability. Chromosoma. https:// doi.org/10.1007/s00412-019-00720-w

140. Govil S, Babbar SB, Gupta SC (1986) Plant regeneration from *in vitro* cultured anthers of black mustard (*Brassica nigra* Koch). Plant Breed 97(1):64–71. https://doi.org/ 10.1111/j.1439-0523.1986.tb01302.x

141. Kurtar ES (2017) Anther culture in red cab-bage (*Brassica oleraceae* L. var. capitata sub-var. rubra): embryogenesis and plantlet initiation. Ekin J Crop Breed Genet 3 (2):82–87

142. Yuan S, Su Y, Liu Y, Li Z, Fang Z, Yang L, Zhuang M, Zhang Y, Lv H, Sun P (2015) Chromosome doubling of microspore-derived plants from cabbage (*Brassica oleracea* var. capitata L.) and broccoli (*Brassica oleracea* var. *italica* L.). Front Plant Sci 6. https://doi.org/10.3389/fpls.2015. 01118

143. da Silva Dias JC (1999) Effect of activated charcoal on *Brassica oleracea* microspore culture embryogenesis. Euphytica 108 (1):65–69. https://doi.org/10.1023/ a:1003634030835

144. Chen W, Zhang Y, Ren J, Ma Y, Liu Z, Hui F (2019) Effects of methylene blue on microspore embryogenesis and plant regeneration in ornamental kale (*Brassica oleracea* var. *acephala*). Sci Hortic 248:1–7. https://doi. org/10.1016/j.scienta.2018.12.048

145. Gu H, Zhao Z, Sheng X, Yu H, Wang J (2014) Efficient doubled haploid production in microspore culture of loose-curd cauliflower (*Brassica oleracea* var. botrytis). Euphytica 195(3):467–475

146. Sato S, Katoh N, Iwai S, Hagimori M (2005) Frequency of spontaneous polyploidization of embryos regenerated from cultured anthers or microspores of *Brassica rapa* var. *pekinensis* L. and *B. oleracea* var. *capitata* L. Breed Sci 55(1):99–102

147. Takahata Y, Keller WA (1991) High frequency embryogenesis and plant regeneration in isolated microspore culture of *Brassica oleracea* L. Plant Sci 74(2):235–242

148. Mousa MA, Haridy AG, Abbas HS, Mohammed MF (2014) Improved androgenesis of broccoli (*Brassica oleracea* var italica) anthers using sucrose and growth regulators. Asian J Crop Sci 6(2):133–141

149. Stipic M, Campion B (1997) An improved protocol for androgenesis in cauliflowers (*Brassica oleracea* var. botrytis). Plant Breed 116(2):153–157

150. Pilih KR, Potokar UK, Bohanec B (2018) Improvements of doubled haploid production protocol for white cabbage (*Brassica oleracea* var. capitata L.). Folia Hortic 30 (1):57–66

151. Arnison PG, Donaldson P, Ho LCC, Keller WA (1990) The influence of various physical parameters on anther culture of broccoli (*Brassica oleracea* var. italica). Plant Cell Tissue Organ Cult 20(3):147–155. https:// doi.org/10.1007/bf00041875

152. Rudolf K, Bohanec B, Hansen M (1999) Microspore culture of white cabbage, *Brassica oleracea* var. capitata L.: genetic improvement of non-responsive cultivars and effect of genome doubling agents. Plant Breed 118(3):237–241

153. Zeng A, Yan Y, Yan J, Song L, Gao B, Li J, Hou X, Li Y (2015) Microspore embryogenesis and plant regeneration in Brussels sprouts (*Brassica oleracea* L. var. gemmifera). Sci Hortic 191:31–37. https://doi. org/10.1016/j.scienta.2015.05.002

154. Bhatia R, Dey SS, Parkash C, Sharma K, Sood S, Kumar R (2018) Modification of important factors for efficient microspore embryogenesis and doubled haploid production in field grown white cabbage (*Brassica oleracea* var. capitata L.) genotypes in India. Sci Hortic 233:178–187. https:// doi.org/10.1016/j.scienta.2018.01.017

155. Bhatia R, Dey SS, Sood S, Sharma K, Sharma VK, Parkash C, Kumar R (2016) Optimizing protocol for efficient microspore embryogenesis and doubled haploid development in different maturity groups of cauliflower (*B. oleracea* var. botrytis L.) in India. Euphytica 212(3):439–454. https://doi.org/10. 1007/s10681-016-1775-2

156. Osolnik B, Bohanec B, Jelaska S (1993) Stimulation of androgenesis in white cabbage (*Brassica oleracea* var. *capitata*) anthers by low temperature and anther dissection. Plant Cell Tissue Organ Cult 32 (2):241–246. https://doi.org/10.1007/ bf00029849

157. Arnison PG, Keller WA (1990) A survey of the anther culture response of *Brassica oleracea* L. cultivars grown under field conditions. Plant Breed 104(2):125–133. https://doi.org/10.1111/j.1439-0523. 1990.tb00414.x

158. Li Q, Shi YT, Wang Y, Liu LJ, Zhang XS, Chen XW, Zhang LZ, Su YB, Zhang TZ (2020) Breeding of cabbage lines resistant to both head splitting and fusarium wilt via an isolated microspore culture system and marker-assisted selection. Euphytica 216 (2). https://doi.org/10.1007/s10681-020-2570-7

159. Singh S, Bhatia R, Kumar R, Sharma K, Dash S, Dey SS (2018) Cytoplasmic male sterile and doubled haploid lines with desirable combining ability enhances the concentration of important antioxidant attributes in *Brassica oleracea*. Euphytica 214(11). https://doi.org/10.1007/s10681-018-2291-3

160. Bhatia R, Dey SS, Sood S, Sharma K, Parkash C, Kumar R (2017) Efficient microspore embryogenesis in cauliflower (*Brassica oleracea* var. botrytis L.) for development of

plants with different ploidy level and their use in breeding programme. Sci Hortic 216:83–92. https://doi.org/10.1016/j.scienta.2016.12.020

161. Huang SN, Liu ZY, Li DY, Yao RP, Feng H (2016) A new method for generation and screening of Chinese cabbage mutants using isolated microspore culturing and EMS mutagenesis. Euphytica 207 (1):23–33. https://doi.org/10.1007/s10681-015-1473-5

162. Na H, Hwang G, Kwak J-H, Yoon MK, Chun C (2011) Microspore derived embryo formation and doubled haploid plant production in broccoli (Brassica oleracea L. var italica) according to nutritional and environmental conditions. Afr J Biotechnol 10 (59):12535–12541

163. Burnett L, Yarrow S, Huang B (1992) Embryogenesis and plant regeneration from isolated microspores of Brassica rapa L ssp Oleifera. Plant Cell Rep 11 (4):215–218

164. Cao MQ, Li Y, Liu F, Doré C (1994) Embryogenesis and plant regeneration of pakchoi (Brassica rapa L. ssp. chinensis) via in vitro isolated microspore culture. Plant Cell Rep 13(8):447–450

165. Ferrie AMR, Epp DJ, Keller WA (1995) Evaluation of Brassica rapa L. genotypes for microspore culture response and identification of a highly embryogenic line. Plant Cell Rep 14(9):580–584

166. Gu HH, Zhou WJ, Hagberg P (2003) High frequency spontaneous production of doubled haploid plants in microspore cultures of Brassica rapa ssp chinensis. Euphytica 134 (3):239–245

167. Jia J, Zhang Y, Feng H (2019) Effects of brassinolide on microspore embryogenesis and plantlet regeneration in pakchoi (Brassica rapa var. multiceps). Sci Hortic 252:354–362. https://doi.org/10.1016/j.scienta.2019.04.004

168. Kitashiba H, Taguchi K, Kaneko I, Inaba K, Yokoi S, Takahata Y, Nishio T (2016) Identification of loci associated with embryo yield in microspore culture of Brassica rapa by segregation distortion analysis. Plant Cell Rep 2016:1–8. https://doi.org/10.1007/s00299-016-2029-4

169. Shumilina DV, Shmykova NA, Bondareva LL, Suprunova TP (2015) Effect of genotype and medium culture content on microspore-derived embryo formation in Chinese cabbage (Brassica rapa ssp. chinensis) cv. Lastochka. Biol Bull 42(4):302–309.

https://doi.org/10.1134/s1062359015040135

170. Jo M, Ham I, Park M, Kim T, Lim Y, Lee E (2012) Seed production ability of doubled haploid plants through microspore culture in Chinese cabbage (Brassica rapa L. ssp. pekinensis) introduced from China. Korean J Hortic Sci Technol 30(5):573–578

171. Shumilina D, Kornyukhin D, Domblides E, Soldatenko A, Artemyeva A (2020) Effects of genotype and culture conditions on microspore embryogenesis and plant regeneration in Brassica Rapa ssp. Rapa L. Plants (Basel) 9(2):278. https://doi.org/10.3390/plants9020278

172. Park S, 장석우 최 (2020) Developing double-haploid inbred lines of 'Wonkyo20051ho' Kimchi Cabbage (Brassica rapa. L) characterized by formation of tight head at low temperatures (저온에서도결구가 잘 형성되 배가 반체 배추 '원교20051호' 개발). Korean J Breed Sci 52(1):41–52. https://doi.org/10.9787/kjbs.2020.52.1.41

173. Lu Y, Dai S, Gu A, Liu M, Wang Y, Luo S, Zhao Y, Wang S, Xuan S, Chen X, Li X, Bonnema G, Zhao J, Shen S (2016) Microspore induced doubled haploids production from ethyl methanesulfonate (EMS) soaked flower buds is an efficient strategy for mutagenesis in Chinese cabbage. Front Plant Sci 7(1780). https://doi.org/10.3389/fpls.2016.01780

174. Zhang L, Zhang Y, Gao Y, Jiang XL, Zhang MD, Wu H, Liu ZY, Feng H (2016) Effects of histone deacetylase inhibitors on microspore embryogenesis and plant regeneration in Pakchoi (Brassica rapa ssp. chinensis L.). Sci Hortic 209:61–66. https://doi.org/10.1016/j.scienta.2016.05.001

175. Zhang Y, Wang A, Liu Y, Wang Y, Feng H (2012) Improved production of doubled haploids in Brassica rapa through microspore culture. Plant Breed 131(1):164–169

176. Lee S, Yoon Y (1987) Anther culture of Brassicoraphanus. Cruciferae Newsl 12:68

177. Shon T-K, Kim S-K, Acquah D, Lee S-C (2004) Haploid plantlet production through somatic embryogenesis in anther-derived callus of Bupleurum falcatum. Plant Prod Sci 7(2):204–211

178. Shon T-K, Yoshida T (1997) Induction of haploid plantlets by anther culture of Bupleurum falcatum L. Jpn J Crop Sci 66 (1):137–138

179. Bajaj Y, Singh H, Gosal S (1980) Haploid embryogenesis in anther cultures of pigeon-

pea (*Cajanus cajan*). Theor Appl Genet 58 (3–4):157–159

180. Fougat R, Pathak A, Bharodia P (1992) Regeneration of haploid callus from anthers of pigeonpea. Gujarat Agric Univ Res J 17 (2):151–152

181. Kaur P, Bhalla J (1998) Regeneration of haploid plants from microspore culture of pigeonpea (*Cajanus cajan* L.). Indian J Exp Biol 36(7):736–738

182. Vishukumar U, Patil M, Nayak S (2000) Anther culture studies in pigeonpea. Karnataka J Agric Sci 13(1):16–19

183. Bajpai R, Chaturvedi R (2018) Haploid embryogenesis in tea. In: Jain S, Gupta P (eds) Step wise protocols for somatic embryogenesis of important woody plants. forestry sciences, vol 85. Springer, Cham, pp 349–368. https://doi.org/10.1007/978-3-319-79087-9_26

184. Pedroso MC, Pais S (1993) Regeneration from anthers of adult *Camellia japonica* L. In Vitro Cell Dev Biol Plant 29 (4):155–159

185. Pedroso MC, Pais MS (1997) Anther and microspore culture in *Camellia japonica*. In: Jain SM, Sopory SK, Veilleux RE (eds) In vitro haploid production in higher plants. Springer, Dordrecht, pp 89–107

186. Pedroso MC, Pais MS (1994) Induction of microspore embryogenesis in *Camellia japonica* cv. Elegans. Plant Cell Tissue Organ Cult 37(2):129–136

187. Raina S, Iyer R (1992) Multicell pollen proembryoid and callus formation in tea. J Plant Crops 9:100–104

188. Shimokado T, Murata T, Miyaji Y (1986) Formation of embryoid by anther culture of tea. Jpn J Breed 36(Suppl 2):282–283

189. Saha D, Bhattacharya N (1992) Stimulating effect of elevated temperature treatments on production of meristemoids from pollen callus of tea, *Camellia sinensis* (L.) O. Kuntze. Indian J Exp Biol 30(2):83–86

190. Seran TH, Hirimburegama K, Hirimburegama W, Shanmugarajah V (1999) Callus formation in anther culture of tea clones, *Camellia sinensis* (L.) Kuntze. J Natl Sci Found 27(3):165–175

191. Chen Z, Liao H (1982) Obtaining plantlet through anther culture of tea plants. Zhongguo Chaye 4:6–7

192. Rekha HR, Rakhi C (2013) Establishment of dedifferentiated callus of haploid origin from unfertilized ovaries of tea (*Camellia sinensis* (L.) O. Kuntze) as a potential source

of total phenolics and antioxidant activity. In Vitro Cell Dev Biol Plant 49(1):60–69

193. Cao H, Biswas MK, Lü Y, Amar MH, Tong Z, Xu Q, Xu J, Guo W, Deng X (2011) Doubled haploid callus lines of Valencia sweet orange recovered from anther culture. Plant Cell Tissue Organ Cult 104 (3):415–423

194. Heidari-Zefreh AA, Shariatpanahi ME, Mousavi A, Kalatejari S (2018) Enhancement of microspore embryogenesis induction and plantlet regeneration of sweet pepper (*Capsicum annuum* L.) using putrescine and ascorbic acid. Protoplasma. https://doi.org/10.1007/s00709-018-1268-3

195. Gémesné Juhász A, Kristóf Z (2016) Highly efficient genome doubling method for haploid paprika (*Capsicum annuum* L.) plants. Paper presented at the proceedings of XVIth EUCARPIA capsicum and eggplant working group meeting, Budapest

196. Ari E, Bedir H, Yildirim S, Yildirim T (2016) Androgenic responses of 64 ornamental pepper (*Capsicum annuum* L.) genotypes to shed-microspore culture in the autumn season. Turk J Biol 40(3):706–717

197. Barroso PA, Rego MM, Rego ER, Soares WS (2015) Embryogenesis in the anthers of different ornamental pepper (*Capsicum annuum* L.) genotypes. Genet Mol Res GMR 14(4):13349–13363. https://doi.org/10.4238/2015.October.26.32

198. Olszewska D, Kisiala A, Niklas-Nowak A, Nowaczyk P (2014) Study of *in vitro* anther culture in 23 selected genotypes of genus *Capsicum*. Turk J Biol 38:118–124. https://doi.org/10.3906/biy-1307-50

199. Parra-Vega V, Renau-Morata B, Sifres A, Seguí-Simarro JM (2013) Stress treatments and in vitro culture conditions influence microspore embryogenesis and growth of callus from anther walls of sweet pepper (*Capsicum annuum* L.). Plant Cell Tissue Organ Cult 112(3):353–360. https://doi.org/10.1007/s11240-012-0242-6

200. Parra-Vega V, González-García B, Seguí-Simarro JM (2013) Morphological markers to correlate bud and anther development with microsporogenesis and microgametogenesis in pepper (*Capsicum annuum* L.). Acta Physiol Plant 35 (2):627–633. https://doi.org/10.1007/s11738-012-1104-x

201. Kim M, Park E-J, An D, Lee Y (2013) High-quality embryo production and plant regeneration using a two-step culture system in isolated microspore cultures of hot pepper (*Capsicum annuum* L.). Plant Cell Tissue

Organ Cult 112(2):191–201. https://doi.org/10.1007/s11240-012-0222-x

202. Ochoa-Alejo N (2012) Anther culture of chilli pepper (*Capsicum* spp.). In: Loyola-Vargas VM, Ochoa-Alejo N (eds) Plant cell culture protocols. Methods in molecular biology, vol 877. Humana, New York, NY, pp 227–231. https://doi.org/10.1007/978-1-61779-818-4_17

203. Lantos C, Juhasz AG, Vagi P, Mihaly R, Kristof Z, Pauk J (2012) Androgenesis induction in microspore culture of sweet pepper (*Capsicum annuum* L.). Plant Biotechnol Rep 6(2):123–132. https://doi.org/10.1007/s11816-011-0205-0

204. Supena EDJ, Custers JBM (2011) Refinement of shed-microspore culture protocol to increase normal embryos production in hot pepper (*Capsicum annuum* L.). Sci Hortic 130(4):769–774. https://doi.org/10.1016/j.scienta.2011.08.037

205. Nowaczyk P, Olszewska D, Kisiała A (2009) Individual reaction of *Capsicum* F_2 hybrid genotypes in anther cultures. Euphytica 168(2):225–233. https://doi.org/10.1007/s10681-009-9909-4

206. Lantos C, Juhász A, Somogyi G, Ötvös K, Vági P, Mihály R, Kristóf Z, Somogyi N, Pauk J (2009) Improvement of isolated microspore culture of pepper (*Capsicum annuum* L.) via co-culture with ovary tissues of pepper or wheat. Plant Cell Tissue Organ Cult 97(3):285–293. https://doi.org/10.1007/s11240-009-9527-9

207. Gémes Juhász A, Kristóf Z, Vági P, Lantos C, Pauk J (2009) In vitro anther and isolated microspore culture as tools in sweet and spice pepper breeding. Acta Hortic 829:61–64

208. Kim M, Jang I-C, Kim J-A, Park E-J, Yoon M, Lee Y (2008) Embryogenesis and plant regeneration of hot pepper (*Capsicum annuum* L.) through isolated microspore culture. Plant Cell Rep 27(3):425–434

209. Ozkum D, Tipirdamaz R (2007) Effects of silver nitrate, activated charcoal and cold treatment on the in vitro androgenesis of pepper (*Capsicum annuum* L.). In: Sivritepe HO, Sivritepe N (eds) Proceedings of the IIIrd Balkan symposium on vegetable and potatoes. Acta Horticulturae, vol 729. Humana, New York, NY, pp 133–136

210. Supena EDJ, Suharsono S, Jacobsen E, Custers JBM (2006) Successful development of a shed-microspore culture protocol for doubled haploid production in Indonesian hot pepper (*Capsicum annuum* L.). Plant Cell Rep 25(1):1–10

211. Supena EDJ, Muswita W, Suharsono S, Custers JBM (2006) Evaluation of crucial factors for implementing shed-microspore culture of Indonesian hot pepper (*Capsicum annuum* L.) cultivars. Sci Hortic 107(3):226–232

212. Mitykó J, Juhász AG (2006) Improvement in the haploid technique routinely used for breeding sweet and spice peppers in Hungary. Acta Agron Hung 54(2):203–219. https://doi.org/10.1556/AAgr.54.2006.2.8

213. Ercan N, Sensoy FA, Sensoy AS (2006) Influence of growing season and donor plant age on anther culture response of some pepper cultivars (*Capsicum annuum* L.). Sci Hortic 110(1):16–20

214. Buyukalaca S, Comlekcioglu N, Abak K, Ekbic E, Kilic N (2004) Effects of silver nitrate and donor plant growing conditions on production of pepper (*Capsicum annuum* L.) haploid embryos via anther culture. Eur J Hortic Sci 69(5):206–209

215. Dolcet-Sanjuan R, Claveria E, Huerta A (1997) Androgenesis in *Capsicum annuum* L – effects of carbohydrate and carbon dioxide enrichment. J Am Soc Hortic Sci 122(4):468–475

216. Regner F (1996) Anther and microspore culture in *Capsicum*. In: Jain SM, Sopory SK, Veilleux RE (eds) *In vitro* haploid production in higher plants, vol 3. Kluwer Academic, Dordrecht, pp 77–89

217. Dumas de Vaulx R, Chambonnet D, Pochard E (1981) Culture *in vitro* d'anthères de piment (*Capsicum annuum* L.): amèlioration des taux d'obtention de plantes chez différents génotypes par des traitments à +35°C. Agronomie 1(10):859–864

218. Sibi M, Dumas de Vaulx R, Chambonnet D (1979) Obtention of haploid plants through in vitro androgenesis in sweet pepper (*Capsicum annuum* L.). Ann Amélior Plant 29(5):583–606

219. George L, Narayanaswamy S (1973) Haploid *Capsicum* through experimental androgenesis. Protoplasma 78:467–470

220. Campos FF, Morgan DTJ (1958) Haploid pepper from a sperm. J Hered 49:135–137

221. Wang Y-Y, Sun C-S, Wang C-C, Chien N-F (1973) The induction of the pollen plantlets of triticale and *Capsicum annuum* from anther culture. Sci Sinica 16:147–151

222. Zamani MJ, Moieni A, Choukan R (2016) Effect of temperature stress on androgenesis induction in bell pepper (*Capsicum annuum*

L.) by anther culture. Int J Adv Biotechnol Res 7(4):1725–1733

223. Keles D, Pinar H, Ata A, Taskin H, Yildiz S, Buyukalaca S (2015) Effect of pepper types on obtaining spontaneous doubled haploid plants via anther culture. HortScience 50 (11):1671–1676. https://doi.org/10.21273/hortsci.50.11.1671

224. Nowaczyk P, Kisiala A, Lszewska D (2005) *In vitro* anther culture of *Capsicum frutescen*s L. red- and yellow-fruited forms. Acta Biol Cracov Ser Bot 47:76–76

225. Wu HN, Zhang SZ (1986) Effect of acridine yellow on development of anthers of *Capsicum frutescens* var. longum cultured *in vitro*. J Agric Sci 2:34–39

226. Nowaczyk P, Kisiala A, Olszewska D (2006) Induced androgenesis of *Capsicum frutescens* L. Acta Physiol Plant 28(1):35–39. https://doi.org/10.1007/s11738-006-0066-2

227. Tsay HS, Su CY (1985) Anther culture of papaya (*Carica papaya* L.). Plant Cell Rep 4 (1):28–30. https://doi.org/10.1007/bf00285498

228. Rimberia FK, Sunagawa H, Urasaki N, Ishimine Y, Adaniya S (2005) Embryo induction via anther culture in papaya and sex analysis of the derived plantlets. Sci Hortic 103(2):199–208

229. Litz R, Conover R (1979) In vitro improvement of *Carica papaya* L. proceedings of the tropical region of the American Society for Horticultural. Society 23:157–159

230. Prasad BR, Khadeer MA, Seeta P, Anwar SY (1991) In vitro induction of androgenic haploids in safflower (*Carthamus tinctorius* L.). Plant Cell Rep 10(1):48–51

231. Smýkalová I, Horáček J, Kubošiová M, Šmirous P Jr, Soukup A, Gasmanová N, Griga M (2012) Induction conditions for somatic and microspore-derived structures and detection of haploid status by isozyme analysis in anther culture of caraway (*Carum carvi* L.). In Vitro Cell Dev Biol Plant 48 (1):30–39. https://doi.org/10.1007/s11627-011-9386-z

232. Smýkalová I, Šmirous P, Kubošiová M, Gasmanová N, Griga M (2009) Doubled haploid production via anther culture in annual, winter type of caraway (*Carum carvi* L.). Acta Physiol Plant 31(1):21

233. Gharyal PK, Rashid A, Maheshwari S (1983) Androgenic response from cultured anthers of a leguminous tree, *Cassia siamea* Lam. Protoplasma 118(1):91–93

234. Bajaj Y, Dhanjy M (1983) Pollen embryogenesis in three ornamental trees – *Cassia fistula*, *Jacaranda acutifolia* and *Poinciana regia*. J Tree Sci 2:16–19

235. Abou-Mandour A, Fischer S, Czygan F-C (1979) Regeneration of intact plants from haploid and diploid callus cells of *Catharanthus roseus*. Zeitschr Pflanzen 91:83–88

236. Kim SW, Song NH, Jung KH, Kwak SS, Liu JR (1994) High frequency plant regeneration from anther-derived cell suspension cultures via somatic embryogenesis in *Catharanthus roseus*. Plant Cell Rep 13 (6):319–322

237. George L (1985) Anther culture of *Catharanthus roseus* L.—development of pollen embryoids. Curr Sci 54(13):641–642

238. Custódio L, Carneiro MF, Romano A (2005) Microsporogenesis and anther culture in carob tree (*Ceratonia siliqua* L.). Sci Hortic 104(1):65–77

239. Wang H, Dong B, Jiang J, Fang W, Guan Z, Liao Y, Chen S, Chen F (2014) Characterization of in vitro haploid and doubled haploid *Chrysanthemum morifolium* plants via unfertilized ovule culture for phenotypical traits and DNA methylation pattern. Front Plant Sci 5:738

240. Watanabe K (1977) Successful ovary culture and production of F hybrids and androgenic haploids in Japanese *Chrysanthemum* species. J Hered 68:317–320

241. Grewal RK, Lulsdorf M, Croser J, Ochatt S, Vandenberg A, Warkentin TD (2009) Doubled-haploid production in chickpea (*Cicer arietinum* L.): role of stress treatments. Plant Cell Rep 28(8):1289–1299. https://doi.org/10.1007/s00299-009-0731-1

242. Khan S, Ghosh P (1983) In vitro induction of androgenesis and organogenesis in *Cicer arietinum* L. Curr Sci 52(18):891–893

243. Bajaj Y, Gosal S (1987) Pollen embryogenesis and chromosomal variation in cultured anthers of chickpea. Int Chickpea Newsl 17:12–13

244. Huda S, Islam R, Bari M, Asaduzzaman M (2001) Anther culture of chickpea. Int Chickpea Pigeonpea Newsl 8:24–26

245. Abdollahi MR, Rashidi S (2018) Production and conversion of haploid embryos in chickpea (*Cicer arietinum* L.) anther cultures using high 2,4-D and silver nitrate containing media. Plant Cell Tissue Organ Cult 133 (1):39–49. https://doi.org/10.1007/s11240-017-1359-4

246. Mallikarjuna N, Jadhav D, Clarke H, Coyne C, Muehlbauer F (2005) Induction of androgenesis as a consequence of wide crossing in chickpea. Int Chickpea Pigeonpea Newsl 12:12–15

247. Panchangam SS, Mallikarjuna N, Gaur PM, Suravajhala P (2014) Androgenesis in chickpea: anther culture and expressed sequence tags derived annotation. NISCAIR-CSIR 52 (2):181–188

248. Reddy V, Reddy G (1997) In vivo production of haploids in chickpea (*Cicer arietinum* L.). J Genet Breed 51:29–32

249. Guedira M, DUBOISTYLSKI T, Vasseur J, Dubois J (1989) Direct somatic embryogenesis from anther cultures of *Cichorium* (*Asteraceae*). Can J Bot 67(4):970–976

250. Theiler-Hedtrich R, Hunter C (1995) Regeneration of dihaploid chicory (*Cichorium intybus* L. var. foliosum Hegi) via microspore culture. Plant Breed 114(1):18–23

251. Van Der Veken J, Eeckhaut T, Baert J, Ruttink T, Maudoux O, Werbrouck S, Van Huylenbroeck J (2019) *Cichorium intybus* L.x *Cicerbita alpina* Walbr.: doubled haploid chicory induction and CENH3 characterization. Euphytica 215(7). https://doi.org/10.1007/s10681-019-2435-0

252. Abdollahi MR, Darbandi M, Hamidvand Y, Majdi M (2015) The influence of phytohormones, wheat ovary co-culture, and temperature stress on anther culture response of watermelon (*Citrullus lanatus* L.). Rev Bras Bot 38(3):447–456. https://doi.org/10.1007/s40415-015-0152-z

253. Taşkın H, Yücel NK, Baktemur G, Çömlekçioğlu S, Büyükalaca S (2013) Effects of different genotypes and gamma ray doses on haploidization with irradiated pollen technique in watermelon (*Citrullus lanatus* L.). Can J Plant Sci 93 (6):1164–1168

254. Chaturvedi H, Sharma A (1985) Androgenesis in *Citrus aurantifolia* (Christm.) swingle. Planta 165(1):142–144

255. Hidaka T, Yamada Y, Shichijo T (1982) Plantlet formation by anther culture of *Citrus aurantium* L. Jpn J Breed 32 (3):247–252

256. Cardoso JC, Abdelgalel AM, Chiancone B, Latado RR, Lain O, Testolin R, Germana MA (2016) Gametic and somatic embryogenesis through in vitro anther culture of different *Citrus* genotypes. Plant Biosyst 150(2):304–312. https://doi.org/10.1080/11263504.2014.987847

257. Ramírez C, Chiancone B, Testillano PS, Garcia-Fojeda B, Germana MA, Risueno MC (2003) First embryogenic stages of *Citrus* microspore-derived embryos. Acta Biol Cracov Ser Bot 45(1):53–58

258. Germanà MA, Wang YY, Barbagallo MG, Iannolino G, Crescimanno FG (1994) Recovery of haploid and diploid plantlets from anther culture of *Citrus clementina* Hort ex Tan and *Citrus reticulata* Blanco. J Hortic Sci 69(3):473–480

259. Germanà M, Crescimanno F, Motisi A (2000) Factors affecting androgenesis in *Citrus clementina* Hort. ex Tan. Adv Hortic Sci 14(2):43–51

260. Germanà M, Crescimanno F, Reforgiato Recupero G, Russo M (1998) Preliminary characterization of several doubled haploids of *Citrus clementina* cv. Nules. Acta Hortic 535:183–190

261. Germanà MA, Chiancone B (2001) Gynogenetic haploids of *Citrus* after in vitro pollination with triploid pollen grains. Plant Cell Tissue Organ Cult 66(1):59–66

262. Germanà M, Chiancone B (2003) Improvement of the anther culture protocol in *Citrus clementina* Hort. ex Tan. Plant Cell Rep 22 (3):181–187

263. Chiancone B, Marli Gniech Karasawa M, Gianguzzi V, Abdelgalel AM, Bárány IV, Testillano PS, Torello Marinoni D, Botta R, Germanà MA (2015) Early embryo achievement through isolated microspore culture in *Citrus clementina* Hort. ex Tan., cvs. 'Monreal Rosso' and 'Nules'. Front Plant Sci 6. https://doi.org/10.3389/fpls.2015.00413

264. Germanà MA, Chiancone B (2003) Improvement of *Citrus clementina* Hort. ex Tan. microspore-derived embryoid induction and regeneration. Plant Cell Rep 22(3):181–187

265. Aleza P, Juárez J, Hernández M, Pina JA, Ollitrault P, Navarro L (2009) Recovery and characterization of a *Citrus clementina* Hort. ex Tan. 'Clemenules' haploid plant selected to establish the reference whole *Citrus* genome sequence. BMC Plant Biol 9 (1):110

266. Germanà MA, Recupero GR (1997) Haploid embryos regeneration from anther culture of 'Mapo' tangelo (*Citrus deliciosa* × *C. paradisi*). Adv Hortic Sci 11(3):147–152

267. Germanà M, Crescimanno F, De Pasquale F, Yu Ying W (1990) Androgenesis in 5 cultivars of *Citrus limon* L. Burm. f. In Vitro Culture, XXIII IHC 300 300:315–324

268. Ling J, Iwamasa M (1988) Nito N Plantlet regeneration by anther culture of Calamondin (*Citrus madurensis* Lour.). In: Goren R, Mendel K (eds) Citriculture: proceedings of the sixth international citrus congress: Middle-East. Rehovot, Balaban Publishers, pp 251–256

269. Yahata M, Nukaya T, Sudo M, Ohta T, Yasuda K, Inagaki H, Mukai H, Harada H, Takagi T, Komatsu H, Kunitake H (2015) Morphological characteristics of a doubled haploid line from 'Banpeiyu' pummelo [*Citrus maxima* (Burm.) Merr.] and its reproductive function. Hortic J 84:30–36

270. Karasawa K (1971) On tte occurrence of haploid seedlings in *Citrus natsudaidai* Hayata, vol 1. Sakushingakuin Junior College for Women Bull, Biological Institute, Utsunomiya, pp 1–2

271. Jedidi E, Kamiri M, Poullet T, Ollitrault P, Froelicher Y (2015) Efficient haploid production on 'Wilking'mandarin by induced gynogenesis. Acta Hortic 1065:60

272. Starrantino A, Caponnetto P (1989) Effect of cytokinins on embryogenic callus formation from undeveloped ovules of orange. Acta Hortic 280:191–194

273. Hidaka T (1984) Induction of plantlets from anthers of 'Trovita' orange. J Jpn Soc Hortic Sci 53(1):1–5

274. Koltunow AM, Soltys K, Nito N, McClure S (1995) Anther, ovule, seed, and nucellar embryo development in *Citrus sinensis* cv. Valencia. Can J Bot 73(10):1567–1582

275. Wang SM, Lan H, Cao HB, Xu Q, Chen CL, Deng XX, Guo WW (2015) Recovery and characterization of homozygous lines from two sweet orange cultivars via anther culture. Plant Cell Tissue Organ Cult 123 (3):633–644. https://doi.org/10.1007/s11240-015-0866-4

276. Froelicher Y, Ollitrault P (1998) Effects of the hormonal balance on *Clausena excavata* androgenesis. First international citrus biotechnology symposium 535:139–146

277. Monfort S (1985) Androgenesis of coconut: embryos from anther culture. Z Pflanzen 94 (3):251–254

278. Thanh-Tuyen NT, De Guzman EV (1983) Formation of pollen embryos in cultured anthers of coconut (*Cocos nucifera* L.). Plant Sci Lett 29(1):81–88. https://doi.org/10.1016/0304-4211(83)90026-3

279. Bandupriya H, Fernando S, Vidhanaarachchil Y (2016) Micropropagation and androgenesis in coconut: an assessment of Sri Lankan implication. Cocos 22:31–47

280. Perera PIP, Yakandawala DMD, Hocher V, Verdeil JL, Weerakoon LK (2009) Effect of growth regulators on microspore embryogenesis in coconut anthers. Plant Cell Tissue Organ Cult 96(2):171–180. https://doi.org/10.1007/s11240-008-9473-y

281. Perera PIP, Motha KF, Vidhanaarchchi VRM (2020) Morphological and histological analysis of anther-derived embryos of coconut (*Cocos nucifera* L.). Plant Cell Tissue Org Cult 140(3):685–689. https://doi.org/10.1007/s11240-019-01762-9

282. Neuenschwander B, Baumann T (1995) Increased frequency of dividing microspores and improved maintenance of multicellular microspores of *Coffea arabica* in medium with coconut milk. Plant Cell Tissue Organ Cult 40(1):49–54

283. Couturon E (1982) Obtaining naturally-occurring haploids of coffea – *Canephora pierre* by grafting of embryos. Cafe Cacao Tee 26(3):155–160

284. Lashermes P, Couturon E, Charrier A (1993) Doubled haploids of *Coffea canephora*: development, fertility and agronomic characteristics. Euphytica 74(1–2):149–157

285. Lashermes P, Couturon E, Charrier A (1994) Combining ability of doubled haploids in *Coffea canephora* P. Plant Breed 112 (4):330–337

286. Raghuramulu Y, Prakash N (1996) Haploidy in coffee. In: Jain SM, Sopory SK, Veilleux RE (eds) In vitro haploid production in higher plants. Springer, Dordrech, pp 349–363

287. Ali MA, Jones J (2000) Microspore culture in *Corchorus olitorius*: effect of growth regulators, temperature and sucrose on callus formation. Indian J Exp Biol 38(6):593–597

288. Gniech Karasawa MM, Chiancone B, Gianguzzi V, Abdelgalel AM, Botta R, Sartor C, Germanà MA (2016) Gametic embryogenesis through isolated microspore culture in *Corylus avellana* L. Plant Cell Tissue Organ Cult 124(3):635–647. https://doi.org/10.1007/s11240-015-0921-1

289. Sacristan MD (1971) Karyotypic changes in callus cultures from haploid and diploid plants of *Crepis capillaris* (L.) Wallr. Chromosoma 33(3):273–283

290. Slusarkiewicz-Jarzina A, Zenkteler M (1979) Cytological and embryological studies on haploids (n = 3) of *Crepis capillaris* L. Bull Soc Amis Sci Lett Poznan D Sci Biol 1:65–73

291. Gerassimowa H (1936) Experimentell erhaltene haploide Pflanze von *Crepis tectorum* L. Planta 25:696–702

292. Debata B (1983) In vitro culture of anther of *Crotalaria pallida* Ait. for induction of haploid. Indian J Exp Biol 21:44–46

293. Matsubara S, Dohya N, Murakami K (1994) Callus formation and regeneration of adventitious embryos from carrot, fennel and mitsuba microspores by anther and isolated microspore cultures. Acta Hortic 392:129–138

294. Dumas de Vaulx R (1979) Obtaining haploid plants in melon (*Cucumis melo* L) after pollination by *Cucumis ficifolius.* Compt Rend Hebdomad Sean L Acad Sci D 289 (12):875

295. Dryanovska OA, Ilieva IN (1983) In vitro anther and ovule culture in muskmelon (*Cucumis melo* L.). Compt Rendus Acad Bulgare Sciences 36(8):1107–1110

296. Sauton A, Dumas de Vaulx R (1987) Obtention de plantes haploides chez melon (*Cucumis melo* L.) par gynogenese indute par du pollen irraidié. Agronomie 7:141–148

297. Savin F, Decomble V, Le Couviour M, Hallard J (1988) The X-ray detection of haploid embryos arisen in muskmelon (*Cucumis melo* L.) seeds, and resulting from a parthenogenetic development induced by irradiated pollen. Rep Cucurbit Genet Coop 11:39–42

298. Cuny F, de Vaulx RD, Longhi B, Siadous R (1992) Analyse des plantes de melon (*Cucumis melo* L) issues de croisements avec du pollen irradié à différentes doses. Agronomie 12:623–630

299. Ficcadenti N, Sestili S, Annibali S, Di Marco M, Schiavi M (1999) *In vitro* gynogenesis to induce haploid plants in melon *Cucumis melo* L. Genet Breed 53:255–257

300. Gonzalo MJ, Claveria E, Monforte AJ, Dolcet-Sanjuan R (2011) Parthenogenic haploids in melon: generation and molecular characterization of a doubled haploid line population. J Am Soc Hortic Sci 136 (2):145–154

301. Yetisir H, Sari N (2003) A new method for haploid muskmelon (*Cucumis melo* L.) dihaploidization. Sci Hortic 98(3):277–283. https://doi.org/10.1016/S0304-4238(02)00226-1

302. Lotfi M, Alan AR, Henning MJ, Jahn MM, Earle ED (2003) Production of haploid and doubled haploid plants of melon (*Cucumis melo* L.) for use in breeding for multiple virus resistance. Plant Cell Rep 21 (11):1121–1128

303. Lim W, Earle ED (2009) Enhanced recovery of doubled haploid lines from parthenogenetic plants of melon (*Cucumis melo* L.). Plant Cell Tissue Organ Cult 98 (3):351–356. https://doi.org/10.1007/ s11240-009-9563-5

304. Sari N, Solmaz I, Yetisir H, Ekiz H, Yucel S (2010) New Fusarium wilt resistant melon (*Cucumis melo* var. cantalupensis) varieties developed by dihaploidization. International Society for Horticultural Science (ISHS), Leuven, pp 267–272. https://doi.org/10. 17660/ActaHortic.2010.871.35

305. Sauton A, Institut National de la Recherche Agronomique (1988) Doubled haploid production in melon (*Cucumis melo* L). Cucurbitaceae 88: proceedings of the Eucarpia meeting on curcurbit genetics and breeding. Institut national de la recherche agronomique, Paris

306. Yashiro K, Hosoya K, Kuzuya M, Tomita K, Ezura H (2002) Efficient production of doubled haploid melon plants by modified colchicine treatment of parthenogenetic haploids. In: Nishimura S, Ezura H, Matsuda T, Tazuke A (eds) Proceedings of the iind international symposium on cucurbits. Acta Horticulturae, vol 588. Springer, New York, NY, pp 335–338. https://doi. org/10.17660/ActaHortic.2002.588.54

307. Le Deunff E, Sauton A (1994) Effect of parthenocarpy on ovule development in cucumber (*Cucumis sativus* L.) after pollination with normal and irradiated pollen. Sex Plant Reprod 7(4):221–228

308. Ebrahimzadeh H, Soltanloo H, Shariatpanahi ME, Eskandari A, Ramezanpour SS (2018) Improved chromosome doubling of parthenogenetic haploid plants of cucumber (*Cucumis sativus* L.) using colchicine, trifluralin, and oryzalin. Plant Cell Tissue Organ Cult 135(3):407–417. https://doi.org/10. 1007/s11240-018-1473-y

309. Ebrahimzadeh H, Shariatpanahi ME, Ahmadi B, Soltanloo H, Lotfi M, Zarifi E (2018) Efficient parthenogenesis induction and in vitro haploid plant regeneration in cucumber (*Cucumis sativus* L.) using putrescine, spermidine, and cycocel. J Plant Growth Regul 37(4):1127–1134. https:// doi.org/10.1007/s00344-018-9803-1

310. Asadi A, Zebarjadi A, Abdollahi MR, Seguí-Simarro JM (2018) Assessment of different anther culture approaches to produce doubled haploids in cucumber (*Cucumis sativus* L.). Euphytica 214(11):216. https://doi. org/10.1007/s10681-018-2297-x

311. Abdollahi MR, Najafi S, Sarikhani H, Moosavi SS (2016) Induction and development of anther-derived gametic embryos in

cucumber (*Cucumis sativus* L.) by optimizing the macronutrient and agar concentrations in culture medium. Turk J Biol 40 (3):571–579

312. Tantasawat PA, Sorntip A, Pornbungkerd P (2015) Effects of exogenous application of plant growth regulators on growth, yield, and in vitro gynogenesis in cucumber. HortScience 50(3):374–382. https://doi.org/10.21273/hortsci.50.3.374

313. Plapung P, Khamsukdee S, Potapohn N, Smitamana P (2014) Screening for cucumber mosaic resistant lines from the ovule culture derived double haploid cucumbers. Am J Agric Biol Sci 9(3):261–269

314. Hamidvand Y, Abdollahi MR, Chaichi M, Moosavi SS (2013) The effect of plant growth regulators on callogenesis and gametic embryogenesis from anther culture of cucumber (*Cucumis sativus* L.). Int J Agric Crop Sci 5(10):1089

315. Zhan Y, J-f C, Malik AA (2009) Embryoid induction and plant regeneration of cucumber (*Cucumis sativus* L.) through microspore culture. Acta Hortic Sin 36 (2):221–226

316. Diao W-P, Jia Y-Y, Song H, Zhang X-Q, Lou Q-F, Chen J-F (2009) Efficient embryo induction in cucumber ovary culture and homozygous identification of the regenetants using SSR markers. Sci Hortic 119 (3):246–251. https://doi.org/10.1016/j.scienta.2008.08.016

317. Suprunova T, Shmykova N (2008) *In vitro* induction of haploid plants in unpollinated ovules, anther and microspore culture of *Cucumis sativus*. Cucurbitaceae 2008: proceedings of the IXth Eucarpia meeting on genetics and breeding of Cucurbitaceae 2008:371–374

318. Song H, Lou QF, Luo XD, Wolukau JN, Diao WP, Qian CT, Chen JF (2007) Regeneration of doubled haploid plants by androgenesis of cucumber (*Cucumis sativus* L.). Plant Cell Tissue Organ Cult 90 (3):245–254. https://doi.org/10.1007/s11240-007-9263-y

319. Claveria E, Garcia-Mas J, Dolcet-Sanjuan R (2005) Optimization of cucumber doubled haploid line production using in vitro rescue of in vivo induced parthenogenic embryos. J Am Soc Hortic Sci 130(4):555–560

320. Dolcet-Sanjuan R, Claveria E, Garcia-Mas J (2004) Cucumber (*Cucumis sativus* L.) dihaploid line production using in vitro rescue of *in vivo* induced parthenogenic embryos. Acta Hortic 725:837–844

321. Ashok Kumar HG, Murthy HN (2004) Effect of sugars and amino acids on androgenesis of *Cucumis sativus*. Plant Cell Tissue Organ Cult 78(3):201–208. https://doi.org/10.1023/b:ticu.0000025637.56693.68

322. Gémes-Juhász A, Balogh P, Ferenczy A, Kristóf Z (2002) Effect of optimal stage of female gametophyte and heat treatment on in vitro gynogenesis induction in cucumber (*Cucumis sativus* L.). Plant Cell Rep 21 (2):105–111. https://doi.org/10.1007/s00299-002-0482-8

323. Çaglar G, Abak K (1999) Progress in the production of haploid embryos, plants and doubled haploids in cucumber (*C. sativus* L.) by gamma irradiated pollen, in Turkey. Acta Hortic 492:317–322

324. Truong-Andre I (1988) *In vitro* haploid plants derived from pollination by irradiated pollen on cucumber. In: Eucarpia meeting on cucurbit genetics and breeding, Montfavet (France), 31 May to 2 Jun. INRA, Montfavet

325. Juhasz AG, Venczel G, Balogh P (1997) Haploid plant induction in zucchini (*Cucurbita pepo* L convar giromontiina Duch) and in cucumber (*Cucumis sativus* L) lines through in vitro gynogenesis. In: Altman A, Ziv M (eds) Horticultural biotechnology in vitro culture and breeding. Acta Horticulturae, vol 447. Springer, New York, NY, pp 623–624

326. Li JW, Si SW, Cheng JY, Li JX, Liu JQ (2013) Thidiazuron and silver nitrate enhanced gynogenesis of unfertilized ovule cultures of *Cucumis sativus*. Biol Plant 57 (1):164–168. https://doi.org/10.1007/s10535-012-0269-x

327. Gemes-Juhasz A, Balogh P, Ferenczy A, Kristóf Z (2002) Effect of optimal stage of female gametophyte and heat treatment on in vitro gynogenesis induction in cucumber (*Cucumis sativus* L.). Plant Cell Rep 21 (2):105–111

328. Sorntip A, Poolsawat O, Kativat C, Tantasawat PA (2017) Gynogenesis and doubled haploid production from unpollinated ovary culture of cucumber (*Cucumis sativus* L.). Can J Plant Sci 98(2):353–361

329. Amirian R, Hojati Z, Azadi P (2020) Male flower induction significantly affects androgenesis in cucumber (*Cucumis sativus* L.). J Hortic Sci Biotechnol 95(2):183–191. https://doi.org/10.1080/14620316.2019.1655488

330. Sorntip A, Poolsawat O, Kativat C, Tantasawat PA (2018) Gynogenesis and doubled

haploid production from unpollinated ovary culture of cucumber (*Cucumis sativus* L.). Can J Plant Sci 98(2):353–361. https://doi.org/10.1139/cjps-2017-0112

331. Golabadi M, Ghanbari S, Keighobadi K, Ercisli S (2017) Embryo and callus induction by different factors in ovary culture of cucumber. J Appl Bot Food Qual 90:68–75. https://doi.org/10.5073/jabfq.2017.090.0101

332. Galazka J, Slomnicka R (2015) From pollination to DH lines. Verification and optimization of protocol for production of doubled haploids in cucumber. Acta Sci Pol-Hortorum Culttus 14(3):81–92

333. Kurtar ES, Ahmet B, Ozbakir OM (2018) Production of callus mediated gynogenic haploids in winter squash (*Cucurbita maxima* Duch.) and pumpkin (*Cucurbita moschata* Duch.). Czech J Genet Plant Breed 54(1):9–16

334. Kurtar ES (2018) The effects of anti-mitotic agents on dihaploidization and fertility in winter squash (*Cucurbita maxima* Duch.) and pumpkin (*Cucurbita moschata* Duch.) androgenic haploids. Acta Sci Pol-Hortorum Cultus 17(5):3–14. https://doi.org/10.24326/asphc.2018.5.1

335. Kurtar ES, Balkaya A, Kandemir D (2016) Evaluation of haploidization efficiency in winter squash (*Cucurbita maxima* Duch.) and pumpkin (*Cucurbita moschata* Duch.) through anther culture. Plant Cell Tissue Organ Cult 127(2):497–511. https://doi.org/10.1007/s11240-016-1074-6

336. Kurtar ES, Balkaya A (2010) Production of in vitro haploid plants from in situ induced haploid embryos in winter squash (*Cucurbita maxima* Duchesne ex Lam.) via irradiated pollen. Plant Cell Tissue Organ Cult 102(3):267–277. https://doi.org/10.1007/s11240-010-9729-1

337. Kurtar ES, Balkaya A, Ozbakir M, Ofluoglu T (2009) Induction of haploid embryo and plant regeneration via irradiated pollen technique in pumpkin (*Cucurbita moschata* Duchesne ex. Poir). Afr J Biotechnol 8(21):5944–5951

338. Dumas de Vaulx R, Chambonnet D (1986) Obtention of embryos and plants from in vitro culture of unfertilized ovules of *Cucurbita pepo*. In: Proceedings of the international symposium, EUCARPIA. Walter de Gruyter & Co., Berlin, pp 295–297

339. Metwally EI, Moustafa SA, El-Sawy BI, Shalaby TA (1998) Haploid plantlets derived by anther culture of *Cucurbita pepo*. Plant Cell

Tissue Organ Cult 52(3):171–176. https://doi.org/10.1023/a:1005908326663

340. Metwally E, Moustafa S, El-Sawy B, Haroun S, Shalaby T (1998) Production of haploid plants from in vitro culture of unpollinated ovules of *Cucurbita pepo*. Plant Cell Tissue Organ Cult 52(3):117–121

341. Košmrlj K, Murovec J, Bohanec B (2013) Haploid induction in hull-less seed pumpkin through parthenogenesis induced by X-ray-irradiated pollen. J Am Soc Hortic Sci 138(4):310–316

342. Rakha M, Metwally E, Moustafa S, Etman A, Dewir Y (2012) Evaluation of regenerated strains from six *Cucurbita* interspecific hybrids obtained through anther and ovule *in vitro* cultures. Aust J Crop Sci 6(1):23–30

343. Shalaby TA (2006) Embryogenesis and plantlets regeneration from anther culture of squash plants (*Cucurbita pepo* L.) as affected by different genotypes. J Agric Res Tanta Univ 32(1):173–183

344. Mohamed M, Refaei E (2004) Enhanced haploids regeneration in anther culture of summer squash (*Curcurbita pepo* L.). Cucurbit Genet Coop Rep 27:57–60

345. Kurtar ES, Sarı N, Abak K (2002) Obtention of haploid embryos and plants through irradiated pollen technique in squash (*Cucurbita pepo* L.). Euphytica 127(3):335–344. https://doi.org/10.1023/a:1020343900419

346. Shalaby TA (2007) Factors affecting haploid induction through in vitro gynogenesis in summer squash (*Cucurbita pepo* L.). Sci Hortic 115(1):1–6

347. Pichot C, El Maâtaoui M, Raddi S, Raddi P (2001) Surrogate mother for endangered *Cupressus*. Nature 412(6842):39–39

348. Ishizaka H (1998) Production of microspore-derived plants by anther culture of an interspecific F1 hybrid between *Cyclamen persicum* and *C. purpurascens*. Plant Cell Tissue Organ Cult 54(1):21–28

349. Motzo R, Deidda M (1993) Anther and ovule culture in globe artichoke. J Genet Breed 47(3):263–266

350. Christensen J, Borrino E, Olesen A, Andersen SB (1997) Diploid, tetraploid, and octoploid plants from anther culture of tetraploid orchard grass, *Dactylis glomerata* L. Plant Breed 116(3):267–270

351. Padmanabhan C, Gurunathan M, Pathmanabhan G, Oblisami G (1977) Induction of haploid plants from anther culture in *Datura ferox* L. Madras Agric J 64(8):542–543

352. Meixner M, Frahm C, Pflug P, Schmidt-Rogge T, Schieder O (1997) Genetic manipulation of haploid *Datura innoxia* Mill: analysis of the transgene integration patterns and the ploidy level of transgenic plants obtained after direct or *Agrobacterium*-mediated gene transfer. In: Altman A, Ziv M (eds) Horticultural biotechnology in vitro culture and breeding. Acta Horticulturae, vol 447. Springer, New York, NY, pp 349–354

353. Sharma VK, Jethwani V, Kothari SL (1993) Embryogenesis in suspension cultures of *Datura innoxia* Mill. Plant Cell Rep 12 (10):581–584

354. Sangwan RS, Ducrocq C, Sangwan-Norreel B (1993) Agrobacterium-mediated transformation of pollen embryos in *Datura innoxia* and *Nicotiana tabacum*: production of transgenic haploid and fertile homozygous dihaploid plants. Plant Sci 95(1):99–115

355. Sangwan RS, Mathivet V, Vasseur G (1989) Ultrastructural localization of acid phosphatase during male meiosis and sporogenesis in *Datura*: evidence for digestion of cytoplasmic structures in the vacuoles. Protoplasma 149:38–46

356. Sangwan RS, Sangwan-Norreel BS (1987) Ultrastructural cytology of plastids in pollen grains of certain androgenic and nonandrogenic plants. Protoplasma 138(1):11–22. https://doi.org/10.1007/bf01281180

357. Sangwan RS, Camefort H (1984) Cold treatment-related structural modifications in the embryogenic anthers of *Datura*. Cytologia 49(3):473–487

358. Tyagi AK, Rashid A, Maheshwari SC (1981) Promotive effect of polyvinylpolypyrrolidone on pollen embryogenesis in *Datura innoxia*. Physiol Plant 53(4):405–406. https://doi.org/10.1111/j.1399-3054.1981.tb02722.x

359. Tyagi AK, Rashid A, Maheshwari SC (1981) Sodium chloride-resistant cell line from haploid *Datura innoxia* mill – a resistance trait carried from cell to plantlet and vice versa *in vitro*. Protoplasma 105(3–4):327–332. https://doi.org/10.1007/bf01279229

360. Forche E, Kibler R, Neumann KH (1981) The influence of developmental stages of haploid and diploid callus cultures of *Datura innoxia* on shoot initiation. Z Pflanzenphysiol 101(3):257–262

361. Forche E, Neumann KH (1977) Influence of various cultural factors on development of haploid plants by anther culture of *Datura innoxia* and *Nicotiana tabacum* ssp. Z Pflanzen 79(3):250–255

362. Sopory SK, Maheshwari SC (1976) Morphogenetic potentialities of haploid and diploid vegetative parts of *Datura innoxia*. Z Pflanzenphysiol 77(3):274–277

363. Sunderland N, Collins GB, Dunwell JM (1974) Role of nuclear fusion in pollen embryogenesis of *Datura innoxia* Mill. Planta 117(3):227–241

364. Nitsch C, Norreel B (1973) Effect of thermal shock on embryogenic power of pollen of *Datura innoxia* cultured in anther or isolated from anther. Compt Rend Hebdomad Sean L Acad Sci D 276(3):303–306

365. Guha S, Maheshwari SC (1966) Cell division and differentiation of embryos in the pollen grains of *Datura in vitro*. Nature 1:97–98

366. Guha S, Maheshwari SC (1964) In vitro production of embryos from anthers of *Datura*. Nature 204:497

367. Iqbal MCM, Wijesekara KB (2007) A brief temperature pulse enhances the competency of microspores for androgenesis in *Datura metel*. Plant Cell Tissue Org Cult 89 (2–3):141–149. https://doi.org/10.1007/s11240-007-9222-7

368. Babbar SB, Gupta SC (1990) Phasic requirement of coconut milk for *Datura metel* microspore embryogenesis. Phytomorphology 40(1–2):53–57

369. Babbar SB, Gupta SC (1986) Effect of carbon source on *Datura metel* microspore embryogenesis and the growth of callus raised from microspore-derived embryos. Biochem Physiol Pflanz 181(5):331–338

370. Babbar SB, Gupta SC (1986) Promotory and inhibitory effects of activated charcoal on microspore embryogenesis in *Datura metel*. Physiol Plant 66(4):602–604. https://doi.org/10.1111/j.1399-3054.1986.tb05586.x

371. Babbar SB, Gupta SC (1986) Obligatory and period-specific requirement of iron for microspore embryogenesis in *Datura metel* anther cultures. Bot Mag Tokyo 99 (1054):225–232. https://doi.org/10.1007/bf02488823

372. Babbar SB, Gupta SC (1986) Putative role of ethylene in *Datura metel* microspore embryogenesis. Physiol Plant 68 (1):141–144. https://doi.org/10.1111/j.1399-3054.1986.tb06609.x

373. Babbar S, Gupta S (1984) Pathways in pollen sporophyte development in anther cultures of *Datura metel* and *Petunia hybrida*. Beitr Biol Pflanzen 59:475–488

374. Sangwan RS, Camefort H (1983) The tonoplast, a specific marker of embryogenic

microspores of *Datura* cultured in vitro. Histochemistry 78(4):473–480

375. Scogin R (1976) Isoenzyme patterns in androgenic, haploid *Datura meteloides* (*Solanaceae*). Experientia 32(5):562–563. https://doi.org/10.1007/bf01990161

376. Blakeslee AF, Belling J, Farnham ME, Bergner AD (1922) A haploid mutant in the Jimson weed *Datura stramonium*. Science 55(1433):646–647. https://doi.org/10.1126/science.55.1433.646

377. Tyukavin G, Shmykova N, Monakhova M (1999) Cytological study of embryogenesis in cultured carrot anthers. Russ J Plant Physiol 46(6):767–773

378. Gorécka K (2005) The influence of several factors on the efficiency of androgenesis in carrot. J Appl. Genetics 46(3):265–269

379. Gorecka K, Kowalska U, Krzyzanowska D, Kiszczak W (2010) Obtaining carrot (*Daucus carota* L.) plants in isolated microspore cultures. J Appl Genet 51(2):141–147

380. Gorecka K, Kiszczak W, Krzyzanowska D, Kowalska U, Kapuscinska A (2014) Effect of polyamines on in vitro anther cultures of carrot (*Daucus carota* L.). Turk J Biol 38 (5):593–600

381. Kiełkowska A, Adamus A, Baranski R (2018) Haploid and doubled haploid plant production in carrot using induced parthenogenesis and ovule excision in vitro. In: Loyola-Vargas VM, Ochoa-Alejo N (eds) Plant cell culture protocols. Springer New York, New York, NY, pp 301–315. https://doi.org/10.1007/978-1-4939-8594-4_21

382. Kiełkowska A, Adamus A, Baranski R (2014) An improved protocol for carrot haploid and doubled haploid plant production using induced parthenogenesis and ovule excision in vitro. In Vitro Cell Dev Biol Plant 50 (3):376–383

383. Li J-R, Zhuang F-Y, Ou C-G, Hu H, Zhao Z-W, Mao J-H (2013) Microspore embryogenesis and production of haploid and doubled haploid plants in carrot (*Daucus carota* L.). Plant Cell Tissue Organ Cult 112 (3):275–287

384. Kiszczak W, Kowalska U, Burian M, Gorecka K (2018) Induced androgenesis as a biotechnology method for obtaining DH plants in *Daucus carota* L. J Hortic Sci Biotechnol 93 (6):625–633. https://doi.org/10.1080/14620316.2018.1431058

385. Domblides AS (2017) Anther and ovule in vitro culture in carrot (*Daucus carota* L.). In: Briard M (ed) International symposium on carrot and other Apiaceae, vol 1153.

Acta Horticulturae, vol 1. International Society of Horticultural Science, Leuven, pp 55–60. https://doi.org/10.17660/ActaHortic.2017.1153.9

386. Górecka K, Krzyżanowska D, Kiszczak W, Kowalska U, Górecki R (2009) Carrot doubled haploids. In: Advances in haploid production in higher plants. Springer, New York, NY, pp 231–239

387. Dunemann F, Unkel K, Sprink T (2019) Using CRISPR/Cas9 to produce haploid inducers of carrot through targeted mutations of centromeric histone H3 (CENH3). International Society for Horticultural Science (ISHS), Leuven, pp 211–220. https://doi.org/10.17660/ActaHortic.2019.1264.26

388. Khandakar RK, Jie Y, Sun-Kyung M, Mi-Kyoung W, Choi HG, Ha-Seung P, Jong-Jin C, Soo-Cheon C, Ji-Youn J, Kyu-Min L (2014) Regeneration of haploid plantlet through anther culture of *Chrysanthemum* (*Dendranthema grandiflorum*). Not Bot Horti Agrobot Cluj Napoca 42 (2):482–487

389. Sato S, Katoh N, Yoshida H, Iwai S, Hagimori M (2000) Production of doubled haploid plants of carnation (*Dianthus caryophyllus* L.) by pseudofertilized ovule culture. Sci Hortic 83(3–4):301–310

390. Badea E, Iordan M, Mihalea A (1985) Induction of androgenesis in anther culture of *Digitalis lanata*. Rev Roum Biol Biol Veg 30:63–71

391. Diettrich B, Ernst S, Luckner M (2000) Haploid plants regenerated from androgenic cell cultures of *Digitalis lanata*. Planta Med 66(03):237–240

392. Pèrez-Bermúdez P, Cornejo MJ, Segura J (1985) Pollen plant formation from anther cultures of *Digitalis obscura* L. Plant Cell Tissue Organ Cult 5(1):63–68. https://doi.org/10.1007/bf00033570

393. Corduan G, Spix C (1975) Haploid callus and regeneration of plants from anthers of *Digitalis purpurea* L. Planta 124(1):1–11

394. Sinha R, Das K (1986) Anther-derived callus of *Dolichos biflorus* L, its protoplast culture and their morphogenic potential. Curr Sci 55(9):447–452

395. Zhao FC, Nilanthi D, Yang YS, Wu H (2006) Anther culture and haploid plant regeneration in purple coneflower (*Echinacea purpurea* L.). Plant Cell Tissue Organ Cult 86(1):55–62. https://doi.org/10.1007/s11240-006-9096-0

396. Dunwell JM, Wilkinson MJ, Nelson S, Wening S, Sitorus AC, Mienanti D, Alfiko Y, Croxford AE, Ford CS, Forster BP (2010) Production of haploids and doubled haploids in oil palm. BMC Plant Biol 10 (1):218

397. Singh M (1979) In vitro induction of haploid roots and shoots from female gametophyte of *Ephedra foliata* Boiss. Beitr Biol Pflanzen 55:169–177

398. Tefera H, Zapata-Arias F, Afza R, Kodym A (1999) Response of tef genotypes to anther culture. Agri 14(1):8–9

399. Gugsa L, Sarial AK, Lorz H, Kumlehn J (2006) Gynogenic plant regeneration from unpollinated flower explants of *Eragrostis tef* (Zuccagni) Trotter. Plant Cell Rep 25 (12):1287–1293. https://doi.org/10. 1007/s00299-006-0200-z

400. Li JQ, Wang YQ, Lin LH, Zhou LJ, Luo N, Deng QX, Xian JR, Hou CX, Qiu Y (2008) Embryogenesis and plant regeneration from anther culture in loquat (*Eriobotrya japonica* L.). Sci Hortic 115(4):329–336. https:// doi.org/10.1016/j.scienta.2007.10.007

401. Germanà MA, Chiancone B, Guarda NL, Testillano PS, Risueno MC (2006) Development of multicellular pollen of *Eriobotrya japonica* Lindl. through anther culture. Plant Sci 171(6):718–725

402. Leskovšek L, Jakše M, Bohanec B (2008) Doubled haploid production in rocket (*Eruca sativa* Mill.) through isolated microspore culture. Plant Cell Tissue Organ Cult 93(2):181–189. https://doi.org/10.1007/ s11240-008-9359-z

403. Sommer HE, Wetzstein HY (1984) Hardwoods. Handbook of plant cell. Culture 3:511–540

404. Yang Y, Wei W (1984) Insection of Longan haploid plantlets from pollens cultured in certain proper media. Acta Genet Sin (China) 11(4):288–293

405. Bohanec B (1997) Haploid induction in buckwheat (*Fagopyrum esculentum* Moench). In: In vitro haploid production in higher plants. Springer, New York, NY, pp 163–170

406. Zheleznov A (1976) Methods of obtaining parthenogenetic haploids in buckwheat. Apomiksis Ego Ispol'zovanie Selektsii 1976:65–68

407. Bohanec B, Nešković M, Vujičić R (1993) Anther culture and androgenetic plant regeneration in buckwheat (*Fagopyrum esculentum* Moench). Plant Cell Tissue Organ

Cult 35(3):259–266. https://doi.org/10. 1007/bf00037279

408. Germanà MA (2009) Haploids and doubled haploids in fruit trees. In: Touraev A, Forster BP, Jain SM (eds) Advances in haploid production in higher plants. Springer, Dordrecht, pp 241–263

409. Germanà MA (2006) Doubled haploid production in fruit crops. Plant Cell Tissue Organ Cult 86(2):131–146

410. Canhoto JM, Cruz GS (1993) Induction of pollen callus in anther cultures of *Feijoa sellowiana* Berg. (Myrtaceae). Plant Cell Rep 13 (1):45–48

411. Kasperbauer M, Buckner R (1979) Haploid plants from anthers of *Festuca arundinacea* cultured with nurse tissue. Agronomy Abstracts (USA)

412. Zare A-G, Humphreys MW, Rogers WJ, Collin HA (1999) Androgenesis from a *Lolium multiflorum × Festuca arundinacea* hybrid to generate extreme variation for freezing-tolerance. Plant Breed 118 (6):497–501. https://doi.org/10.1046/j. 1439-0523.1999.00399.x

413. Zwierzykowski Z, Zwierzykowska E, Slusarkiewicz-Jarzina A, Ponitka A (1999) Regeneration of anther-derived plants from pentaploid hybrids of *Festuca arundinacea × Lolium multiflorum*. Euphytica 105 (3):191–195. https://doi.org/10.1023/ a:1003479915606

414. Zwierzykowski Z, Lukaszewski AJ, Lesniewska A, Naganowska B (1998) Genomic structure of androgenic progeny of pentaploid hybrids, *Festuca arundinacea × Lolium multiflorum*. Plant Breed 117 (5):457–462. https://doi.org/10.1111/j. 1439-0523.1998.tb01973.x

415. Rose J, Dunwell J, Sunderland N (1987) Anther culture of *Lolium temulentum, Festuca pratensis* and *Lolium× Festuca* hybrids. I. Influence of pretreatment, culture medium and culture incubation conditions on callus production and differentiation. Ann Bot 60(2):191–201

416. Rose J, Dunwell J, Sunderland N (1987) Anther culture of *Lolium temulentum, Festuca pratensis* and *Lolium× Festuca* hybrids. II. Anther and pollen development in vivo and in vitro. Ann Bot 60(2):203–214

417. Leśniewska A, Ponitka A, Zwierzykowska E, Zwierzykowski Z, James A, Thomas H, Humphreys M (2001) Androgenesis from *Festuca pratensis × Lolium multiflorum* amphidiploid cultivars in order to select

418. Quarta D, Nati D, Paoloni F (1990) Strawberry anther culture. Acta Hortic 300:335–340

419. Li S, Wu W, Zhang Z, Wang D (1988) Study on anther culture of strawberry (*Fragaria ananassa*). Genet Manipulat Crops Newsl 4 (1):52–62

420. Rose J, Jones R, Simpson D (1993) Anther culture and intergeneric hybridization of *Fragaria × ananassa*. Adv Strawberry Res 12:59–64

421. Svensson M, Johansson L (1994) Anther culture of *Fragaria × ananassa*: environmental factors and medium components affecting microspore divisions and callus production. J Hortic Sci 69(3):417–426

422. Owen HR, Miller AR (1996) Haploid plant regeneration from anther cultures of three north american cultivars of strawberry (*Fragaria × ananassa* Duch.). Plant Cell Rep 15 (12):905–909. https://doi.org/10.1007/bf00231585

423. Xue G, Fei K, Hu J (1981) Induction of haploid plantlets of strawberry (*Fragaria orientalis*) by anther culture in vitro. Acta Hortic Sin 8:9–14

424. Jelenkovic G, Wilson M, Harding P (1984) An evaluation of intergeneric hybridization of *Fragaria* spp. × *Potentilla* spp. as a means of haploid production. Euphytica 33 (1):143–152

425. Doi H, Hoshi N, Yamada E, Yokoi S, Nishihara M, Hikage T, Takahata Y (2013) Efficient haploid and doubled haploid production from unfertilized ovule culture of gentians (*Gentiana* spp.). Breed Sci 63 (4):400–406. https://doi.org/10.1270/jsbbs.63.400

426. Doi H, Yokoi S, Hikage T, Nishihara M, K-i T, Takahata Y (2011) Gynogenesis in gentians (*Gentiana triflora, G. scabra*): production of haploids and doubled haploids. Plant Cell Rep 30(6):1099–1106

427. Doi H, Takahashi R, Hikage T, Takahata Y (2010) Embryogenesis and doubled haploid production from anther culture in gentian (*Gentiana triflora*). Plant Cell Tissue Organ Cult 102(1):27–33

428. Preil W, Huhnke W, Engelhardt M, Hoffmann M (1977) Haploids in *Gerbera jamesonii* from in vitro cultured capitulum explants. Z Pflanzen 79(2):167–171

429. Sitbon M (1981) Production of haploid *Gerbera jamesonii* plants by in vitro culture of unfertilized ovules. Agronomie, EDP Sci 1 (9):807–812

430. Meynet J, Sibi M (1984) Haploid plants from in vitro culture of unfertilized ovules in *Gerbera jamesonii*. Z Pflanzen 93 (1):78–85

431. Miyoshi K, Asakura N (1996) Callus induction, regeneration of haploid plants and chromosome doubling in ovule cultures of pot gerbera (*Gerbera jamesonii*). Plant Cell Rep 16(1–2):1–5

432. Tosca A, Arcara L, Frangi P (1999) Effect of genotype and season on gynogenesis efficiency in *Gerbera*. Plant Cell Tissue Organ Cult 59(1):77. https://doi.org/10.1023/a:1006418619992

433. Ahmim M, Vieth J (1986) Production de plantes haploïdes de *Gerbera jamesonii* par culture in vitro d'ovules. Can J Bot 64 (10):2355–2357. https://doi.org/10.1139/b86-309

434. Cappadocia M, Chrétien L, Laublin G (1988) Production of haploids in *Gerbera jamesonii* via ovule culture: influence of fall versus spring sampling on callus formation and shoot regeneration. Can J Bot 66 (6):1107–1110

435. Honkanen J, Aapola A, Seppänen P, Törmälä T, Oy K, de Wit J, Esendam H, Stravers L, Terra Nigra B (1990) Production of doubled haploid *Gerbera* clones. In Vitro Cult XXIII IHC 300:341–346

436. Shan Q, Wang J, Li S, Qu S, Wang G, Yang C, Jiang H (2017) Effect of colchicine on *Gerbera jamesonii* haploid doubling. Southw Chin J Agric Sci 30(10):2230–2234

437. Laurain D, Trémouillaux-Guiller J, Chénieux J-C (1993) Embryogenesis from microspores of *Ginkgo biloba* L., a medicinal woody species. Plant Cell Rep 12 (9):501–505

438. Crane C, Beversdorf W, Bingham E (1982) Chromosome pairing and associations at meiosis in haploid soybean (*Glycine max*). Can J Genet Cytol 24(3):293–300

439. Jian Y, Liu D, Luo X, Zhao G (1986) Studies on induction of pollen plants in *Glycine max* (L.) Merr. J Agric Sci 2:26–30

440. Hildebrand DF, Phillips GC, Collins GB (1986) Soybean [*Glycine max* (L.)Merr.]. In: Crops I. Springer, New York, NY, pp 283–308

441. Kadlec M, Suchomelova J, Smirnov V, Nikolajevna S (1991) Anther culture in soybean. Soybean Genet Newsl 18:121–124

442. Hu C-Y, Yin G-C, Helena M, Zanettini B (1996) Haploid of soybean. In: In Vitro

haploid production in higher plants. Springer, New York, NY, pp 377–395

443. Kaltchuk-Santos E, Mariath JE, Mundstock E, Hu C-y, Bodanese-Zanettini MH (1997) Cytological analysis of early microspore divisions and embryo formation in cultured soybean anthers. Plant Cell Tissue Organ Cult 49(2):107–115

444. Zhuang X, Hu C, Chen Y, Yin G (1991) Embryoids from soybean anther culture. Soybean Genet Newsl 18:265

445. Hai NH, Lal SK, Singh SK, Talukdar A, Vinod (2016) Anther culture of *Glycine max* (Merr.): effect of media on callus induction and organogenesis. Indian J Genet Plant Breed 76(3):319–325. https://doi.org/10.5958/0975-6906.2016.00048.1

446. Mehetre S (1984) Analysis of chromosome pairing in haploids of cotton (*Gossypium* spp.). Indian J Agric Res 18:49–53

447. Bajaj Y, Gill M (1989) Pollen-embryogenesis and chromosomal variation in anther culture of a diploid cotton (*Gossypium arboreum* L.). SABRAO J 21(1):57–63

448. Mehetre S, Thombre M (1981) Meiotic studies in the haploids (2n = 2x = 26) of tetraploid cottons (2n = 4x = 52). Proc Indian Natl Sci Acad B Biol Sci 47 (4):516–518

449. Bajaj Y, Gill MS (1997) In vitro induction of haploidy in cotton. In: In vitro haploid production in higher plants. Springer, New York, NY, pp 165–174

450. Stelly DM, Lee JA, Rooney WL (1988) Proposed schemes for mass-extraction of doubled haploids of cotton. Crop Sci 28 (6):885–890

451. Singh K, Sandhu BS, Gosal SS (1998) Anther culture response in cotton. Ann Biol 14:11–14

452. Contolini CS, Menzel MY (1987) Early development of duplication-deficiency ovules in upland cotton. Crop Sci 27 (2):345–348

453. Meredith WR Jr, Bridge R, Chism J (1970) Relative performance of F1 and F2 hybrids from doubled haploids and their parent varieties in upland cotton, *Gossypium hirsutum* L. Crop Sci 10(3):295–298

454. Pallares P (1984) First results from "in vitro" culture of unfertilized cotton ovules (*Gossypium hirsutum* L.). Coton Fibres Trop 39 (4):145–152

455. Barrow JR (1986) The conditions required to isolate and maintain viable cotton (*Gossypium hirsutum* L.) microspores. Plant Cell Rep 5(6):405–408

456. Zhou S, Qian D, Cao X (1989) Haploid breeding and its cytogenetics in cotton (*Gossypium hirsutum*). In: Mujeeb Kazi A, Sitch LA (eds) Review of Advances in Plant Biotechnology, 1985–1988. CIMMYT, IRRI, Manila, pp 323–324

457. Chaudhari H (1979) The production and performance of doubled haploids of cotton. B Torrey Bot Club 1979:123–130

458. Kavi Kishor P, Reddy T, Sarvesh A, Venkatesham G (1997) Haploidy in niger (*Guizotia abyssinica* Cass). In: In vitro haploid production in higher plants. Springer, New York, NY, pp 37–51

459. Makhmudov T (1978) Utilization of haploids in cotton breeding. Khlopkovodstvo 1978:31–32

460. Mahill JF, Jenkins JN, McCarty J Jr, Parrott W (1984) Performance and stability of doubled haploid lines of upland cotton derived via semigamy. Crop Sci 24(2):271–277

461. Adda S, Reddy T, Kishor PK (1994) Androclonal variation in niger (*Guizotia abyssinica* Cass). Euphytica 79(1–2):59–64

462. Sarvesh A, Reddy T, Kavi Kishor P (1993) Embryogenesis and organogenesis in cultured anthers of an oil yielding crop niger (*Guizotia abyssinica*. Cass). Plant Cell Tissue Organ Cult 35(1):75–80

463. Bhat JG, Murthy HN (2007) Factors affecting in-vitro gynogenic haploid production in niger (*Guizotia abyssinica* (L. f.) Cass.). Plant Growth Regul 52(3):241–248

464. Zhou C, Orndorff K, Allen RD, DeMaggio AE (1986) Direct observations on generative cells isolated from pollen grains of *Haemanthus katherinae* baker. Plant Cell Rep 5 (4):306–309. https://doi.org/10.1007/bf00269829

465. Mezzarobba A, Jonard R (1986) Effects of the stage of isolation and pretreatments on in vitro development of cultivated sunflower anthers (*Helianthus annuus* L.). Compt Rendus Acad Sci III Life Sci 303 (5):181–186

466. Hongyuan Y, Chang Z, Detian C, Hua Y, Yan W, Xiaoming C (1986) In vitro culture of unfertilized ovules in *Helianthus annuus* L. Haploids of higher plants in vitro/edited by Hu Han and Yang Hongyuan. pp 182–191

467. Gelebart P, San L (1987) Production of haploid plants in sunflower (*Helianthus annuus* L.) by in vitro culture of non fertilized ovaries and ovules. Agronomie (France) 7:81–86

468. Gürel A, Nichterlein K, Friedt W (1991) Shoot regeneration from anther culture of

sunflower (*Helianthus annuus*) and some interspecific hybrids as affected by genotype and culture procedure. Plant Breed 106 (1):68–76

469. Coumans M, Zhong D (1995) Doubled haploid sunflower (*Helianthus annuus*) plant production by androgenesis: fact or artifact? Part 2. In vitro isolated microspore culture. Plant Cell Tissue Organ Cult 41 (3):203–309

470. Nurhidayah T, Horn R, Röcher T, Friedt W (1996) High regeneration rates in anther culture of interspecific sunflower hybrids. Plant Cell Rep 16(3–4):167–173

471. Badigannavar AM, Kuruvinashetti M (1996) Callus induction and shoot bud formation from cultured anthers in sunflower (*Helianthus annuus* L.). Helia Novi Sad 19:39–46

472. Todorova M, Ivanov P, Shindrova P, Christov M, Ivanova I (1997) Doubled haploid production of sunflower (*Helianthus annuus* L.) through irradiated pollen-induced parthenogenesis. Euphytica 97 (3):249–254. https://doi.org/10.1023/a:1002966824988

473. Friedt W, Nurhidayah T, Röcher T, Köhler H, Bergmann R, Horn R (1997) Haploid production and application of molecular methods in sunflower (*Helianthus annuus* L.). In: In vitro haploid production in higher plants. Springer, New York, NY, pp 17–35

474. Saji K, Sujatha M (1998) Embryogenesis and plant regeneration in anther culture of sunflower (*Helianthus annuus* L.). Euphytica 103(1):1–7

475. Todorova M, Ivanov P (1999) Induced parthenogenesis in sunflower: effect of pollen donor. Helia (Yugoslavia) 22(31):49–56

476. Thengane SR, Joshi MS, Khuspe SS, Mascarenhas AF (1994) Anther culture in *Helianthus annuus* L., influence of genotype and culture conditions on embryo induction and plant regeneration. Plant Cell Rep 13 (3–4):222–226

477. Zhong D, Michauxferriere N, Coumans M (1995) Assay for doubled haploid sunflower (*Helianthus annuus*) plant production by androgenesis – fact or artifact. 1 In vitro anther culture. Plant Cell Tissue Organ Cult 41(2):91–97

478. Nenova N, Christov M, Ivanov P (1992) Anter culture regeneration of F1 hybrids of *Helianthus annuus* × *Helianthus smitii* and *Helianthus annuus* × *Heliantus eggerttii*. In: Proceedings of the XIII international

sunflower conference, Pisa. Springer, New York, NY, pp 1509–1514

479. Nenova N, Cristov M, Ivanov P (2000) Anther culture regeneration from some wild *Helianthus* species. Helia 23(32):65–72

480. Zhou C (1989) Cell divisions in pollen protoplast culture of *Hemerocallis fulva* L. Plant Sci 62(2):229–235. https://doi.org/10.1016/0168-9452(89)90085-X

481. Nomizu T, Niimi Y, D-s H (2004) Haploid plant regeneration via embryogenesis from anther cultures of *Hepatica nobilis*. Plant Cell Tissue Organ Cult 79(3):307–313

482. Zhenghua C, Wenbin L, Lihua Z, Xuen X, Shijie Z (1988) Production of haploid plantlets in cultures of unpolinated ovules of *Hevea brasiliensis* Muell.-Arg. In: Somatic cell genetics of woody plants. Springer, New York, NY, pp 39–44

483. Jayasree PK, Asokan M, Sobha S, Ammal LS, Rekha K, Kala R, Jayasree R, Thulaseedharan A (1999) Somatic embryogenesis and plant regeneration from immature anthers of *Hevea brasiliensis* (Muell.) Arg. Curr Sci 76:1242–1245

484. Chen Z, Qian C, Qin M, Xu X, Xiao Y (1982) Recent advances in anther culture of *Hevea brasiliensis* (Muell.-Arg.). Theor Appl Genet 62(2):103–108

485. Susanto D, Ibrahim AM, Hussin ZESM (2013) Pollen and anther cultures as potential means in production of haploid kenaf (*Hibiscus cannabinus* L.). Int J Adv Sci Eng Informat Technol 3(1):38–40

486. Ibrahim AM, Kayat FB, Susanto D, Ariffullah M, Kashiani P (2015) Callus induction from ovules of kenaf (*Hibiscus cannabinus* L.). Biotechnology 14(2):72–78

487. Mahmood Ibrahim A, Binti Kayat F, Ermiena Surya Mat Hussin Z, Susanto D, Ariffulah M (2014) Determination of suitable microspore stage and callus induction from anthers of kenaf (*Hibiscus cannabinus* L.). Sci World J 2014:Article ID 284342

488. Ma'arup R, Aziz MA, Osman M (2012) Development of a procedure for production of haploid plants through microspore culture of roselle (*Hibiscus sabdariffa* L.). Sci Hortic 145:52–61. https://doi.org/10.1016/j.scienta.2012.07.028

489. Bicknell RA, Borst NK (1996) Isolation of reduced genotypes of *Hieracium pilosella* using anther culture. Plant Cell Tissue Organ Cult 45(1):37–41. https://doi.org/10.1007/bf00043426

490. Gudu S, Procunier J, Ziauddin A, Kasha K (1993) Anther culture derived homozygous

lines in *Hordeum bulbosum*. Plant Breed 110 (2):109–115

491. Kihara M, Fukuda K, Funatsuki H, Kishinami I, Aida Y (1994) Plant regeneration through anther culture of three wild species of *Hordeum* (*H. murinum*, *H. marinum* and *H. bulbosum*). Plant Breed 112(3):244–247

492. Jørgensen RB, BOTHMER RV (1988) Haploids of *Hordeum vulgare* and *H. marinum* from crosses between the two species. Hereditas 108(2):207–212

493. Gaj M, Gaj M (1985) Dihaploids of *Hordeum murinum* L. and *H. secalinum* Schreb. from interspecific crosses with *H. bulbosum* L. Barley Genet Newsl 15:33–34

494. Wang XH, Lazzeri PA, Lörz H (1993) Regeneration of haploid, dihaploid and diploid plants from anther- and embryo-derived cell suspensions of wild barley (*Hordeum murinum* L.). J Plant Physiol 141 (6):726–732. https://doi.org/10.1016/S0176-1617(11)81582-8

495. Simpson E, Snape J (1980) Haploid production in *Hordeum spontaneum* × *H. bulbosum* crosses. Barley Genet Newsl 10:66–67

496. Piccirilli M, Arcioni S (1991) Haploid plants regenerated via anther culture in wild barley (*Hordeum spontaneum* C. Kock). Plant Cell Rep 10:273–276

497. Kintzios S, Fischbeck G (1994) Anther culture response of *Hordeum spontaneum*-derived winter barley lines. Plant Cell Tissue Organ Cult 37(2):165–170

498. Makowska K, Kałużniak M, Oleszczuk S, Zimny J, Czaplicki A, Konieczny R (2017) Arabinogalactan proteins improve plant regeneration in barley (*Hordeum vulgare* L.) anther culture. Plant Cell Tissue Organ Cult 131(2):247–257. https://doi.org/10.1007/s11240-017-1280-x

499. Lu R, Chen Z, Gao R, He T, Wang Y, Xu H, Guo G, Li Y, Liu C, Huang J (2016) Genotypes-independent optimization of nitrogen supply for isolated microspore cultures in barley. Biomed Res Int 2016:8. https://doi.org/10.1155/2016/1801646

500. Sriskandarajah S, Sameri M, Lerceteau-Köhler E, Westerbergh A (2015) Increased recovery of green doubled haploid plants from barley anther culture. Crop Sci 55 (6):2806–2812. https://doi.org/10.2135/cropsci2015.04.0245

501. Lippmann R, Friedel S, Mock H-P, Kumlehn J (2015) The low molecular weight fraction of compounds released from immature wheat pistils supports barley pollen embryogenesis. Front Plant Sci 6. https://doi.org/10.3389/fpls.2015.00498

502. Esteves P, Clermont I, Marchand S, Belzile F (2014) Improving the efficiency of isolated microspore culture in six-row spring barley: II. Exploring novel growth regulators to maximize embryogenesis and reduce albinism. Plant Cell Rep. https://doi.org/10.1007/s00299-014-1563-1

503. Castillo AM, Nielsen NH, Jensen A, Vallés MP (2014) Effects of n-butanol on barley microspore embryogenesis. Plant Cell Tissue Organ Cult 117(3):411–418. https://doi.org/10.1007/s11240-014-0451-2

504. Pulido A, Bakos F, Castillo A, Valles MP, Barnabas B, Olmedilla A (2006) Influence of Fe concentration in the medium on multicellular pollen grains and haploid plants induced by mannitol pretreatment in barley (*Hordeum vulgare* L.). Protoplasma 228 (1–3):101–106. https://doi.org/10.1007/s00709-006-0178-y

505. Pretova A, Obert B, Bartosova Z (2006) Haploid formation in maize, barley, flax, and potato. Protoplasma 228 (1–3):107–114. https://doi.org/10.1007/s00709-006-0170-6

506. Oleszczuk S, Sowa S, Zimny J (2006) Androgenic response to preculture stress in microspore cultures of barley. Protoplasma 228(1–3):95–100. https://doi.org/10.1007/s00709-006-0179-x

507. Kruczkowska H, Pawlowska H, Skucinska B (2005) Effect of 2,4-D concentration on the androgenic response in anther culture of barley. Cereal Res Commun 33(4):727–732

508. Wojnarowiez G, Caredda S, Devaux P, Sangwan R, Clément C (2004) Barley anther culture: assessment of carbohydrate effects on embryo yield, green plant production and differential plastid development in relation with albinism. J Plant Physiol 161 (6):747–755

509. Shim YS, Kasha KJ (2003) The influence of pretreatment on cell stage progression and the time of DNA synthesis in barley (*Hordeum vulgare* L.) uninucleate microspores. Plant Cell Rep 21(11):1065–1071

510. Li HC, Devaux P (2003) High frequency regeneration of barley doubled haploid plants from isolated microspore culture. Plant Sci 164(3):379–386. https://doi.org/10.1016/s0168-9452(02)00424-7

511. Hayes P, Corey A, DeNoma J (2003) Doubled haploid production in barley using the *Hordeum bulbosum* (L.) technique. In: Maluszynski M, Kasha KJ, Forster BP,

Szarejko I (eds) Doubled haploid production in crop plants. A manual. Kluwer Academic, Dordretch, pp 5–14

512. Devaux P (2003) The *Hordeum bulbosum* (L.) method. In: Maluszynski M, Kasha KJ, Forster BP, Szarejko I (eds) Doubled haploid production in crop plants. A manual. Kluwer Academic, Dordretch, pp 15–19

513. Castillo AM, Cistué L, Vallés MP, Sanz JM, Romagosa I, Molina-Cano JL (2001) Efficient production of androgenic doubled-haploid mutants in barley by the application of sodium azide to anther and microspore cultures. Plant Cell Rep 20(2):105–111. https://doi.org/10.1007/s002990000289

514. Castillo AM, Valles MP, Cistue L (2000) Comparison of anther and isolated microspore cultures in barley. Effects of culture density and regeneration medium. Euphytica 113(1):1–8

515. Hu TC, Kasha KJ (1999) A cytological study of pretreatments used to improve isolated microspore cultures of wheat (*Triticum aestivum* L.) cv. Chris. Genome 42(3):432–441

516. Davies PA, Morton S (1998) A comparison of barley isolated microspore and anther culture and the influence of cell culture density. Plant Cell Rep 17(3):206–210

517. Cistué L, Ramos A, Castillo AM (1998) Influence of anther pretreatment and culture medium composition on the production of barley doubled haploids from model and low responding cultivars. Plant Cell Tissue Organ Cult 55(3):159–166

518. Salmenkallio-Marttila M, Kurten U, Kauppinen V (1995) Culture conditions for efficient induction of green plants from isolated microspores of barley. Plant Cell Tissue Organ Cult 43(1):79–81

519. Cistué L, Ramos A, Castillo AM, Romagosa I (1994) Production of large number of doubled haploid plants from barley anthers pretreated with high concentrations of mannitol. Plant Cell Rep 13(12):709–712

520. Hoekstra S, Vanzijderveld MH, Heidekamp F, Vandermark F (1993) Microspore culture of *Hordeum vulgare* L. – the influence of density and osmolality. Plant Cell Rep 12(12):661–665

521. Hoekstra S, Vanzijderveld MH, Louwerse JD, Heidekamp F, Vandermark F (1992) Anther and microspore culture of *Hordeum vulgare* L. cv Igri. Plant Sci 86(1):89–96

522. Olsen FL (1991) Isolation and cultivation of embryogenic microspores from barley (*Hordeum vulgare* L.). Hereditas 115(3):255–266

523. Ziauddin A, Simion E, Kasha KJ (1990) Improved plant regeneration from shed microspore culture in barley (*Hordeum vulgare* L) cv. Igri. Plant Cell Rep 9(2):69–72

524. Sunderland N, Xu ZH (1982) Shed pollen culture in *Hordeum vulgare*. J Exp Bot 33:1086–1095

525. San Noeum LH (1976) Haploides d'*Hordeum vulgare* L. par culture *in vitro* d'ovaries non fécondés. Ann Amélior Plant 26:751–754

526. Kasha KJ, Kao KN (1970) High frequency haploid production in barley (*Hordeum vulgare* L.). Nature 225:874–876

527. Sanei M, Pickering R, Kumke K, Nasuda S, Houben A (2011) Loss of centromeric histone H3 (CENH3) from centromeres precedes uniparental chromosome elimination in interspecific barley hybrids. Proc Natl Acad Sci 108(33):E498–E505. https://doi.org/10.1073/pnas.1103190108

528. Roberts-Oehlschlager SL, Dunwell JM (1990) Barley anther culture: pretreatment on mannitol stimulates production of microspore-derived embryos. Plant Cell Tissue Organ Cult 20(3):235–240. https://doi.org/10.1007/bf00041887

529. Orlowska R, Pachota KA, Machczynska J, Niedziela A, Makowska K, Zimny J, Bednarek PT (2020) Improvement of anther cultures conditions using the Taguchi method in three cereal crops. Electron J Biotechnol 43:8–15. https://doi.org/10.1016/j.ejbt.2019.11.001

530. Echavarri B, Cistue L (2016) Enhancement in androgenesis efficiency in barley (*Hordeum vulgare* L.) and bread wheat (*Triticum aestivum* L.) by the addition of dimethyl sulfoxide to the mannitol pretreatment medium. Plant Cell Tissue Organ Cult 125(1):11–22. https://doi.org/10.1007/s11240-015-0923-z

531. Wernicke W, Lorz H, Thomas E (1979) Plant regeneration from leaf protoplasts of haploid *Hyoscyamus muticus* L. produced via anther culture. Plant Sci Lett 15(3):239–249. https://doi.org/10.1016/0304-4211(79)90116-0

532. Chand S, Basu P (1998) Embryogenesis and plant regeneration from callus cultures derived from unpollinated ovaries of *Hyoscyamus muticus* L. Plant Cell Rep 17(4):302–305

533. Strauss A, Bucher F, King PJ (1981) Isolation of biochemical mutants using haploid mesophyll protoplasts of *Hyoscyamus muticus*. Planta 153(1):75–80

534. Fankhauser H, Bucher F, King PJ (1984) Isolation of biochemical mutants using haploid mesophyll protoplasts of *Hyoscyamus muticus*. Planta 160(5):415–421

535. Reynolds TL (1985) Ultrastructure of anomalous pollen development in embryogenic anther cultures of *Hyoscyamus niger*. Am J Bot 72(1):44–51. https://doi.org/10.2307/2443567

536. Corduan G (1975) Regeneration of anther-derived plants of *Hyoscyamus niger* L. Planta 127(1):27–36. https://doi.org/10.1007/bf00388860

537. Raghavan V (1975) Role of the generative cell in androgenesis in henbane. Science 191:388–389

538. Raghavan V (1978) Origin and development of pollen embryoids and pollen calluses in cultured anther segments of *Hyoscyamus niger* (henbane). Am J Bot 65:984–1002

539. Dodds JH, Reynolds TL (1980) A scanning electron-microscope study of pollen embryogenesis in *Hyoscyamus niger*. Z Pflanzenphysiol 97(3):271–276

540. Reynolds TL (1984) An ultrastructural and stereological analysis of pollen grains of *Hyoscyamus niger* during normal ontogeny and induced embryogenic development. Am J Bot 71(4):490–504

541. Raghavan V, Nagmani R (1989) Cytokinin effects on pollen embryogenesis in cultured anthers of *Hyoscyamus niger*. Can J Bot 67(1):247–257

542. Garcia RB, Cisneros A, Schneider B, Tel-Zur N (2009) Gynogenesis in the vine cacti *Hylocereus* and *Selenicereus* (Cactaceae). Plant Cell Rep 28(5):719–726

543. Schulte J, Büter B, Schaffner W, Berger K (1996) Gametic embryogenesis in *Hypericum* spp. In: Pank F (ed) International symposium on breeding research on medicinal and aromatic plants, Quedlinburg, Germany. BREEDMAP 6, Quedlinburg, pp 307–310

544. Canhoto JM, Ludovina M, Guimaraes S, Cruz GS (1990) In vitro induction of haploid, diploid and triploid plantlets by anther culture of *Iochroma warscewiczii* Regel. Plant Cell Tissue Organ Cult 21(2):171–177

545. Tsay H, Lai P, Chen L (1982) Organ regeneration from anther callus of sweet potato. J Agric Res China 31(2):123–126

546. Mukherjee A, Unnikrishnan M, Nair N (1991) Callus induction, embryogenesis and regeneration from sweet potato anther. J Root Crops 17:302–304

547. Madan NS, Arockiasamy S, Narasimham JV, Patil M, Yepuri V, Sarkar P (2019) Anther culture for the production of haploid and doubled haploids in *Jatropha curcas* L. and its hybrids. Plant Cell Tissue Organ Cult 138(1):181–192. https://doi.org/10.1007/s11240-019-01616-4

548. Grouh MSH, Vahdati K, Lotfi M, Hassani D, Biranvand NP (2011) Production of haploids in Persian walnut through parthenogenesis induced by gamma-irradiated pollen. J Am Soc Hortic Sci 136(3):198–204

549. Piosik Ł, Zenkteler E, Zenkteler M (2016) Development of haploid embryos and plants of *Lactuca sativa* induced by distant pollination with *Helianthus annuus* and *H. tuberosus*. Euphytica 208(3):439–451. https://doi.org/10.1007/s10681-015-1578-x

550. von Aderkas P, Bonga JM (1988) Formation of haploid embryoids of *Larix decidua*: early embryogenesis. Am J Bot 75(5):690–700. https://doi.org/10.1002/j.1537-2197.1988.tb13491.x

551. Nagmani R, Bonga J (1985) Embryogenesis in subcultured callus of *Larix decidua*. Can J For Res 15(6):1088–1091

552. von Aderkas P, Klimaszewska K, Bonga JM (1990) Diploid and haploid embryogenesis in *Larix leptolepis*, L. decidua, and their reciprocal hybrids. Can J Forest Res 20(1):9–14. https://doi.org/10.1139/x90-002

553. Ochatt S, Pech C, Grewal R, Conreux C, Lulsdorf M, Jacas L (2009) Abiotic stress enhances androgenesis from isolated microspores of some legume species (Fabaceae). J Plant Physiol 166(12):1314–1328. https://doi.org/10.1016/j.jplph.2009.01.011

554. Croser J, Lülsdorf M, Davies P, Clarke H, Bayliss K, Mallikarjuna N, Siddique K (2006) Toward doubled haploid production in the Fabaceae: progress, constraints, and opportunities. Crit Rev Plant Sci 25(2):139–157

555. Tomasi P, Dierig DA, Backhaus RA, Pigg KB (1999) Floral bud and mean petal length as morphological predictors of microspore cytological stage in *Lesquerella*. HortScience 34(7):1269–1270

556. Han D-S, Niimi Y, Nakano M (1997) Regeneration of haploid plants from anther cultures of the Asiatic hybrid lily 'Connecticut King'. Plant Cell Tissue Organ Cult 47(2):153–158

557. Han DS, Niimi Y, Nakano M (1999) Production of doubled haploid plants through colchicine treatment of anther-derived haploid calli in the Asiatic hybrid lily 'Connecticut King'. J Jpn Soc Hortic Sci 68 (5):979–983

558. Zhu-ping G, Kuo-chang C (1983) In vitro induction of haploid plantlets from unpollinated young ovaries of lily and its embryo logical observations. J Integr Plant Biol 25 (1):73–88

559. Han D-S, Niimi Y (2004) Production of haploid and doubled haploid plants from anther-derived callus of *Lilium formosanum*. IX international symposium on flower bulbs 673:389–393

560. Arzate-Fernández A-M, Nakazaki T, Yamagata H, Tanisaka T (1997) Production of doubled-haploid plants from *Lilium longiflorum* Thunb. anther culture. Plant Sci 123 (1):179–187. https://doi.org/10.1016/S0168-9452(96)04573-6

561. Qu Y, Mok MC, Mok DW, Stang JR (1988) Phenotypic and cytological variation among plants derived from anther cultures of *Lilium longiflorum*. In Vitro Cell Dev Biol 24(5):471–476

562. Vassileva-Dryanovska OA (1966) The induction of embryos and tetraploid endosperm nuclei with irradiated pollen in *Lilium*. Hereditas 55(2–3):160–165. https://doi.org/10.1111/j.1601-5223.1966.tb02044.x

563. Prakash J, Giles K (1986) Production of doubled haploids in oriental lilies. In: Horn W, Jensen CJ, Oldenbach W, Schieder O (eds) Genetic manipulation in plant breeding. Walter de Gruyter and Co, Berlin, pp 335–337

564. Van den Bulk R, Van Tuyl J (1997) In vitro induction of haploid plants from the gametophytes of lily and tulip. In: Jain SM, Sopory SK, Veilleux RE (eds) In vitro haploid production in higher plants. Springer, Dordrecht, pp 73–88

565. Obert B, Zackova Z, Samaj J, Pret'ova A (2009) Doubled haploid production in Flax (*Linum usitatissimum* L.). Biotechnol Adv 27(4):371–375. https://doi.org/10.1016/j.biotechadv.2009.02.004

566. Chen Y, Dribnenki P (2004) Effect of medium osmotic potential on callus induction and shoot regeneration in flax anther culture. Plant Cell Rep 23(5):272–276. https://doi.org/10.1007/s00299-004-0831-x

567. Nichterlein K, Umbach H, Friedt W (1991) Genotypic and exogenous factors affecting shoot regeneration from anther callus of linseed (*Linum usitatissimum* L.). Euphytica 58(2):157–164. https://doi.org/10.1007/bf00022816

568. Nichterlein K, Friedt W (1993) Plant regeneration from isolated microspores of linseed (*Linum usitatissimum* L.). Plant Cell Rep 12 (7):426–430. https://doi.org/10.1007/bf00234706

569. Fu L, Tang D (1983) Induction of pollen plants of litchi tree (*Litchi chinensis* Sonn.). Acta Genet Sin 10(5):369–374

570. Hussain W, Richardson K, Faville M, Woodfield D (2006) Production of haploids and double haploids in annual (*Lolium multiflorum*) and perennial (*L. perenne*) ryegrasses. Proceedings of the 13th Australasian plant breeding conference 12

571. Pašakinskienė I, Anamthawat-Jónsson K, Humphreys MW, Jones RN (1997) Novel diploids following chromosome elimination and somatic recombination in *Lolium multiflorum* × *Festuca arundinacea* hybrids. Heredity 78(5):464–469. https://doi.org/10.1038/hdy.1997.74

572. Humphreys M, Zare A, Pašakinskienė I, Thomas H, Rogers W, Collin H (1998) Interspecific genomic rearrangements in androgenic plants derived from a *Lolium multiflorum* × *Festuca arundinacea* (2n = 5x = 35) hybrid. Heredity 80 (1):78–82

573. Guo YD, Mizukami Y, Yamada T (2005) Genetic characterization of androgenic progeny derived from *Lolium perenne* × *Festuca pratensis* cultivars. New Phytol 166 (2):455–464

574. Tomes DT, Peterson RL (1981) Isolation of a dwarf plant responsive to exogenous GA3 from anther cultures of birdsfoot trefoil. Can J Bot 59(7):1338–1342. https://doi.org/10.1139/b81-180

575. Séguin-Swartz G, Grant W (1995) Evidence for androgenesis in the genus *Lotus* (Fabaceae). Lotus Newsl 26:4–8

576. Negri V, Veronesi F (1989) Evidence for the existence of 2n gametes in *Lotus tenuis* Wald. et Kit. (2n = 2x = 12): their relevance in evolution and breeding of *Lotus corniculatus* L. (2n = 4x = 24). Theor Appl Genet 78 (3):400–404

577. Bayliss K, Wroth J, Cowling W (2004) Pro-embryos of *Lupinus* spp. produced from isolated microspore culture. Aust J Agric Res 55(5):589–593

578. Simioniuc D, Burlacu-Arsene M-C, Morariu A, Lipsa F (2010) Induction of the embryogenesis process in anther and microspores cultures at the *Lupinus albus* species. Lucrări Ştiinţifice, Universitatea de Stiinte Agricole Şi Medicină Veterinară "Ion Ionescu de la Brad" Iaşi, Seria. Agronomie 53(1):60–63

579. Ormerod AJ, Caligari PDS (1994) Anther and microspore culture of *Lupinus albus* in liquid culture medium. Plant Cell Tissue Organ Cult 36(2):227–236. https://doi.org/10.1007/bf00037724

580. Kozak K, Galek R, Waheed MT, Sawicka-Sienkiewicz E (2012) Anther culture of *Lupinus angustifolius*: callus formation and the development of multicellular and embryo-like structures. Plant Growth Regul 66(2):145–153. https://doi.org/10.1007/s10725-011-9638-2

581. Sator C, Mix G, Menge U (1982) Investigations on anther culture of *Lupinus polyphyllus*. Landbauforschung Voelkenrode 32:37–42

582. Fan Y, Zang S, Zhao J (1982) Induction of haploid plants in *Lycium chinense* Mill. and *Lycium barbarum* by anther culture. Hereditas 5:25–26

583. Hofer M, Touraev A, Heberle-Bors E (1999) Induction of embryogenesis from isolated apple microspores. Plant Cell Rep 18(12):1012–1017

584. Höfer M (2004) In vitro androgenesis in apple—improvement of the induction phase. Plant Cell Rep 22(6):365–370. https://doi.org/10.1007/s00299-003-0701-y

585. Höfer M (2005) Regeneration of androgenic embryos in apple (*Malus × domestica* Borkh.) via anther and microspore culture. Acta Physiol Plant 27(4B):709–716

586. Kadota M, Han D-S, Niimi Y (2002) Plant regeneration from anther-derived embryos of apple and pear. HortScience 37 (6):962–965

587. Höfer M, Hanke V (1989) Induction of androgenesis in vitro in apple and sweet cherry. I International symposium on in vitro culture and horticultural breeding 280:333–336

588. Monika H (1994) In vitro androgenesis in apple: induction, regeneration and ploidy level. In: Schmidt H, Kellerhals M (eds) Progress in temperate fruit breeding. Developments in plant breeding, vol 1. Springer, Dordrecht, pp 399–402. https://doi.org/10.1007/978-94-011-0467-8_80

589. Zhang YX, Lespinasse Y (1991) Pollination with gamma-irradiated pollen and development of fruits, seeds and parthenogenetic plants in apple. Euphytica 54(1):101–109. https://doi.org/10.1007/bf00145636

590. Zhang CF, Sato S, Tsukuni T, Sato M, Okada H, Yamamoto T, Wada M, Matsumoto S, Yoshikawa N, Mimida N, Takagishi K, Watanabe M, Cao QF, Komori S (2017) Elucidating cultivar differences in plant regeneration ability in an apple anther culture. Hortic J 86(1):1–10. https://doi.org/10.2503/hortj.MI-094

591. Hofer M, Flachowsky H (2015) Comprehensive characterization of plant material obtained by in vitro androgenesis in apple. Plant Cell Tissue Organ Cult 122 (3):617–628. https://doi.org/10.1007/s11240-015-0794-3

592. Wu J (1981) Obtaining haploid plantlets of crab apple from anther culture in vitro. Acta Hortic Sin 8:36

593. Cheema GS, Mehra PN (1981) Anther culture of a cactus: *Mammillaria elongata* var. *tenuis* (DC) Schumann. Natl Cactus Succulent J 36(1):8–11

594. Liu M-C, Chen W-H (1978) Organogenesis and chromosome number in callus derived from cassava anthers. Can J Bot 56 (10):1287–1290

595. Abraham A, Krishnan P, Seeni S (1995) Induction of androgenesis, callus formation and root differentiation in anther culture of cassava (*Manihot esculenta* Crantz). Indian J Exp Biol 33:186–189

596. Perera PIP, Ordoñez CA, Dedicova B, Ortega PEM (2014) Reprogramming of cassava (*Manihot esculenta*) microspores towards sporophytic development. AoB Plants 6. https://doi.org/10.1093/aobpla/plu022

597. Perera PI, Ordoñez CA, Lopez-Lavalle LA, Dedicova B (2013) A milestone in the doubled haploid pathway of cassava: a milestone in the doubled haploid pathway of cassava (*Manihot esculenta* Crantz): cellular and molecular assessment of anther-derived structures. Protoplasma. https://doi.org/10.1007/s00709-013-0543-6

598. Bingham ET (1969) Haploids from cultivated Alfalfa, *Medicago sativa* L. Nature 221(5183):865–866. https://doi.org/10.1038/221865a0

599. Bingham E, Gillies C (1971) Chromosome pairing, fertility, and crossing behavior of haploids of tetraploid alfalfa, *Medicago sativa* L. Can J Genet Cytol 13(2):195–202

600. Tanner G, Moore A, Larkin P (1988) Reducing the ploidy of lucerne by anther culture or induced parthenogenesis. In: McWhirter KS, Downes RW, Read BJ (eds) Ninth Australian plant breeding conference, Wagga Wagga. Agricultural Research Organising Committee, Wagga Wagga, NSW, p 136

601. Tanner G, Piccirilli M, Moore A, Larkin P, Arcioni S (1990) Initiation of non-physiological division and manipulation of developmental pathway in cultured microspores of *Medicago* sp. Protoplasma 158(3):165–175

602. Ray I, Bingham E (1989) Breeding diploid alfalfa for regeneration from tissue culture. Crop Sci 29(6):1545–1548

603. Zagorska N, Dimitrov B (1995) Induced androgenesis in alfalfa (*Medicago sativa* L.). Plant Cell Rep 14(4):249–252. https://doi.org/10.1007/bf00233643

604. Skinner DZ, Liang GH (1996) Haploidy in alfalfa. In: Jain SM, Sopory SK, Veilleux RE (eds) In vitro haploid production in higher plants. Springer, Dordrecht, pp 365–375

605. Zagorska N, Dimitrov B, Gadeva P, Robeva P (1997) Regeneration and characterization of plants obtained from anther cultures in *Medicago sativa* L. In Vitro Cell Dev Biol Plant 33(2):107–110

606. Yi D, Sun J, Su Y, Tong Z, Zhang T, Wang Z (2019) Doubled haploid production in alfalfa (*Medicago sativa* L.) through isolated microspore culture. Sci Rep 9(1):9458. https://doi.org/10.1038/s41598-019-45946-x

607. Bingham E, Binek A (1969) Comparative morphology of haploids from cultivated alfalfa, *Medicago sativa* L. 1. Crop Sci 9 (6):749–751

608. Bingham T (1971) Isolation of haploids of tetraploid alfalfa. Crop Sci 11(3):433–435

609. Veuskens J, Ye D, Oliveira M, Ciupercescu DD, Installé P, Verhoeven HA, Negrutiu I (1992) Sex determination in the dioecious *Melandrium album*: androgenic embryogenesis requires the presence of the X chromosome. Genome 35(1):8–16. https://doi.org/10.1139/g92-002

610. Mól R (1992) *In vitro* gynogenesis in *Melandrium album*: from parthenogenetic embryos to mixoploid plants. Plant Sci 81 (2):261–269

611. Paulíková D, Vagera J (1993) *In vitro* induced androgenesis in *Melandrium album*. Biol Plant 35(4):645–647

612. Van Eck J, Kitto S (1990) Callus initiation and regeneration in *Mentha*. HortScience 25 (7):804–806

613. Murovec J, Bohanec B (2013) Haploid induction in *Mimulus aurantiacus* Curtis obtained by pollination with gamma irradiated pollen. Sci Hortic 162:218–225

614. Nguyen ML, Ta THT, Huyen TNBT, Voronina AV (2019) Anther-derived callus formation in bitter melon (*Momordica charantia* L.) as influenced by microspore development stage and medium composition. Sel'skokhozyaistvennaya Biol 54 (1):140–148. https://doi.org/10.15389/agrobiology.2019.1.140rus

615. Dennis Thomas T, Bhatnagar AK, Razdan MK, Bhojwani SS (1999) A reproducible protocol for the production of gynogenic haploids of mulberry, *Morus alba* L. Euphytica 110(3):169–173. https://doi.org/10.1023/a:1003797328246

616. Jain A, Sarkar A, Datra R (1996) Induction of haploid callus and embryogenesis in in vitro cultured anthers of mulberry (*Morus indica*). Plant Cell Tissue Organ Cult 44(2):143–147

617. Lin S, Ji D, Qin J (1987) In vitro production of haploid plants from mulberry (*Morus*) anther culture. Sci China Ser B Chem Biol Agric Med Earth Sci 30(8):853–863

618. Sita GL, Ravindran S (1991) Gynogenic plants from ovary cultures of mulberry (*Morus indica*). In: Horticulture—new technologies and applications. Springer, New York, NY, pp 225–229

619. Perea Dallos M (1997) Pollen and anther culture in *Musa* spp. II International symposium on banana: I international symposium on banana in the subtropics 490:493–500

620. Assani A, Bakry F, Kerbellec F, Haicour R, Wenzel G, Foroughi-Wehr B (2003) Production of haploids from anther culture of banana [*Musa balbisiana* (BB)]. Plant Cell Rep 21(6):511–516

621. Imelda M, Sastrapradja S, Lubis S (1988) Anther culture of rambután (*Nephelium* sp). Ann Bogorienses N Ser 1(1):7–9

622. Collins GB, Sunderland N (1974) Pollen-derived haploids of *Nicotiana knightiana*, *N. raimondii*, and *N. attenuata*. J Exp Bot 25(6):1030–1039. https://doi.org/10.1093/jxb/25.6.1030

623. Kehr AE (1951) Monoploidy in *Nicotiana*. J Hered 42(2):107–112

624. Kostoff D (1929) An androgenic *Nicotiana* haploid. Z Zellforsch 9:640–642

625. Clausen RE, Mann MC (1924) Inheritance in *Nicotiana tabacum*: V. The occurrence of haploid plants in interspecific progenies. Proc Natl Acad Sci U S A 10(4):121

626. Kostoff D (1942) The problem of haploidy (cytogenetic studies in *Nicotiana* haploids and their bearing on some other cytogenetic problems). Bib Genet 13:1–148

627. Bourgin JP, Nitsch JP (1967) Obtention de *Nicotiana* haploids à partir d'etamines cultivées *in vitro*. Ann Physiol Veg 9:377–382

628. Nitsch JP, Nitsch C (1969) Haploid plants from pollen grains. Science 163 (3862):85–87

629. Sunderland N, Wicks FM (1969) Cultivation of haploid plants from tobacco pollen. Nature 224(5225):1227–1229. https://doi. org/10.1038/2241227b0

630. Sunderland N, Wicks FM (1971) Embryoid formation in pollen grains of *Nicotiana tabacum*. J Exp Bot 22(70):213–226

631. Sunderland N, Dunwell JM (1974) Anther and pollen culture. In: Street HE (ed) Plant tissue and cell culture. Blackwell Scientific Publications, Oxford, pp 223–265

632. Heberle-Bors E (1982) In vitro pollen embryogenesis in *Nicotiana tabacum* L. and its relation to pollen sterility, sex balance, and floral induction of the pollen donor plants. Planta 156:396–401

633. Harada H, Kyo M, Imamura J (1988) The induction of embryogenesis in *Nicotiana* inmature pollen culture. In: Bock G, Marsh J (eds) Applications of plant cell and tissue culture. John Wiley and Sons, Chichester, pp 59–74

634. Garrido D, Charvat B, Benito-Moreno RM, Alwen A, Vicente O, Heberle-Bors E (1991) Pollen culture for haploid plant formation in tobacco. In: Negrutiu I, Gharti-Chhetri G (eds) A laboratory guide for cellular and molecular plant biology. Birkhaüser-Verlag, Basel, pp 59–69

635. Touraev A, Ilham A, Vicente O, Heberle-Bors E (1996) Stress-induced microspore embryogenesis in tobacco: an optimized system for molecular studies. Plant Cell Rep 15 (8):561–565

636. Atanassov A, Djilianov D (1997) Androgenesis in vitro in tobacco. Biotechnol Biotechnol Eq 11(1–2):3–11

637. Touraev A, Heberle-Bors E (2003) Anther and microspore culture in tobacco. In: Maluszynski M, Kasha KJ, Forster BP, Szarejko I (eds) Doubled haploid production in crop plants. Kluwer Academic, Dordrecht, pp 223–228

638. Sood S, Dwivedi S, Reddy TV, Prasanna PS, Sharma N (2013) Improving androgenesis-mediated doubled haploid production efficiency of FCV tobacco (*Nicotiana tabacum* L.) through in vitro colchicine application. Plant Breed 132(6):764–771. https://doi. org/10.1111/pbr.12114

639. De Oliveira E (2016) Optimization of doubled haploid production in burley tobacco (*Nicotiana Tabacum* L.). University of Kentucky, Lexington, KY

640. Yamaji N, Kyo M (2006) Two promoters conferring active gene expression in vegetative nuclei of tobacco immature pollen undergoing embryogenic dedifferentiation. Plant Cell Rep 25(8):749–757. https://doi. org/10.1007/s00299-005-0076-3

641. Garrido D, Eller N, Heberle-Bors E, Vicente O (1993) De novo transcription of specific mRNAs during the induction of tobacco pollen embryogenesis. Sex Plant Reprod 6:40–45

642. Orcen N, Emiroglu U (2014) Cytological characterization of tobacco plantlets obtained from androgenic haploids through chromosome doubling. Fresenius Environ Bull 23(2):378–381

643. Schedel S, Pencs S, Hensel G, Muller A, Rutten T, Kumlehn J (2017) RNA-guided Cas9-induced mutagenesis in tobacco followed by efficient genetic fixation in doubled haploid plants. Front Plant Sci 7. https:// doi.org/10.3389/fpls.2016.01995

644. Floss DM, Kumlehn J, Conrad U, Saalbach I (2009) Haploid technology allows for the efficient and rapid generation of homozygous antibody-accumulating transgenic tobacco plants. Plant Biotechnol J 7 (7):593–601

645. Martinez LD, de Halac IN (1995) Organogenesis of anther-derived calluses in long-term cultures of *Oenothera hookeri* de Vries. Plant Cell Tissue Organ Cult 42(1):91–96

646. Perri E, Parlati M, Mulé R, Fodale A (1993) Attempts to generate haploid plants from in vitro cultures of *Olea europaea* L. anthers. Acta Hortic 356:47–50

647. Solis M-T, Pintos B, Prado M-J, Bueno M-A, Raska I, Risueno M-C, Testillano PS (2008) Early markers of in vitro microspore reprogramming to embryogenesis in olive (*Olea europaea* L.). Plant Sci 174 (6):597–605. https://doi.org/10.1016/j. plantsci.2008.03.014

648. Bouamama-Gzara B, Zemni H, Zoghlami N, Gandoura S, Mliki A, Arnold M, Ghorbel A (2020) Behavior of

Opuntia ficus-indica (L.) Mill. heat-stressed microspores under in vitro culture conditions as evidenced by microscopic analysis. In Vitro Cell Dev Biol Plant 56(1):122–133. https://doi.org/10.1007/s11627-019-10032-4

649. Tang K, Sun X, He Y, Zhang Z (1998) Anther culture response of wild *Oryza* species. Plant Breed 117(5):443–446

650. Gueye T, Ndir KN (2010) In vitro production of double haploid plants from two rice species (*Oryza sativa* L. and *Oryza glaberrima* Steudt.) for the rapid development of new breeding material. Sci Res Essays 5 (7):709–713

651. Wakasa K, Watanabe Y (1979) Haploid plant of *Oryza perennis* (spontanea type) induced by anther culture. Jpn J Breed 29 (2):146–150

652. Myint A, de Fossard RA (1974) Induction of haploid callus from rice anthers and regeneration of plants. In: Kasha KJ (ed) Haploids in higher plants: advances and potential. University of Guelph, Guelph, ON, p 139

653. Miah MAA, Earle ED, Khush GS (1985) Inheritance of callus formation ability in anther cultures of rice, *Oryza sativa* L. Theor Appl Genet 70(2):113–116. https://doi.org/10.1007/bf00275308

654. Cho MS, Zapata FJ (1988) Callus formation and plant-regeneration in isolated pollen culture of rice (*Oryza Sativa* L. cv. Taipei 309). Plant Sci 58(2):239–244

655. Yamagishi M, Yano M, Fukuta Y, Fukui K, Otani M, Shimada T (1996) Distorted segregation of RFLP markers in regenerated plants derived from anther culture of an F1 hybrid of rice. Genes Genet Syst 71:37–41

656. Yamagishi M, Otani M, Higashi M, Fukuta Y, Fukui K, Shimada T (1998) Chromosomal regions controlling anther culturability in rice (*Oryza sativa* L.). Euphytica 103(2):227–234. https://doi.org/10.1023/a:1018328708322

657. Grewal D, Manito C, Bartolome V (2011) Doubled haploids generated through anther culture from crosses of elite *Indica* and *Japonica* cultivars and/or lines of rice: large-scale production, agronomic performance, and molecular characterization. Crop Sci 51(6):2544–2553. https://doi.org/10.2135/cropsci2011.04.0236

658. Premvaranon P, Vearasilp S, S-n T, Karladee D, Gorinstein S (2011) In vitro studies to produce double haploid in *Indica* hybrid rice. Biologia 66(6):1074

659. Hooghvorst I, Ramos-Fuentes E, López-Cristofannini C, Ortega M, Vidal R, Serrat X, Nogués S (2018) Antimitotic and hormone effects on green double haploid plant production through anther culture of Mediterranean japonica rice. Plant Cell Tissue Organ Cult 134:205–215

660. López-Cristoffanini C, Serrat X, Ramos-Fuentes E, Hooghvorst I, Lla ó R, López-Carbonell M, Nogués S (2018) An improved anther culture procedure for obtaining new commercial Mediterranean temperate *Japonica* rice (*Oryza sativa*) genotypes. Plant Biotechnol 35(2):161–166. https://doi.org/10.5511/plantbiotechnology.18.0409a

661. Ferreres I, Ortega M, López-Cristoffanini C, Nogués S, Serrat X (2019) Colchicine and osmotic stress for improving anther culture efficiency on long grain temperate and tropical japonica rice genotypes. Plant Biotechnol (Tokyo) 36(4):269–273. https://doi.org/10.5511/plantbiotechnology.19.1022a

662. Sahoo SA, Jha Z, Verulkar SB, Srivastava AK, Suprasanna P (2019) High-throughput cell analysis based protocol for ploidy determination in anther-derived rice callus. Plant Cell Tissue Organ Cult 137(1):187–192. https://doi.org/10.1007/s11240-019-01561-2

663. Usenbekov BN, Kaykeev DT, Yhanbirbaev EA, Berkimbaj H, Tynybekov BM, Satybaldiyeva GK et al (2014) Doubled haploid production through culture of anthers in rice. Indian J Genet Plant Breed 74 (1):90–92

664. Alsabah R, Purwoko BS, Dewi IS, Wahyu Y (2019) Selection index for selecting promising double haploid lines of Black Rice. Sabrao J Breed Genet 51(4):430–441

665. Mayakaduwa D, Silva TD (2019) Flow citometry detection of haploids, diploids and mixoploids among the anther-derived plants in indica rice (*Oryza sativa* L.). J Anim Plant Sci 29(5):1344–1351

666. Samal P, Pote TD, Krishnan SG, Singh AK, Salgotra RK, Rathour R (2019) Integrating marker-assisted selection and doubled haploidy for rapid introgression of semi-dwarfing and blast resistance genes into a Basmati rice variety 'Ranbir Basmati'. Euphytica 215(9). https://doi.org/10.1007/s10681-019-2473-7

667. Kaushal L, Ulaganathan K, Shenoy V, Balachandran SM (2018) Geno- and phenotyping of submergence tolerance and elongated uppermost internode traits in doubled

haploids of rice. Euphytica 214(12). https://doi.org/10.1007/s10681-018-2305-1

668. Naik N, Rout P, Umakanta N, Verma RL, Katara JL, Sahoo KK, Singh ON, Samantaray S (2017) Development of doubled haploids from an elite indica rice hybrid (BS6444G) using anther culture. Plant Cell Tissue Organ Cult 128(3):679–689. https://doi.org/10.1007/s11240-016-1149-4

669. Rout P, Naik N, Ngangkham U, Verma RL, Katara JL, Singh ON, Samantaray S (2016) Doubled haploids generated through anther culture from an elite long duration rice hybrid, CRHR32: method optimization and molecular characterization. Plant Biotechnol 33(3):177–186. https://doi.org/10.5511/plantbiotechnology.16.0719a

670. Nguyen H, Chen XY, Jiang M, Wang Q, Deng L, Zhang WZ, Shu QY (2016) Development and molecular characterization of a doubled haploid population derived from a hybrid between *Japonica* rice and wide compatible *Indica* rice. Breed Sci 66(4):552–559. https://doi.org/10.1270/jsbbs.15141

671. Cha-Um S, Srianan B, Pichakum A, Kirdmanee C (2009) An efficient procedure for embryogenic callus induction and double haploid plant regeneration through anther culture of Thai aromatic rice (*Oryza sativa* L. subsp *indica*). In Vitro Cell Dev Biol Plant 45(2):171–179. https://doi.org/10.1007/s11627-009-9203-0

672. Yao L, Zhang Y, Liu C, Liu Y, Wang Y, Liang D, Liu J, Sahoo G, Kelliher T (2018) OsMATL mutation induces haploid seed formation in *indica* rice. Nat Plants 4:530–533. https://doi.org/10.1038/s41477-018-0193-y

673. Hooghvorst I, Ribas P, Nogués S (2020) Chromosome doubling of androgenic haploid plantlets of rice (*Oryza sativa*) using antimitotic compounds. Plant Breed. https://doi.org/10.1111/pbr.12824

674. Woo S-C, Ko S-W, Wong C-K, Wu X (1983) Anther culture of pollen plants derived from cross *Oryza sativa* L. × *O. glaberrima* Steud. Bot Bull Acad Sin 24:53–58

675. Lee B, Ko J, Kim Y (1992) Studies on the thidiazuron treatment of anther culture in *Paeonia albiflora*. J Korean Soc Hortic Sci 33(5):384–395

676. Lee H-Y, Khorolragchaa A, Sun M-S, Kim Y-J, Kim Y-J, Kwon W-S, Yang D-C (2013) Plant regeneration from anther culture of *Panax ginseng*. Korean J Plant Resour 26(3):383–388

677. Du L, Shao Q, Li A (1986) Somatic embryogenesis and plant regeneration from anther culture of *Panax quinquifolius* (Ginseng). Int Plant Biotechnol Netw 6(9):2

678. Dieu P, Dunwell JM (1988) Anther culture with different genotypes of opium poppy (*Papaver somniferum* L.): effect of cold treatment. Plant Cell Tissue Organ Cult 12(3):263–271. https://doi.org/10.1007/bf00034367

679. Gerstel D, Mishanec W (1950) On the inheritance of apomixis in *Parthenium argentatum*. Bot Gaz 112(1):96–106

680. Tsay H, Hsu J, Yang T, Yang C (1984) Anther culture of passion fruit (*Passiflora edulis*). J Agric Res China 33:126–131

681. Rêgo M, Rêgo E, Bruckner C, Otoni W, Pedroza C (2011) Variation of gynogenic ability in passion fruit (*Passiflora edulis* Sims.) accessions. Plant Breed 130(1):86–91

682. El-Nil MA, Hildebrandt A (1973) Origin of androgenetic callus and haploid geranium plants. Can J Bot 51(11):2107–2109

683. Kato M, Suga T, Tokumasu S (1980) Effect of 2, 4-D and NAA on callus formation and haploid production in anther culture of *Pelargonium roseum*. Mem Coll Agric Ehime Univ 24(2):199–207

684. Pol'kheim F (1972) On the problem of selecting for breeding mutation chimeras and mutants in haploids of *Pelargonium zonale* Kleiner Liebling and *Thuja gigantea gracilis*. Eksperimental'nyi Mutagenez Selektsii 1972:199–221

685. Rao PL, De Deepesh N (1987) Haploid plants from in vitro anther culture of the leguminous tree, *Peltophorum pterocarpum* (DC) K. Hayne (Copper pod). Plant Cell Tissue Organ Cult 11(3):167–177

686. Nitsch C, Andersen S, Godard M, Neuffer M, Sheridan W (1982) Production of haploid plants of *Zea mays* and *Pennisetum* through androgenesis. In: Earle ED, Demarly Y (eds) Variability in plants regenerated from tissue culture. Praeger Publishers, New York, NY, pp 69–91

687. Le Thi K, Lespinasse R, Siljak-Yakovlev S, Robert T, Khalfallah N, Sarr A (1994) Karyotypic modifications in androgenetic plantlets of pearl millet, *Pennisetum glaucum* (L.) R. Brunken: occurrence of B chromosomes. Caryologia 47(1):1–10

688. Choi B-H, Park K-Y, Park R-K (1997) Haploidy in pearl millet [*Pennisetum glaucum* (L.) R. Br.]. In: In vitro haploid production

in higher plants. Springer, New York, NY, pp 171–179

689. Caredda S, Clément C (1999) Androgenesis and albinism in Poaceae: influence of genotype and carbohydrates. In: Anther and pollen. Springer, New York, NY, pp 211–228

690. Sastry PS, Mallikarjuna N (2014) Induction of androgenesis in pearl millet. Univ J Agric Res 2(06):216–223

691. Haydu Z, Vasil I (1981) Somatic embryogenesis and plant regeneration from leaf tissues and anthers of *Pennisetum purpureum* Schum. Theor Appl Genet 59(5):269–273

692. Robert T, San A, Pernes J (1989) Haploid selections in pearl millet (*Pennisetum typhoides* (Bunn.) Stapf et Hubb.): temperature effect. Genome 32:946–952

693. Ha DBD, Pernes J (1982) Androgenesis in pearl millet: I. Analysis of plants obtained from microspore culture. Z Pflanzenphysiol 108(4):317–327

694. DeVerna J, Collins G (1984) Maternal haploids of *Petunia axillaris* (Lam.) BSP via culture of placenta attached ovules. Theor Appl Genet 69(2):187–192

695. Raquin C (1983) Utilization of different sugars as carbon source for *in vitro* anther culture of *Petunia*. Z Pflanzenphysiol 111 (5):453–457

696. Mitchell AZ, Hanson MR, Skvirsky RC, Ausubel FM (1980) Anther culture of *Petunia* – genotypes with high-frequency of callus, root, or plantlet formation. Z Pflanzenphysiol 100(2):131–146

697. Jain SM, Bhalla-Sarin N (1997) Haploidy in *Petunia*. In: Jain SM, Sopory SK, Veilleux RE (eds) In vitro haploid production in higher plants, vol 29. Current plant science and biotechnology in agriculture. Springer Netherlands, Dordrecht, pp 53–71. https://doi.org/10.1007/978-94-017-1856-1_4

698. Singh I, Cornu A (1976) Research into androgenetic *Petunia* haploids with gynogenetic cytoplasmic pollen sterility. Ann Amélior Plant 26:565–568

699. Raquin C, Cornu A, Farcy E, Maizonnier D, Pelletier G, Vedel F (1989) Nucleus substitution between *Petunia* species using gamma-ray-induced androgenesis. Theor Appl Genet 78(3):337–341

700. Malhotra K, Maheshwari S (1977) Enhancement by cold treatment of pollen embryoid development in *Petunia hybrida*. Z Pflanzenphysiol 85(2):177–180

701. Babbar SB, Gupta SC (1980) Chilling induced androgenesis in anthers of *Petunia hybrida* without any culture medium. Z Pflanzenphysiol 100(3):279–283

702. Raquin C (1982) Genetic control of embryo production and embryo quality in anther culture of *Petunia*. Theor Appl Genet 63 (2):151–154. https://doi.org/10.1007/bf00303698

703. Raquin C, Amssa M, Henry Y, Debuyser J, Essad S (1982) Origin of polyhaploid plants obtained through *in vitro* anther culture – cytophotometrical analysis of *Petunia* and wheat microspore *in situ* and *in vitro*. Z Pflanzen 89(4):265–277

704. Gupta PP (1983) Microspore-derived haploid, diploid and triploid plants in *Petunia violacea* Lindl. Plant Cell Rep 2 (5):255–256. https://doi.org/10.1007/BF00269154

705. Peters J, Crocomo O, Sharp W, Paddock E, Tegenkamp I, Tegenkamp T (1977) Haploid callus cells from anthers of *Phaseolus vulgaris*. Phytomorphology 27:79–85

706. Munoz L, Baudoin J (1994) Influence of the cold pretreatment and the carbon source on callus induction from anthers in *Phaseolus*. Ann Rep Bean Improv Coop 37:129–130

707. Pulli S, Guo Y-D (2003) Anther culture and isolated microspore culture in timothy. In: Maluszynski M, Kasha KJ, Forster BP, Szarejko I (eds) Doubled haploid production in crop plants: a manual. Springer Netherlands, Dordrecht, pp 173–177. https://doi.org/10.1007/978-94-017-1293-4_27

708. Guo YD, Pulli S (2000) An efficient androgenic embryogenesis and plant regeneration method through isolated microspore culture in timothy (*Phleum pratense* L.). Plant Cell Rep 19(8):761–767. https://doi.org/10.1007/s002990000193

709. Razdan A, Razdan MK, Rajam MV, Raina SN (2008) Efficient protocol for in vitro production of androgenic haploids of *Phlox drummondii*. Plant Cell Tissue Organ Cult 95(2):245. https://doi.org/10.1007/s11240-008-9431-8

710. Bapat V, Wenzel G (1982) In vitro haploid plantlet induction in *Physalis ixocarpa* Brot. through microspore embryogenesis. Plant Cell Rep 1(4):154–156

711. Escobar-Guzmán R, Hernández-Godínez F, Martínez de la Vega O, Ochoa-Alejo N (2009) *In vitro* embryo formation and plant regeneration from anther culture of different cultivars of Mexican husk tomato (*Physalis ixocarpa* Brot.). Plant Cell Tissue Organ Cult 96(2):181–189. https://doi.org/10.1007/s11240-008-9474-x

712. Garcia-Arias F, Sánchez-Betancourt E, Núñez V (2018) Fertility recovery of anther-derived haploid plants in Cape gooseberry (*Physalis peruviana* L.). Agron Colomb 36(3):201–209. https://doi.org/10.15446/agron.colomb.v36n3.73108

713. Baldursson S, Nørgaard J, Krogstrup P (1993) Factors influencing haploid callus initiation and proliferation in megagametophyte cultures of Sitka spruce (*Picea sitchensis*). Silvae Genet 42:79–79

714. Ribalta FM, Croser JS, Ochatt SJ (2012) Flow cytometry enables identification of sporophytic eliciting stress treatments in gametic cells. J Plant Physiol 169 (0):104–110. https://doi.org/10.1016/j.jplph.2011.08.013

715. Gupta S (1976) Morphogenetic response of haploid callus tissue of *Pisum sativum* (var. B22). Indian Agric 194(4):11–21

716. Hidaka T, Yamada Y, Shichijo T (1979) In vitro differentiation of haploid plants by anther culture in *Poncirus trifoliata* (L.) Raf. Jpn J Breed 29(3):248–254

717. Hyun S, Kim J, Noh E, Park J (1986) Induction of haploid plants of *Populus* species. In: Withers LA, Alderson PG (eds) Plant tissue culture and its agricultural applications. Butterworths, London, pp 413–418

718. Baldursson S, Krogstrup P, Norgaard JV, Andersen SB (1993) Microspore embryogenesis in anther culture of 3 species of *Populus* and regeneration of dihaploid plants of *Populus trichocarpa*. Can J Forest Res 23 (9):1821–1825

719. Li Y, Li H, Chen Z, Ji L-X, Ye M-X, Wang J, Wang L, An X-M (2013) Haploid plants from anther cultures of poplar (*Populus* × *beijingensis*). Plant Cell Tissue Organ Cult 114(1):39–48. https://doi.org/10.1007/s11240-013-0303-5

720. Uddin MR, Meyer MM Jr, Jokela JJ (1988) Plantlet production from anthers of Eastern cottonwood (*Populus deltoïdes*). Can J For Res 18(7):937–941. https://doi.org/10.1139/x88-142

721. Kiss J, Kondrák M, Törjék O, Kiss E, Gyulai G, Mázik-Tökei K, Heszky L (2001) Morphological and RAPD analysis of poplar trees of anther culture origin. Euphytica 118 (2):213–221

722. Mofidabadi A, Kiss J, Mazik-Tokei K, Gergacz E, Heszky L (1995) Callus induction and haploid plant regeneration from anther culture of two poplar species. Silvae Genet 44(2–3):141–145

723. Kim J, Noh E, Park J (1983) Haploid plantlets formation through anther culture of *Populus glandulosa*. Res Rep Inst For Genet 19:93–98

724. Kim J, Moon H, Park J (1986) Haploid plantlet induction through anther culture of *Populus maximowiczii*. Res Rep Inst For Genet 1986:116–121

725. Stoehr M, Zsuffa L (1990) Genetic evaluation of haploid clonal lines of a single donor plant of *Populus maximowiczii*. Theor Appl Genet 80(4):470–474

726. Wang C, Chu C, Sun C (1975) Induction of *Populus nigra* pollen-plants. Acta Bot Sin 17:56–59

727. Deutsch F, Kumlehn J, Ziegenhagen B, Fladung M (2004) Stable haploid poplar callus lines from immature pollen culture. Physiol Plant 120(4):613–622

728. Yang JL, Li K, Li CY, Li JX, Zhao B, Zheng W, Gao YC, Li CH (2018) In vitro anther culture and Agrobacterium-mediated transformation of the AP1 gene from *Salix integra* Linn. in haploid poplar (*Populus simonii* × *P. nigra*). J For Res 29 (2):321–330. https://doi.org/10.1007/s11676-017-0453-0

729. Wu K, Nagarajan P (1990) Poplars (*Populus* spp.): *in vitro* production of haploids. In: Bajaj YPS (ed) Haploids in crop improvement I. Springer, Berlin, pp 237–249. https://doi.org/10.1007/978-3-642-61499-6_10

730. Andersen SB (2003) Doubled haploid production in poplar. In: Maluszynski M, Kasha KJ, Forster BP, Szarejko I (eds) Doubled haploid production in crop plants: a manual. Springer Netherlands, Dordrecht, pp 293–296. https://doi.org/10.1007/978-94-017-1293-4_43

731. Ho R, Raj Y (1985) Haploid plant production through anther culture in poplars. For Ecol Manag 13(3–4):133–142

732. Illies ZM (1974) Induction of haploid parthenogenesis in *Populus tremula* by male gametes inactivated with toluidine blue. In: Kasha KJ (ed) Haploids in higher plants: advances and potential. University of Guelph, Guelph, ON, p 136

733. Stettler R, Bawa K (1971) Experimental induction of haploid parthenogenesis in Black Cottonwood (*Populus trichocarpa* T. & G. ex Hook.). Aspen Bibliogr 20:343

734. Okura E (1933) A haploid plant in *Portulacea grandiflora* Hook. Jpn J Genet 8:251–260

735. Jia Y, Zhang Q-X, Pan H-T, Wang S-Q, Liu Q-L, Sun L-X (2014) Callus induction and haploid plant regeneration from baby primrose (*Primula forbesii* Franch.) anther culture. Sci Hortic 176:273–281. https://doi.org/10.1016/j.scienta.2014.07.018

736. Peixe A, Barroso J, Potes A, Pais MS (2004) Induction of haploid morphogenic calluses from in vitro cultured anthers of *Prunus armeniaca* cv. 'Harcot'. Plant Cell Tissue Organ Cult 77(1):35–41. https://doi.org/10.1023/B:TICU.0000016498.95516.e6

737. Long CM, Mulinix CA, Iezzoni AF (1994) Production of a microspore-derived callus population from sweet cherry. HortScience 29(11):1346–1348

738. Cimò G, Marchese A, Germanà MA (2017) Microspore embryogenesis induced through in vitro anther culture of almond (*Prunus dulcis* Mill.). Plant Cell Tissue Organ Cult 128(1):85–95. https://doi.org/10.1007/s11240-016-1086-2

739. Todorovic R, Mišic P, Petrovic D, Mirkovic M (1990) Anther culture of peach cultivars 'Cresthaven' and 'Vesna'. Acta Hortic 300:331–334

740. Hammerschlag FA (1983) Factors influencing the frequency of callus formation among cultured peach anthers. HortScience 18:210–211

741. Pooler MR, Scorza R (1995) Aberrant transmission of RAPD markers in haploids, doubled haploids, and F1 hybrids of peach: observations and speculation on causes. Sci Hortic 64(4):233–241. https://doi.org/10.1016/0304-4238(95)00846-2

742. Toyama TK (1974) Haploidy in peach. HortScience 9:187–188

743. Livingston GK (1971) Experimental studies on the induction of haploid parthenogenesis in Douglas fir and the effects of radiation on the germination and growth of Douglas fir pollen. Dissert Abstr Int B 32:4331–4332

744. Durzan DJ (2011) Female parthenogenetic apomixis and androsporogenesis in Douglas-fir embryonal initials in an artificial sporangium. Sex Plant Reprod 24(4):283–296. https://doi.org/10.1007/s00497-011-0171-2

745. Babbar SB, Gupta SC (1986) Induction of androgenesis and callus formation in *in vitro* cultured anthers of a myrtaceous fruit tree (*Psidium guajava* L). Bot Mag Tokyo 99(1053):75–83. https://doi.org/10.1007/bf02488624

746. Pal A (1983) Isolated microspore culture of the winged bean, *Psophocarpus tetragonolobus* (L) DC-growth, development and chromosomal status. Indian J Exp Biol 21:597–603

747. Usha Rao I, Rao R, Narasimham M (1986) Induction of androgenesis in the vitro in grown anthers of winged bean (*Psophocarpus tetragonolobus*). Phytomorphology 36(1–2):111–116

748. Chand S, Sahrawat A (2007) Embryogenesis and plant regeneration from unpollinated ovary culture of *Psoralea corylifolia*. Biol Plant 51(2):223–228

749. Asker S (1983) A monoploid of *Potentilla argentea*. Hereditas 99:303–304

750. Moriguchi T, Omura M, Matsuta N, Kozaki I (1987) In vitro adventitious shoot formation from anthers of pomegranate. HortScience 22(5):947–948

751. Zhong CH, Aruga Y, Yan X (2019) Morphogenesis and spontaneous chromosome doubling during the parthenogenetic development of haploid female gametophytes in *Pyropia haitanensis* (Bangiales, Rhodophyta). J Appl Phycol 31(4):2729–2741. https://doi.org/10.1007/s10811-019-01769-x

752. Braniste N, Popescu A (1984) Coman T Producing and multiplication of *Pyrus communis* haploid plants Symposium on production and preservation of pears 161. pp 147–162

753. Sniezko R, Visser T (1987) Embryo development and fruit-set in pear induced by untreated and irradiated pollen. Euphytica 36(1):287–294

754. Bouvier L, Zhang Y-X, Lespinasse Y (1993) Two methods of haploidization in pear, *Pyrus communis* L.: greenhouse seedling selection and in situ parthenogenesis induced by irradiated pollen. Theor Appl Genet 87(1–2):229–232 ·

755. Pintos B, Sánchez N, Bueno MA, Navarro RM, Jorrín J, Manzanera JA, Gómez-Garay A (2013) Induction of *Quercus ilex* L. haploid and doubled-haploid embryos from anther cultures by temperature-stress. Silvae Genet 62(1–6):210–217

756. Bueno MA, Gómez A, Boscaiu M, Manzanera JA, Vicente O (1997) Stress induced haploid plant production from anther cultures of *Quercus suber*. Physiol Plant 99:335–341

757. Bueno MA, Gómez A, Sepúlveda F, Seguí-Simarro JM, Testillano PS, Manzanera JA, Risueño MC (2003) Microspore-derived embryos from *Quercus suber* anthers mimic zygotic embryos and maintain haploidy in

long-term anther culture. J Plant Physiol 160(8):953–960

758. Pintos B, Manzanera JA, Bueno MA (2007) Antimitotic agents increase the production of doubled-haploid embryos from cork oak anther culture. J Plant Physiol 164 (12):1595–1604

759. Takahata Y, Komatsu H, Kaizuma N (1996) Microspore culture of radish (*Raphanus sativus* L.): influence of genotype and culture conditions on embryogenesis. Plant Cell Rep 16(3–4):163–166

760. Chung YS, Lee YG, Silva RR, Park S, Park MY, Lim YP, Choi SC, Kim C (2018) Potential SNPs related to microspore culture in *Raphanus sativus* based on a single-marker analysis. Can J Plant Sci 98(5):1072–1083. https://doi.org/10.1139/cjps-2017-0333

761. Han N, Na H, Kim J (2018) Identification and variation of major aliphatic glucosinolates in doubled haploid lines of radish (*Raphanus sativus* L.). Hortic Sci Technol 36(2):302–311

762. Kim K, Kang Y, Lee S-J, Choi S-H, Jeon D-H, Park M-Y, Park S, Lim YP, Kim C (2020) Quantitative trait loci (QTLs) associated with microspore culture in *Raphanus sativus* L. (Radish). Genes 11(3). https://doi.org/10.3390/genes11030337

763. Kozar EV, Domblides EA, Soldatenko AV (2020) Factors affecting DH plants in vitro production from microspores of European radish. Vavilovskii Zhurnal Genet Sel 24 (1):31–39. https://doi.org/10.18699/vj20.592

764. Tuncer B (2017) Callus formation from isolated microspore culture in radish (*Raphanus sativus* L.). J Anim Plant Sci 27 (1):277–282

765. Sankina A, Sankin L (1988) Characteristics of meiosis in the remote currant hybrid *Ribes nigrum* × *Ribes Holland* Red at the amphihaploid and amphidiploid levels. Cytol Genet 22(6):12–16

766. Jelenkovic G, Shifriss O, Harrington E (1980) Association and distribution of meiotic chromosomes in a haploid of *Ricinus communis* L. Cytologia 45(3):571–577. https://doi.org/10.1508/cytologia.45.571

767. Tabaeezadeh Z, Khosh-Khui M (1981) Anther culture of *Rosa*. Sci Hortic 15 (1):61–66. https://doi.org/10.1016/0304-4238(81)90062-5

768. Wissemann V, Möllers C, Hellwig F (1998) Microspore culture in the genus *Rosa*, further investigations. Angew Bot 72(1–2):7–9

769. Meynet J, Barrade R, Duclos A, Siadous R (1994) Dihaploid plants of roses (*Rosa* × *hybrida*, cv 'Sonia') obtained by parthenogenesis induced using irradiated pollen and in vitro culture of immature seeds. Agronomie 2:169–175

770. Liu M, Chen W, Yang L (1980) Anther culture in sugarcane. I Structure of anther and its pollen grain development stages. Taiwan Sugar 27(3):86–91

771. Fitch MM, Moore PH (1996) Haploids of sugarcane. In: In vitro haploid production in higher plants. Springer, New York, NY, pp 1–16

772. Fitch MM, Moore PH (1984) Production of haploid *Saccharum spontaneum* L.-comparison of media for cold incubation of panicle branches and for float culture of anthers. J Plant Physiol 117(2):169–178

773. Hinchee MA, Cruz AD, Maretzki A (1984) Developmental and biochemical characteristics of cold-treated anthers of *Saccharum spontaneum*. J Plant Physiol 115 (4):271–284

774. Bugara A, Rusina L, Reznikova S (1986) Embryoidogenesis in anther culture of *Salvia sclarea*. Fiziol Biokhim Kult Rast 18:381–386

775. Radojevic L, Vapa L, Borojevic K, Joksimovic J (1985) Plant regeneration in *Saintpaulia ionantha* Wendl. anther cultures. Savrem Poljopr 33:485–491

776. Hughes KW, Bell SL, Caponetti JD (1975) Anther-derived haploids of the African violet. Can J Bot 53(14):1442–1444

777. Weatherhead MA, Grout BWW, Short KC (1982) Increased haploid production in *Saintpaulia ionantha* by anther culture. Sci Hortic 17(2):137–144. https://doi.org/10.1016/0304-4238(82)90006-1

778. Bhaskaran S, Smith RH, Finer JJ (1983) Ribulose bisphosphate carboxylase activity in anther-derived plants of *Saintpaulia ionantha* Wendl. Shag. Plant Physiol 73 (3):639–642

779. Uno Y, Koda-Katayama H, Kobayashi H (2016) Application of anther culture for efficient haploid production in the genus *Saintpaulia*. Plant Cell Tissue Organ Cult 125 (2):241–248. https://doi.org/10.1007/s11240-016-0943-3

780. Kernan Z, Ferrie AMR (2006) Microspore embryogenesis and the development of a double haploidy protocol for cow cockle (*Saponaria vaccaria*). Plant Cell Rep 25 (4):274–280. https://doi.org/10.1007/s00299-005-0064-7

781. Chakravarty B, Sen S (1989) Regeneration through somatic embryogenesis from anther explants of *Scilla indica* (Roxb.) Baker. Plant Cell Tissue Organ Cult 19(1):71–75

782. Zieliński K, Krzewska M, Żur I, Juzoń K, Kopeć P, Nowicka A, Moravčiková J, Skrzypek E, Dubas E (2020) The effect of glutathione and mannitol on androgenesis in anther and isolated microspore cultures of rye (*Secale cereale* L.). Plant Cell Tissue Organ Cult 140(3):577–592. https://doi.org/10.1007/s11240-019-01754-9

783. Sharma P, Chaithhary HK, Manoj NV, Singh K, Relan A, Sood VK (2019) Haploid induction in triticale × wheat and wheat × rye derivatives following imperata cylindrica-mediated chromosome elimination approach. Cereal Res Commun 47(4):701–713. https://doi.org/10.1556/0806.47.2019.46

784. Ma R, Guo YD, Pulli S (2004) Comparison of anther and microspore culture in the embryogenesis and regeneration of rye (*Secale cereale*). Plant Cell Tissue Organ Cult 76(2):147–157

785. Ranaweera K, Pathirana R (1992) Optimization of media and conditions for callus induction from anthers of sesame cultivar MI 3. J Natl Sci Found 20(2):309–316

786. Govil CM, Singh VRR (1982) Induction of haploids in anther culture of *Sesamum indicum*. In: Proceedings of the 5th international congress plant tissue and cell culture, pp 545–546

787. Yifter M, Sbhatu DB, Mekbib F, Abraha E (2013) *In vitro* regeneration of four ethiopian varieties of sesame (*Sesamum indicum* L.) using anther culture. Asian J Plant Sci 12:214–218

788. Ban Y, Kokubu T, Miyaji Y (1971) Production of haploid plant by anther-culture of *Setaria italica*. Kagoshima Univ Fac Agr Bull 21:77–81

789. Šafářová D, Kopecký D, Vagera J (2005) The effect of a short heat treatment on the in vitro induced androgenesis in *Silene latifolia* ssp. alba. Biol Plant 49(2):261–264

790. Jain R, Brune U, Friedt W (1989) Plant regeneration from in vitro cultures of cotyledon explants and anthers of *Sinapis alba* and its implications on breeding of crucifers. Euphytica 43(1–2):153–163

791. Tsay HS, Yeh CC, Hsu JY (1990) Embryogenesis and plant regeneration from anther culture of bamboo (*Sinocalamus latiflora* (Munro) McClure). Plant Cell Rep 9(7):349–351. https://doi.org/10.1007/bf00232396

792. Khoshoo T (1957) A polyhaploid plant of the tetraploid race of *Sisymbrium irio*. J Hered 48(5):239–242

793. Rokka VM, Ishimaru CA, Lapitan NLV, Pehu E (1998) Production of androgenic dihaploid lines of the disomic tetraploid potato species *Solanum acaule* ssp. acaule. Plant Cell Rep 18(1–2):89–93

794. Lysenko EG, Sidorov VA (1985) The obtaining of *S. bulbocastanum* androgenic haploids and mesophyll protoplast culture. Tsitologiya I. Genetika 19(6):433–436

795. Reynolds TL (1987) The roles of auxin and ethylene during pollen embryogenesis in *Solanum carolinense* L. Am J Bot 74(5):623–624

796. Reynolds TL (1984) Callus formation and organogenesis in anther cultures of *Solanum carolinense* L. J Plant Physiol 117(2):157–161

797. Reynolds TL (1990) Interactions between calcium and auxin during pollen androgenesis in anther cultures of *Solanum carolinense* L. Plant Sci 72(1):109–114. https://doi.org/10.1016/0168-9452(90)90192-q

798. Reynolds TL (1986) Pollen embryogenesis in anther cultures of *Solanum carolinense* L. Plant Cell Rep 5(4):273–275

799. Reynolds TL (1987) A possible role for ethylene during iaa-induced pollen embryogenesis in anther cultures of *Solanum carolinense* L. Am J Bot 74(6):967–969. https://doi.org/10.2307/2443878

800. Reynolds TL (1989) Ethylene effects on pollen callus formation and organogenesis in anther cultures of *Solanum carolinense* L. Plant Sci 61(1):131–136. https://doi.org/10.1016/0168-9452(89)90127-1

801. Hermsen JGT (1969) Induction of haploids and aneuhaploids in colchicine-induced tetraploid *Solanum chacoense* Bitt. Euphytica 18(2):183–189. https://doi.org/10.1007/bf00035690

802. Cappadocia M, Ahmim M (1988) Comparison of two culture methods for the production of haploids by anther culture in *Solanum chacoense*. Can J Bot 66(5):1003–1005. https://doi.org/10.1139/b88-144

803. Birhman RK, Rivard SR, Cappadocia M (1994) Restriction fragment length polymorphism analysis of anther-culture-derived *Solanum chacoense*. HortScience 29(3):206–208

804. Rivard SR, Sabaelleil MK, Landry BS, Cappadocia M (1994) RFLP analyses and segregation of molecular markers in plants produced by in vitro anther culture, selfing, and reciprocal crosses of 2 lines of self incompatible *Solanum chacoense*. Genome 37(5):775–783

805. Veilleux RE, Shen LY, Paz MM (1995) Analysis of the genetic composition of anther-derived potato by randomly amplified polymorphic DNA and simple sequence repeats. Genome 38(6):1153–1162

806. Boluarte-Medina T, Veilleux RE (2002) Phenotypic characterization and bulk segregant analysis of anther culture response in two backcross families of diploid potato – RAPD markers for androgenesis in potato. Plant Cell Tissue Org Cult 68(3):277–286. https://doi.org/10.1023/a:1013973323546

807. Zenkteler M (1973) In vitro development of embryos and seedlings from pollen grains of *Solanum dulcamara*. Z Pflanzenphysiol 69(2):189–192. https://doi.org/10.1016/S0044-328X(73)80038-8

808. Binding H, Mordhorst G (1984) Haploid *Solanum dulcamara* L.: shoot culture and plant regeneration from isolated protoplasts. Plant Sci Lett 35(1):77–79. https://doi.org/10.1016/0304-4211(84)90161-5

809. Hernández Amasifuen AD, Díaz Pillasca HB (2019) Inducción in vitro de callo embriogénico a partir del cultivo de anteras en "papa amarilla" *Solanum goniocalyx* Juz. & Bukasov (Solanaceae). Arnaldoa 26(1):277–286. https://doi.org/10.22497/arnaldoa.261.26111

810. Pacheco-Sanchez M, Lozoya-Saldana H, Colinas-Leon MT (2003) Growth regulators and cold pretreatment on in vitro androgenesis of *Solanum iopetalum* L. Agrociencia 37(3):257–265

811. Niazian M, Shariatpanahi ME, Abdipour M, Oroojloo M (2019) Modeling callus induction and regeneration in an anther culture of tomato (*Lycopersicon esculentum* L.) using image processing and artificial neural network method. Protoplasma. https://doi.org/10.1007/s00709-019-01379-x

812. Juliao SA, Carvalho CR, Dias Koehlers T, Ribeiro da Silva C (2015) Multiploidy occurrence in tomato calli from anther culture. Afr J Biotechnol 14(40):2846–2855. https://doi.org/10.5897/AJB2015.14525

813. Corral-Martínez P, Nuez F, Seguí-Simarro JM (2011) Genetic, quantitative and microscopic evidence for fusion of haploid nuclei and growth of somatic calli in cultured *ms10^{35}* tomato anthers. Euphytica 178(2):215–228. https://doi.org/10.1007/s10681-010-0303-z

814. Motallebi-Azar A, Panahandeh J (2010) Effects of colchicine and cold duration pretreatments on androgenesis responses of tomato (*Lycopersicon esculentum* Mill) via anther culture. Russ Agric Sci 36(5):338–341. https://doi.org/10.3103/s106836741005006x

815. Farooq AM, Tabassum B, Nasir IA, Husnain T (2010) Androgenesis induction, callogenesis, regeneration and cytogenetic studies of tomato haploid. J Agric Res 48(4):457–470

816. Seguí-Simarro JM, Nuez F (2007) Embryogenesis induction, callogenesis, and plant regeneration by *in vitro* culture of tomato isolated microspores and whole anthers. J Exp Bot 58(5):1119–1132

817. Seguí-Simarro JM, Nuez F (2006) Androgenesis induction from tomato anther cultures: callus characterization. Acta Hortic 725:855–861

818. Motallebi-Azar A, Khosroshahli M, Valizadeh M, Massiha S, Moeini A (2006) Effect of genotype, and cold and heat pretreatment on callus and shoot induction in tomato anther culture. Iran J Agric Sci 37:899–909

819. Seguí-Simarro JM, Nuez F (2005) Meiotic metaphase I to telophase II is the most responsive stage of microspore development for induction of androgenesis in tomato (*Solanum lycopersicum*). Acta Physiol Plant 27(4B):675–685

820. Bal U, Abak K (2005) Induction of symmetrical nucleus division and multicellular structures from the isolated microspores of *Lycopersicon esculentum* Mill. Biotechnol Biotechnol Eq 19(1):35–42

821. Zagorska NA, Shtereva LA, Kruleva MM, Sotirova VG, Baralieva DL, Dimitrov BD (2004) Induced androgenesis in tomato (*Lycopersicon esculentum* Mill.). III. Characterization of the regenerants. Plant Cell Rep 22(7):449–456

822. Shtereva L, Atanassova B (2001) Callus induction and plant regeneration via anther culture in mutant tomato (*Lycopersicon esculentum* Mill.) lines with anther abnormalities. Israel J Plant Sci 49(3):203–208

823. Zagorska NA, Shtereva A, Dimitrov BD, Kruleva MM (1998) Induced androgenesis in tomato (*Lycopersicon esculentum* Mill.) – I. Influence of genotype on androgenetic ability. Plant Cell Rep 17(12):968–973

824. Shtereva LA, Zagorska NA, Dimitrov BD, Kruleva MM, Oanh HK (1998) Induced androgenesis in tomato (*Lycopersicon esculentum* Mill). II. Factors affecting induction of androgenesis. Plant Cell Rep 18 (3–4):312–317

825. Evans DA, Morrison RA (1989) Tomato anther culture. USA Patent

826. Varghese TM, Gulshan Y (1986) Production of embryoids and calli from isolated microspores of tomato (*Lycopersicon esculentum* Mill.) in liquid media. Biol Plant 28 (2):126–129

827. Gulshan TMV, Sharma DR (1981) Studies on anther cultures of tomato – *Lycopersicon esculentum* Mill. Biol Plant 23(6):414–420

828. Zamir D, Jones RA, Kedar N (1980) Anther culture of male sterile tomato (*Lycopersicon esculentum* Mill.) mutants. Plant Sci Lett 17:353–361

829. Gresshoff PM, Doy CH (1972) Development and differentiation of haploid *Lycopersicon esculentum* (tomato). Planta 107 (2):161–170

830. Gavrilenko T, Thieme R, Rokka VM (2001) Cytogenetic analysis of *Lycopersicon esculentum* (+) *Solanum etuberosum* somatic hybrids and their androgenetic regenerants. Theor Appl Genet 103(2–3):231–239

831. Calabuig-Serna A, Porcel R, Corral-Martínez P, Seguí-Simarro JM (2020) Anther culture in eggplant (*Solanum melongena* L.). In: Bayer M (ed) Plant embryogenesis: methods and protocols, vol. 2122. Methods in molecular biology. Springer US, New York, NY, pp 283–293. https://doi.org/10.1007/978-1-0716-0342-0_20

832. Rivas-Sendra A, Campos-Vega M, Calabuig-Serna A, Seguí-Simarro JM (2017) Development and characterization of an eggplant (*Solanum melongena*) doubled haploid population and a doubled haploid line with high androgenic response. Euphytica 213(4):89. https://doi.org/10.1007/s10681-017-1879-3

833. Rotino GL (2016) Anther culture in eggplant (*Solanum melongena* L.). In: Germana MA, Lambardi M (eds) In vitro embryogenesis in higher plants, vol 1359. Methods in molecular biology. Springer, New York, NY, pp 453–466. https://doi.org/10.1007/978-1-4939-3061-6_25

834. Rivas-Sendra A, Corral-Martínez P, Camacho-Fernández C, Seguí-Simarro JM (2015) Improved regeneration of eggplant doubled haploids from microspore-derived calli through organogenesis. Plant Cell Tissue

Organ Cult 122(3):759–765. https://doi.org/10.1007/s11240-015-0791-6

835. Corral-Martínez P, Seguí-Simarro JM (2014) Refining the method for eggplant microspore culture: effect of abscisic acid, epibrassinolide, polyethylene glycol, naphthaleneacetic acid, 6-benzylaminopurine and arabinogalactan proteins. Euphytica 195(3):369–382. https://doi.org/10.1007/s10681-013-1001-4

836. Salas P, Rivas-Sendra A, Prohens J, Seguí-Simarro JM (2012) Influence of the stage for anther excision and heterostyly in embryogenesis induction from eggplant anther cultures. Euphytica 184 (2):235–250. https://doi.org/10.1007/s10681-011-0569-9

837. Corral-Martínez P, Seguí-Simarro JM (2012) Efficient production of callus-derived doubled haploids through isolated microspore culture in eggplant (*Solanum melongena* L.). Euphytica 187(1):47–61. https://doi.org/10.1007/s10681-012-0715-z

838. Salas P, Prohens J, Seguí-Simarro JM (2011) Evaluation of androgenic competence through anther culture in common eggplant and related species. Euphytica 182 (2):261–274. https://doi.org/10.1007/s10681-011-0490-2

839. Başay S, Şeniz V, Ellialtioğlu Ş (2011) Obtaining dihaploid lines by using anther culture in the different eggplant cultivars. J Food Agric Environ 9(2):188–190

840. Toppino L, Mennella G, Rizza F, D'Alessandro A, Sihachakr D, Rotino GL (2008) ISSR and isozyme characterization of androgenetic dihaploids reveals tetrasomic inheritance in tetraploid somatic hybrids between *Solanum melongena* and *Solanum aethiopicum* group Gilo. J Hered 99 (3):304–315. https://doi.org/10.1093/jhered/esm122

841. Alpsoy HC, Seniz V (2007) Researches on the in vitro androgenesis and obtaining haploid plants in some eggplant genotypes. Acta Hortic 729:137–141

842. Rotino GL, Sihachakr D, Rizza F, Vale G, Tacconi MG, Alberti P, Mennella G, Sabatini E, Toppino L, D'Alessandro A, Acciarri N (2005) Current status in production and utilization of dihaploids from somatic hybrids between eggplant (*Solanum melongena* L.) and its wild relatives. Acta Physiol Plant 27(4B):723–733

843. Rizza F, Mennella G, Collonnier C, Shiachakr D, Kashyap V, Rajam MV,

Prestera M, Rotino GL (2002) Androgenic dihaploids from somatic hybrids between *Solanum melongena* and *S. aethiopicum* group *Gilo* as a source of resistance to *Fusarium oxysporum* f. sp. *melongenae*. Plant Cell Rep 20(11):1022–1032

844. Miyoshi K (1996) Callus induction and plantlet formation through culture of isolated microspores of eggplant (*Solanum melongena* L). Plant Cell Rep 15 (6):391–395

845. Matsubara S, Hu KL, Murakami K (1992) Embryoid and callus formation from pollen grains of eggplant and pepper by anther culture. J Jpn Soc Hortic Sci 61(1):69–77

846. Rotino GL, Restaino F, Gjomarkaj M, Massimo M, Falavigna A, Schiavi M, Vicini E (1991) Evaluation of genetic variability in embryogenetic and androgenetic lines of eggplant. Acta Hortic 300:357–362

847. Sanguineti MC, Tuberosa R, Conti S (1990) Field evaluation of androgenetic lines of eggplant. Acta Hortic 280:177–182

848. Chambonnet D (1988) Production of haploid eggplant plants. Bulletin interne de la Station d'Amélioration des Plantes Maraichères d'Avignon-Montfavet, France, pp 1–10

849. Tuberosa R, Sanguineti MC, Conti S (1987) Anther culture of eggplant *Solanum melongena* L. lines and hybrids. Genét Agrár 41 (3):267–274

850. Rotino GL, Falavigna A, Restaino F (1987) Production of anther-derived plantlets of eggplant. Capsicum Newsl 6:89–90

851. Borgel A, Arnaud M (1986) Progress in eggplant breeding, use of haplomethod. Capsicum Newsl 5:65–66

852. Misra NR, Varghese TM, Maherchandani N, Jain RK (1983) Studies on induction and differentiation of androgenic callus of *Solanum melongena* L. In: Sen SK, Giles KL (eds) Plant cell culture in crop improvement. Plenum, New York, NY, pp 465–468

853. Dumas de Vaulx R, Chambonnet D (1982) Culture *in vitro* d'anthères d'aubergine (*Solanum melongena* L.): stimulation de la production de plantes au moyen de traitements à 35°C associés à de faibles teneurs en substances de croissance. Agronomie 2 (10):983–988

854. Isouard G, Raquin C, Demarly Y (1979) Obtention de plantes haploides et diploides par culture in vitro d'anthères d'aubergine (*Solanum melongena* L.). C R Acad Sci Paris 288:987–989

855. S-r G (1979) Plantlets from isolated pollen cultures of eggplant (*Solanum melongena* L.). Acta Bot Sin 21:30–36

856. Breeding RGoH (1978) Induction of haploid plants of *Solanum melongena*. In: Proceedings of the symposium on plant tissue culture. Science Press, Peking, pp 227–232

857. Raina SK, Iyer RD (1973) Differentiation of diploid plants from pollen callus in anther cultures of *Solanum melongena* L. Z Pflanzen 70(4):275–280

858. Sree Ramulu K (1982) Genetic instability at the S-locus of *Lycopersicon peruvianum* plants regenerated from *in vitro* culture of anthers: generation of new S-specificities and S-allele reversions. Heredity 49(3):319–330

859. Nishiyama I, Uematsu S (1967) Radiobiological studies in plants—XIII. Embryogenesis following X-irradiation of pollen in *Lycopersicum pimpinellifolium*. Radiat Bot 7(6):481–489

860. Sharma S, Sarkar D, Pandey SK (2010) Phenotypic characterization and nuclear microsatellite analysis reveal genomic changes and rearrangements underlying androgenesis in tetraploid potatoes (*Solanum tuberosum* L.). Euphytica 171(3):313–326. https://doi.org/10.1007/s10681-009-9983-7

861. Teparkum S, Veilleux RE (1998) Indifference of potato anther culture to colchicine and genetic similarity among anther-derived monoploid regenerants determined by RAPD analysis. Plant Cell Tissue Org Cult 53(1):49–58. https://doi.org/10.1023/a:1006099423651

862. Teten Snider K, Veilleux RE (1994) Factors affecting variability in anther culture and in conversion of androgenic embryos of *Solanum phureja*. Plant Cell Tissue Organ Cult 36(3):345–354. https://doi.org/10.1007/bf00046092

863. Owen HR, Veilleux RE, Haynes FL, Haynes KG (1988) Photoperiod effects on 2n pollen production, response to anther culture, and net photosynthesis of a diplandrous clone of *Solanum phureja*. Am Potato J 65 (3):131–139. https://doi.org/10.1007/bf02871602

864. Pehu E, Veilleux RE, Hilu KW (1987) Cluster analysis of anther-derived plants of *Solanum phureja* (Solanaceae) based on morphological characteristics. Am J Bot 74 (1):47–52. https://doi.org/10.2307/2444330

865. Veilleux RE, Booze-Daniels J, Pehu E (1985) Anther culture of a 2n pollen producing clone of *Solanum phureja* Juz. &

Buk. Can J Genet Cytol 27(5):559–564. https://doi.org/10.1139/g85-082

866. Sinha S, Roy RP, Jha KK (1979) Callus formation and shoot bud differentiation in anther culture of *Solanum surattense*. Can J Bot 57(22):2524–2527

867. Jaiswal VS, Narayan P (1981) Induction of pollen embryoids in *Solanum torvum* Swartz. Curr Sci 50(22):998–999

868. Asakaviciute R (2008) Androgenesis in anther culture of Lithuanian spring barley (*Hordeum vulgare* L.) and potato (*Solanum tuberosum* L.) cultivars. Turk J Biol 32 (3):155–160

869. Iovene M, Aversano R, Savarese S, Caruso I, Di Matteo A, Cardi T, Frusciante L, Carputo D (2012) Interspecific somatic hybrids between *Solanum bulbocastanum* and *S. tuberosum* and their haploidization for potato breeding. Biol Plant 56(1):1–8

870. Rokka VM (2009) Potato haploids and breeding. In: Touraev A, Forster BP, Jain SM (eds) Advances in haploid production in higher plants. Springer, Dordrecht, pp 199–208

871. Tai GCC, Xiong XY (2003) Haploid production of potatoes by anther culture. In: Maluszynski M, Kasha KJ, Forster BP, Szarejko I (eds) Doubled haploid production in crop plants. A manual. Kluwer Academic, Dordretch, pp 229–234

872. Rokka VM (2003) Anther culture through direct embryogenesis in a genetically diverse range of potato (*Solanum*) species and their interspecific and intergeneric hybrids. In: Maluszynski M, Kasha KJ, Forster BP, Szarejko I (eds) Doubled haploid production in crop plants. A manual. Kluwer Academic, Dordretch, pp 235–240

873. Gavrilenko T, Larkka J, Pehu E, Rokka VM (2002) Identification of mitotic chromosomes of tuberous and non-tuberous *Solanum* species (*Solanum tuberosum* and *Solanum brevidens*) by GISH in their interspecific hybrids. Genome 45(2):442–449. https://doi.org/10.1139/g01-136

874. Rihova L, Tupy J (1999) Manipulation of division symmetry and developmental fate in cultures of potato microspores. Plant Cell Tissue Organ Cult 59(2):135–145

875. Rokka VM, Tauriainen A, Pietila L, Pehu E (1998) Interspecific somatic hybrids between wild potato *Solanum acaule* Bitt. and anther-derived dihaploid potato (*Solanum tuberosum* L.). Plant Cell Rep 18 (1–2):82–88

876. Rokka VM, Pietila L, Pehu E (1996) Enhanced production of dihaploid lines via anther culture of tetraploid potato (*Solanum tuberosum* L ssp *tuberosum*) clones. Am Potato J 73(1):1–12. https://doi.org/10.1007/bf02849299

877. Bugárová Z, Pret'ová A (1996) Isolated microspore cultures in *Solanum tuberosum* L. cultivars. Biologia 51:411–416

878. Rokka VM, Valkonen JPT, Pehu E (1995) Production and characterization of haploids derived from somatic hybrids between *Solanum brevidens* and *S. tuberosum* through anther culture. Plant Sci 112(1):85–95

879. Sopory SK (1977) Development of embryoids in isolated pollen culture of dihaploid *Solanum tuberosum*. Z Pflanzenphysiol 84:453–457

880. Pham GM, Braz GT, Conway M, Crisovan E, Hamilton JP, Laimbeer FPE, Manrique-Carpintero N, Newton L, Douches DS, Jiang JM, Veilleux RE, Buell CR (2019) Genome-wide inference of somatic translocation events during potato dihaploid production. Plant Genome 12(2). https://doi.org/10.3835/plantgenome2018.10.0079

881. Amundson KR, Ordoñez B, Santayana M, Tan EH, Henry IM, Mihovilovich E, Bonierbale M, Comai L (2020) Genomic outcomes of haploid induction crosses in potato (*Solanum tuberosum* L.). Genetics 214(2):369–380. https://doi.org/10.1534/genetics.119.302843

882. Debata BK, Patnaik SN (1988) Induction of androgenesis in anther cultures of *Solanum viarum* Dunal. J Plant Physiol 133 (1):124–125

883. Arrillaga I, Lerma V, Pérez-Bermúdez P, Segura J (1995) Callus and somatic embryogenesis from cultured anthers of service tree (*Sorbus domestica* L.). HortScience 30 (5):1078–1079

884. Rose JB, Dunwell JM, Sunderland N (1986) Anther culture of *Sorghum bicolor* (L.) Moench. I. Effect of panicle pretreatment, anther incubation temperature and 2,4-D concentration. Plant Cell Tissue Organ Cult 6(1):15–22

885. Rose JB, Dunwell JM, Sunderland N (1986) Anther culture of *Sorghum bicolor* (L.) Moench. II. Pollen development in vivo and in vitro. Plant Cell Tissue Organ Cult 6 (1):23–31

886. Elkonin L, Gudova T, Ishin A, Tyrnov U (1993) Diploidization in haploid tissue

cultures of sorghum. Plant Breed 110 (3):201–206

887. Kumaravadivel N, Rangasamy SRS (1994) Plant regeneration from sorghum anther cultures and field evaluation of progeny. Plant Cell Rep 13(5):286–290

888. Sairam R, Seetharama N (1996) Androgenic response of cultured anthers and microspores of sorghum. Int Sorghum Millets Newsl 37:69–71

889. Liang GH, Gu X, Yue G, Shi Z, Kofoid K (1997) Haploidy in sorghum. In: In vitro haploid production in higher plants. Springer, New York, NY, pp 149–161

890. Can ND, Yoshida T (1999) Combining ability of callus induction and plant regeneration in sorghum anther culture. Plant Prod Sci 2 (2):125–128

891. Can ND, Nakamura S, Haryanto TAD, Yoshida T (1998) Effects of physiological status of parent plants and culture medium composition on the anther culture of sorghum. Plant Prod Sci 1(3):211–215

892. Hussain T, Franks C (2019) Discovery of *Sorghum* haploid induction system. In: Zhao Z-Y, Dahlberg J (eds) Sorghum: methods and protocols. Springer New York, New York, NY, pp 49–59. https://doi.org/10.1007/978-1-4939-9039-9_4

893. Schertz K (1963) Chromosomal, morphological, and fertility characteristics of haploids and their derivatives in *Sorghum vulgare* Pers. Crop Sci 3(5):445–447

894. Eeckhaut T, Werbrouck S, Dendauw J, Van Bockstaele E, Debergh P (2001) Induction of homozygous *Spathiphyllum wallisii* genotypes through gynogenesis. Plant Cell Tissue Organ Cult 67(2):181–189

895. Keles D, Ozcan C, Pinar H, Ata A, Denli N, Yucel NK, Taskin H, Buyukalaca S (2016) First report of obtaining haploid plants using tissue culture techniques in spinach. HortScience 51(6):742–749. https://doi.org/10.21273/hortsci.51.6.742

896. Uskutoglu T, Uskutoglu D, Turgut K (2019) Effects on pre-treatment and different tissue culture media for androgenesis in *Stevia rebaudiana* Bertoni. Sugar Tech 21 (6):1016–1023. https://doi.org/10.1007/s12355-019-00722-z

897. Wolff DW, Veilleux RE, Jensen CJ (1986) Evaluation of anther-derived *Streptocarpus X hybridus* and their progeny. Plant Cell Tissue Organ Cult 6(2):167–172. https://doi.org/10.1007/bf00180800

898. Frederiksen S (1989) Chromosome elimination in a hybrid between *Taeniatherum*

caput-medusae and *Hordeum bulbosum*. Hereditas 110(1):87–88

899. Kumar KR, Singh KP, Bhatia R, Raju DVS, Panwar S (2019) Optimising protocol for successful development of haploids in marigold (*Tagetes* spp.) through *in vitro* androgenesis. Plant Cell Tissue Organ Cult 138 (1):11–28. https://doi.org/10.1007/s11240-019-01598-3

900. Mehraj U, Panwar S, Singh KP, Namita PR, Solanke AU, Mallick N, Kumar S (2019) Assessment of clonal fidelity of doubled haploid line of marigold (*Tagetes erecta*) using microsatellite markers. Indian J Agric Sci 89 (7):102–106

901. Mehraj U, Panwar S, Singh KP, Namita PR, Solanke AU, Mallick N, Kumar S (2019) In vitro regeneration of double haploid line of African marigold (*Tagetes erecta*) derived from ovule culture using non-axillary explants. Indian J Agric Sci 89(6):969–974

902. Kumar KR, Singh KP, Jan PK, Raju DVS, Kumar P, Bhatia R, Panwar S (2018) Influence of growth regulators on callus induction and plant regeneration from anthers of *Tagetes* spp. Indian J Agric Sci 88 (6):970–977

903. Mehraj U, Panwar S, Singh KP, Namita PR, Solanke AU, Mallick N (2018) Development of protocol for in vitro rooting and hardening of doubled haploid line of *Tagetes erecta* L. derived through ovule culture. Indian J Hortic 75(4):651–655. https://doi.org/10.5958/0974-0112.2018.00108.1

904. Dublin P (1978) Diploidised haploids and production of fertile homozygous genotpyes in cultivated cocoa trees (*Theobroma cacao*). Cafe Cacao Tee 22:275–284

905. Sounigo O, Lachenaud P, Bastide P, Cilas C, N Goran J, Lanaud C (2003) Assessment of the value of doubled haploids as progenitors in cocoa (*Theobroma cacao* L.) breeding. J Appl Genet 44(3):339–354

906. Falque M, Kodia A, Sounigo O, Eskes A, Charrier A (1992) Gamma-irradiation of cacao (*Theobroma cacao* L.) pollen: effect on pollen grain viability, germination and mitosis and on fruit set. Euphytica 64 (3):167–172

907. Falque M (1994) Pod and seed development and phenotype of the M1 plants after pollination and fertilization with irradiated pollen in cacao (*Theobroma cacao* L.). Euphytica 75(1–2):19–25

908. Mokhtarzadeh A, Constantin MJ (1978) Plant regeneration from hypocotyl-and

anther-derived callus of berseem clover. Crop Sci 18(4):567–572

909. Butterfass T (1969) The distribution of plastids in mitosis of guard cell mother cells of haploid *Trifolium hybridum* L. Planta 84 (3):230–234. https://doi.org/10.1007/bf00388108

910. Ponitka A, Slusarkiewicz-Jarzina A (1987) Induction of gynogenesis in selected plant species from the family Papilionaceae. Genet Polonica 28(3):239–242

911. Singh AK, Zhang P, Dong CM, Li JB, Trethowan R, Sharp P (2019) Molecular cytogenetic characterization of stem rust and stripe rust resistance in wheat (*Thinopyrum bessarabicum*) derived doubled haploid lines. Mol Breed 39(9). https://doi.org/10.1007/s11032-019-1034-z

912. Vassileva-Dryanovska OA (1966) Fertilization with irradiated pollen in *Tradescantia*. Radiat Bot 6(5):469–479. https://doi.org/10.1016/S0033-7560(66)80079-0

913. Vassileva-Dryanovska OA (1966) Development of embryo and endosperm produced after irradiation of pollenin *Tradescantia*. Hereditas 55:129–148

914. Immonen S, Robinson J (2000) Stress treatments and ficoll for improving green plant regeneration in triticale anther culture. Plant Sci 150(1):77–84. https://doi.org/10.1016/S0168-9452(99)00169-7

915. Żur I, Dubas E, Krzewska M, Zieliński K, Fodor J, Janowiak F (2019) Glutathione provides antioxidative defence and promotes microspore-derived embryo development in isolated microspore cultures of triticale (×*Triticosecale* Wittm.). Plant Cell Rep 38 (2):195–209. https://doi.org/10.1007/s00299-018-2362-x

916. Würschum T, Tucker MR, Maurer HP, Leiser WL (2015) Ethylene inhibitors improve efficiency of microspore embryogenesis in hexaploid triticale. Plant Cell Tissue Organ Cult 122(3):751–757. https://doi.org/10.1007/s11240-015-0808-1

917. Oleszczuk S, Rabiza-Swider J, Zimny J, Lukaszewski AJ (2011) Aneuploidy among androgenic progeny of hexaploid triticale (X *Triticosecale Wittmack*). Plant Cell Rep 30(4):575–586. https://doi.org/10.1007/s00299-010-0971-0

918. Tuvesson S, Ljungberg A, Johansson N, Karlsson KE, Suijs LW, Josset JP (2000) Large-scale production of wheat and triticale double haploids through the use of a single-anther culture method. Plant Breed 119 (6):455–459

919. Krzewska M, Czyczyło-Mysza I, Dubas E, Gołębiowska-Pikania G, Żur I (2015) Identification of QTLs associated with albino plant formation and some new facts concerning green versus albino ratio determinants in triticale (×*Triticosecale Wittm.*) anther culture. Euphytica 206(1):263–278. https://doi.org/10.1007/s10681-015-1509-x

920. Zur I, Dubas E, Golemiec E, Szechynska-Hebda M, Golebiowska G, Wedzony M (2009) Stress-related variation in antioxidative enzymes activity and cell metabolism efficiency associated with embryogenesis induction in isolated microspore culture of triticale (x *Triticosecale* Wittm.). Plant Cell Rep 28(8):1279–1287. https://doi.org/10.1007/s00299-009-0730-2

921. Żur I, Dubas E, Krzewska M, Janowiak F, Hura K, Pociecha E, Bączek-Kwinta R, Płażek A (2014) Antioxidant activity and ROS tolerance in triticale (×*Triticosecale Wittm.*) anthers affect the efficiency of microspore embryogenesis. Plant Cell Tissue Organ Cult 119(1):79–94. https://doi.org/10.1007/s11240-014-0515-3

922. Żur I, Dubas E, Golemiec E, Szechyńska-Hebda M, Janowiak F, Wędzony M (2008) Stress-induced changes important for effective androgenic induction in isolated microspore culture of triticale (×*Triticosecale Wittm.*). Plant Cell Tissue Organ Cult 94 (3):319–328. https://doi.org/10.1007/s11240-008-9360-6

923. Żur I, Dubas E, Krzewska M, Sánchez-Díaz RA, Castillo AM, Vallés MP (2014) Changes in gene expression patterns associated with microspore embryogenesis in hexaploid triticale (×*Triticosecale Wittm.*). Plant Cell Tissue Organ Cult 116(2):261–267. https://doi.org/10.1007/s11240-013-0399-7

924. Sinha R, Eudes F (2015) Dimethyl tyrosine conjugated peptide prevents oxidative damage and death of triticale and wheat microspores. Plant Cell Tissue Organ Cult 122:227–237. https://doi.org/10.1007/s11240-015-0763-x

925. González JM, López LA, Bernard S, Jouvé N (1993) Prolamin analysis of progenies from androgenetic plants of triticale. Plant Breed 111(1):42–48

926. Asif M, Eudes F, Randhawa H, Amundsen E, Yanke J, Spaner D (2013) Cefotaxime prevents microbial contamination and improves microspore embryogenesis in wheat and triticale. Plant Cell Rep. https://doi.org/10.1007/s00299-013-1476-4

927. Arzani A, Darvey NL (2001) The effect of colchicine on triticale anther-derived plants: microspore pretreatment and haploid plant treatment using a hydroponic recovery system. Euphytica 122(2):235–241

928. Slusarkiewicz-Jarzina A, Ponitka A (2003) Efficient production of spontaneous and induced doubled haploid triticale plants derived from anther culture. Cereal Res Commun 31(3–4):289–296

929. Asif M, Eudes F, Randhawa H, Amundsen E, Spaner D (2014) Phytosulfokine alpha enhances microspore embryogenesis in both triticale and wheat. Plant Cell Tissue Organ Cult 116(1):125–130. https://doi.org/10.1007/s11240-013-0379-y

930. Asif M, Eudes F, Goyal A, Amundsen E, Randhawa H, Spaner D (2013) Organelle antioxidants improve microspore embryogenesis in wheat and triticale. In Vitro Cell Dev Biol Plant 49(5):489–497. https://doi.org/10.1007/s11627-013-9514-z

931. Nowicka A, Juzon K, Krzewska M, Dziurka M, Dubas E, Kopec P, Zielinski K, Zur I (2019) Chemically-induced DNA de-methylation alters the effectiveness of microspore embryogenesis in triticale. Plant Sci 287. https://doi.org/10.1016/j.plantsci.2019.110189

932. Yerzhebayeva RS, Abdurakhmanova MA, Bastaubayeva SO, Tadjibayev D (2019) Efect of zeatin on in vitro embryogenesis and plant regeneration from anther culture of hexaploid triticale (*Triticosecale* Wittmack) (ЭМБРИОГЕНЕЗ И РЕГЕНЕРАЦИЯ РАСТЕНИЙ В КУЛЬТУРЕ ПЫЛЬНИКОВ ГЕКСАПЛОИДНОЙ ТРИТИКАЛЕ (*Triticosecale* Wittmack) ПОД ВЛИЯНИЕМ ЦИТОКИНИНА ЗЕАТИНА). Sel'skokhozyaistvennaya Biol 54(5):934–945

933. Tyrka M, Oleszczuk S, Rabiza-Swider J, Wos H, Wedzony M, Zimny J, Ponitka A, Slusarkiewicz-Jarzina A, Metzger RJ, Baenziger PS, Lukaszewski AJ (2018) Populations of doubled haploids for genetic mapping in hexaploid winter triticale. Mol Breed 38(4). https://doi.org/10.1007/s11032-018-0804-3

934. Zaitseva OI (2017) In vitro androgenesis in triticale. Eur Biotech J 1(1):99–100. https://doi.org/10.24190/issn2564-615x/2017/01.20

935. Wurschum T, Tucker MR, Maurer HP, Leiser WL (2015) Ethylene inhibitors improve efficiency of microspore embryogenesis in hexaploid triticale. Plant Cell Tissue Org Cult 122(3):751–757. https://doi.org/10.1007/s11240-015-0808-1

936. Würschum T, Tucker MR, Reif JC, Maurer HP (2012) Improved efficiency of doubled haploid generation in hexaploid triticale by in vitro chromosome doubling. BMC Plant Biol 12(1):109

937. Oleszczuk S, Sowa S, Zimny J (2004) Direct embryogenesis and green plant regeneration from isolated microspores of hexaploid triticale (× *Triticosecale Wittmack*) cv. Bogo Plant Cell Rep 22(12):885–893. https://doi.org/10.1007/s00299-004-0796-9

938. Moradi P, Hagh NA, Bozorgipour R, Sharma B (2009) Development of yellow rust resistant doubled haploid lines of wheat through wheat × maize crosses. Int J Plant Prod 3(3):77–88

939. Wang HM, Enns JL, Nelson KL, Brost JM, Orr TD, Ferrie AMR (2019) Improving the efficiency of wheat microspore culture methodology: evaluation of pretreatments, gradients, and epigenetic chemicals. Plant Cell Tissue Organ Cult 139(3):589–599. https://doi.org/10.1007/s11240-019-01704-5

940. Zhang W, Wang K, Lin ZS, Du LP, Ma HL, Xiao LL, Ye XG (2014) Production and identification of haploid dwarf male sterile wheat plants induced by corn inducer. Bot Stud 55(1):26

941. Reynolds TL (2000) Effects of calcium on embryogenic induction and the accumulation of abscisic acid, and an early cysteine-labeled metallothionein gene in androgenic microspores of *Triticum aestivum*. Plant Sci 150(2):201–207. https://doi.org/10.1016/S0168-9452(99)00187-9

942. Liu W (2004) Transformation of microspores for generating doubled haploid transgenic wheat (*Triticum aestivum* L.). Washington State University, Washington, DC

943. Chauhan H, Khurana P (2010) Use of doubled haploid technology for development of stable drought tolerant bread wheat (*Triticum aestivum* L.) transgenics. Plant Biotechnol J 9(3):408–417. https://doi.org/10.1111/j.1467-7652.2010.00561.x

944. Letarte J, Simion E, Miner M, Kasha KJ (2006) Arabinogalactans and arabinogalactan-proteins induce embryogenesis in wheat (*Triticum aestivum* L.) microspore culture. Plant Cell Rep 24:691–698

945. Redha A, Attia T, Büter B, Saisingtong S, Stamp P, Schmid JE (1998) Improved

production of doubled haploids by colchicine application to wheat (*Triticum aestivum* L.) anther culture. Plant Cell Rep 17 (12):974–979. https://doi.org/10.1007/s002990050520

946. Shariatpanahi ME, Belogradova K, Hessamvaziri L, Heberle-Bors E, Touraev A (2006) Efficient embryogenesis and regeneration in freshly isolated and cultured wheat (*Triticum aestivum* L.) microspores without stress pretreatment. Plant Cell Rep 25 (12):1294–1299

947. Soriano M, Cistue L, Castillo AM (2008) Enhanced induction of microspore embryogenesis after n-butanol treatment in wheat (*Triticum aestivum* L.) anther culture. Plant Cell Rep 27(5):805–811. https://doi.org/10.1007/s00299-007-0500-y

948. Karsai I, Bedo Z, Hayes PM (1994) Effect of induction medium Ph and maltose concentration on in-vitro androgenesis of hexaploid winter triticale and wheat. Plant Cell Tissue Organ Cult 39(1):49–53

949. Tuvesson IKD, Öhlund RCV (1993) Plant regeneration through culture of isolated microspores of *Triticum aestivum* L. Plant Cell Tissue Organ Cult 34(2):163–167

950. Adamski T, Krystkowiak K, Kuczyńska A, Mikołajczak K, Ogrodowicz P, Ponitka A, Surma M, Ślusarkiewicz-Jarzina A (2014) Segregation distortion in homozygous lines obtained via anther culture and maize doubled haploid methods in comparison to single seed descent in wheat (*Triticum aestivum* L.). Electron J Biotechnol 17(1):6–13

951. Tayeng T, Chaudhary HK, Kishore N (2012) Enhancing doubled haploid production efficiency in wheat (*Triticum aestivum* L. em. Thell) by in vivo colchicine manipulations in *Imperata cylindrica*-mediated chromosome elimination approach. Plant Breed 131(5):574–578

952. Weigt D, Kiel A, Siatkowski I, Zyprych-Walczak J, Tomkowiak A, Kwiatek M (2020) Comparison of the androgenic response of spring and winter wheat (*Triticum aestivum* L.). Plants 9(1):49. https://doi.org/10.3390/plants9010049

953. Weigt D, Niemann J, Siatkowski I, Zyprych-Walczak J, Olejnik P, Kurasiak-Popowska D (2019) Effect of zearalenone and hormone regulators on microspore embryogenesis in anther culture of wheat. Plan Theory 8 (11):487

954. Barakat MN, Al-Doss AA, Ghazy AI, Moustafa KA, Elshafei AA, Ahmed EI (2018) Doubled haploid wheat lines with high molecular weight glutenin alleles derived from microspore cultures. New Zeal J Crop Hortic Sci 46(3):198–211. https://doi.org/10.1080/01140671.2017.1368674

955. Lantos C, Pauk J (2016) Anther culture as an effective tool in winter wheat (*Triticum aestivum* L.) breeding. Genetika 52 (8):910–918

956. Santra M, Wang H, Seifert S, Haley S (2017) Doubled haploid laboratory protocol for wheat using wheat–maize wide hybridization. In: Bhalla PL, Singh MB (eds) Wheat biotechnology: methods and protocols. Springer New York, New York, NY, pp 235–249. https://doi.org/10.1007/978-1-4939-7337-8_14

957. Santra M, Ankrah N, Santra DK, Kidwell KK (2012) An improved wheat microspore culture technique for the production of doubled haploid plants. Crop Sci 52 (5):2314–2320

958. Castillo AM, Sánchez-Díaz RA, Valles MP (2015) Effect of ovary induction on bread wheat anther culture: ovary genotype and developmental stage, and candidate gene association. Front Plant Sci 6. https://doi.org/10.3389/fpls.2015.00402

959. Lazaridou T, Pankou C, Xynias I, Roupakias D (2016) Effect of D genome in wheat anther culture response after cold and mannitol pretreatment. Acta Biol Cracov Ser Bot 58(1):95–102. https://doi.org/10.1515/abcsb-2016-0006

960. Kim K-M, Baenziger PS (2005) A simple wheat haploid and doubled haploid production system using anther culture. In Vitro Cell Dev Biol Plant 41(1):22–27. https://doi.org/10.1079/ivp2004594

961. Daniel G, Baumann A, Schmucker S (2005) Production of wheat doubled haploids (*Triticum aestivum L.*) by wheat × maize crosses using colchicine enriched medium for embryo regeneration. Cereal Res Commun 33(2):461–468. https://doi.org/10.1556/crc.33.2005.2-3.107

962. Gaines EF, Aase HC (1926) A haploid wheat plant. Am J Bot 13(6):373–385. https://doi.org/10.2307/2435439

963. Tan B, Halloran G (1982) Pollen dimorphism and the frequency of inductive anthers in anther culture of *Triticum monococcum*. Biochem Physiol Pflanz 177(2):197–202

964. Katayama Y (1934) Haploid formation by X-rays in *Triticum monococcum*. Cytologia 5(2):235–237

965. Lantos C, Bóna L, Nagy É, Békés F, Pauk J (2018) Induction of in vitro androgenesis in anther and isolated microspore culture of

different spelt wheat (*Triticum spelta* L.) genotypes. Plant Cell Tissue Organ Cult 133(3):385–393. https://doi.org/10.1007/s11240-018-1391-z

966. Lantos C, Jenes B, Bona L, Cserhati M, Pauk J (2016) High frequency of Double Haploid plant reproduction in spelt wheat. Acta Biol Cracov Ser Bot 58(2):107–112. https://doi.org/10.1515/abcsb-2016-0014

967. Castillo AM, Allue S, Costar A, Alvaro F, Valles MP (2019) Doubled haploid production from Spanish and Central European spelt by anther culture. J Agric Sci Technol 21(5):1313–1324

968. Cistué L, Soriano M, Castillo AM, Valles MP, Sanz JM, Echavarri B (2006) Production of doubled haploids in durum wheat (*Triticum turgidum* L.) through isolated microspore culture. Plant Cell Rep 25(4):257–264

969. Mahato A, Chaudhary HK (2019) Auxin induced haploid induction in wide crosses of durum wheat. Cereal Res Commun 47(3):552–565. https://doi.org/10.1556/0806.47.2019.31

970. Slama-Ayed O, Slim-Amara H (2007) Production of doubled haploids in durum wheat (*Triticum durum* Desf.) through culture of unpollinated ovaries. Plant Cell Tissue Organ Cult 91(2):125–133

971. Labbani Z, Richard N, De Buyser J, Picard E (2005) Plantes chlorophylliennes de blé dur obtenues par culture de microspores isolées: importance des prétraitements. C R Biol 328(8):713–723. https://doi.org/10.1016/j.crvi.2005.05.009

972. Fedak G (1983) Haploids in *Triticum ventricosum* via intergeneric hybridization with *Hordeum bulbosum*. Can J Genet Cytol 25(2):104–106

973. Barcelo P, Cabrera A, Hagel C, Lörz H (1994) Production of doubled haploid plants from *Tritordeum* anther culture. Theor Appl Genet 87(6):741–745

974. Van den Bulk R, De Vries-Van Hulten H, Custers J, Dons J (1994) Induction of embryogenesis in isolated microspores of tulip. Plant Sci 104(1):101–111

975. Custers J, Ennik E, Eikelboom W, Dons J, van Lookeren Campagne M (1996) Embryogenesis from isolated microspores of tulip; towards developing F1 hybrid varieties. VII international symposium on flowerbulbs 430:259–266

976. Redenbaugh MK, Westfall RD, Karnosky DF (1981) Dihaploid callus production from *Ulmus americana* anthers. Bot Gaz 142(1):19–26. https://doi.org/10.1086/337191

977. Smagula J, Lyrene P (1984) Blueberry. Handbook of plant cell. Culture 3:383–401

978. Chong-Perez B, Carrasco B, Silva H, Herrera F, Quiroz K, Garcia-Gonzales R (2018) Regeneration of highland papaya (*Vasconcellea pubescens*) from anther culture. Appl Plant Sci 6(9). https://doi.org/10.1002/aps3.1182

979. Hesemann C (1980) Haploid cells in calli from anther culture of *Vicia faba* [broad bean]. Zeitschr Pflanzen 84:18–27

980. Gosal S, Bajaj YS (1988) Pollen embryogenesis and chromosomal variation in anther of three food legumes – *Cicer arietinumPisum sativum and Vigna mungo*. SABRAO J 20:51–58

981. Arya I, Chandra N (1989) Organogenesis in anther-derived callus culture of cowpea [*Vigna unguiculata* (l.) Walp]. Curr Sci 58(5):257–259

982. Wijowska M, Kuta E, Przywara L (1999) In vitro culture of unfertilized ovules of *Viola odorata* L. Acta Biol Cracov Ser Bot 41:95–101

983. Salunkhe C, Rao P, Mhatre M (1999) Plantlet regeneration via somatic embryogenesis in anther callus of *Vitis latifolia* L. Plant Cell Rep 18(7–8):670–673

984. Mozsar J, Süle S (1994) A rapid method for somatic embryogenesis and plant regeneration from cultured anthers of *Vitis riparia*. Vitis 33(4):245–246

985. Altamura M, Cersosimo A, Majoli C, Crespan M (1992) Histological study of embryogenesis and organogenesis from anthers of *Vitis rupestris* du Lot cultured *in vitro*. Protoplasma 171(3–4):134–141

986. Mauro MC, Nef C, Fallot J (1986) Stimulation of somatic embryogenesis and plant regeneration from anther culture of *Vitis vinifera* cv. Cabernet-Sauvignon. Plant Cell Rep 5(5):377–380

987. Emershad RL, Ramming DW, Serpe MD (1989) In ovulo embryo development and plant formation from stenospermic genotypes of *Vitis vinifera*. Am J Bot 76(3):397–402

988. Cersosimo A, Crespan M, Paludetti G, Altamura M (1989) Embryogenesis, organogenesis and plant regeneration from anther culture in *Vitis*. I International symposium on in vitro culture and horticultural breeding 280:307–314

989. Gresshoff PM, Doy CH (1974) Derivation of a haploid cell line from *Vitis vinifera* and

importance of stage of meiotic development of anthers for haploid culture of this and other genera. Z Pflanzenphysiol 73 (2):132–141

990. Bensaad Z, Hennerty M, Roche T (1996) Effects of cold pretreatment, carbohydrate source and gelling agents on somatic embryogenesis from anthers of *Vitis vinifera* L. cvs. 'Regina' and 'Reichensteiner'. International symposium on plant production in closed ecosystems 440:504–509

991. Zhang X, Wu Q, Li X, Zheng S, Wang S, Guo L, Zhang L, Custers JB (2011) Haploid plant production in *Zantedeschia aethiopica* 'Hong Gan' using anther culture. Sci Hortic 129(2):335–342

992. Kelliher T, Starr D, Wang W, McCuiston J, Zhong H, Nuccio ML, Martin B (2016) Maternal haploids are preferentially induced by CENH3-tailswap transgenic complementation in maize. Front Plant Sci 7(414). https://doi.org/10.3389/fpls.2016.00414

993. Molenaar WS, de Oliveira Couto EG, Piepho H-P, Melchinger AE (2019) Early diagnosis of ploidy status in doubled haploid production of maize by stomata length and flow cytometry measurements. Plant Breed 138(3):266–276. https://doi.org/10.1111/pbr.12694

994. Chaikam V, Molenaar W, Melchinger AE, Boddupalli PM (2019) Doubled haploid technology for line development in maize: technical advances and prospects. Theor Appl Genet 132(12):3227–3243. https://doi.org/10.1007/s00122-019-03433-x

995. Molenaar WS, Schipprack W, Melchinger AE (2018) Nitrous oxide-induced chromosome doubling of maize haploids. Crop Sci 58:650–659

996. Hu H, Schrag TA, Peis R, Unterseer S, Schipprack W, Chen S, Lai J, Yan J, Prasanna BM, Nair SK, Chaikam V, Rotarenco V, Shatskaya OA, Zavalishina A, Scholten S, Schön C-C, Melchinger AE (2016) The genetic basis of haploid induction in maize identified with a novel genome-wide association method. Genetics 202(4):1267–1276. https://doi.org/10.1534/genetics.115.184234

997. Liu Z, Wang Y, Ren J, Mei M, Frei UK, Trampe B, Lübberstedt T (2016) Maize doubled haploids. In: Janick J (ed) Plant breeding reviews, vol 40. John Wiley & Sons, Hoboken, NJ, pp 123–166

998. Geiger HH, Gordillo GA (2009) Doubled haploids in hybrid maize breeding. Maydica 54(4):485–499

999. Prigge V, Xu X, Li L, Babu R, Chen S, Atlin GN, Melchinger AE (2012) New insights into the genetics of in vivo induction of maternal haploids, the backbone of doubled haploid technology in maize. Genetics 190 (2):781–793. https://doi.org/10.1534/genetics.111.133066

1000. Prigge V, Melchinger AE (2012) Production of haploids and doubled haploids in maize. In: Loyola-Vargas VM, Ochoa-Alejo N (eds) Plant cell culture protocols, vol 877. Methods in molecular biology. Humana, New York, NY, pp 161–172. https://doi.org/10.1007/978-1-61779-818-4_13

1001. Nageli M, Schmid JE, Stamp P, Buter B (1999) Improved formation of regenerable callus in isolated microspore culture of maize: impact of carbohydrates, plating density and time of transfer. Plant Cell Rep 19 (2):177–184

1002. Kermicle JL (1974) Origin of androgenetic haploids and diploids induced by the indeterminate gametophyte (ig) mutation in maize. In: Kasha KJ (ed) Haploids in higher plants: advances and potential. University of Guelph, Guelph, ON, p 137

1003. Barnabas B, Obert B, Kovacs G (1999) Colchicine, an efficient genome-doubling agent for maize (*Zea mays* L.) microspores cultured in anthero. Plant Cell Rep 18 (10):858–862

1004. Gilles LM, Khaled A, Laffaire J-B, Chaignon S, Gendrot G, Laplaige J, Bergès H, Beydon G, Bayle V, Barret P, Comadran J, Martinant J-P, Rogowsky PM, Widiez T (2017) Loss of pollen-specific phospholipase NOT LIKE DAD triggers gynogenesis in maize. EMBO J 36 (6):707–717. https://doi.org/10.15252/embj.201796603

1005. Kelliher T, Starr D, Richbourg L, Chintamanani S, Delzer B, Nuccio ML, Green J, Chen Z, McCuiston J, Wang W, Liebler T, Bullock P, Martin B (2017) MATRILINEAL, a sperm-specific phospholipase, triggers maize haploid induction. Nature 542(7639):105–109. https://doi.org/10.1038/nature20827

1006. Belicuas PR, Guimaraes CT, Paiva LV, Duarte JM, Maluf WR, Paiva E (2007) Androgenetic haploids and SSR markers as tools for the development of tropical maize hybrids. Euphytica 156(1–2):95–102. https://doi.org/10.1007/s10681-007-9356-z

1007. Brettel R, Thomas E, Wernicke W (1981) Production of haploid maize plants by anther culture. Maydica 26:101–111

1008. Tang F, Tao Y, Zhao T, Wang G (2006) In vitro production of haploid and doubled haploid plants from pollinated ovaries of maize (*Zea mays*). Plant Cell Tissue Organ Cult 84(2):233–237

1009. Battistelli G, Von Pinho R, Justus A, Couto E, Balestre M (2013) Production and identification of doubled haploids in tropical maize. Genet Mol Res 12 (4):4230–4242

1010. Wu P, Ren J, Li L, Chen S (2014) Early spontaneous diploidization of maternal maize haploids generated by in vivo haploid induction. Euphytica 200:127–138

1011. Nair SK, Chaikam V, Gowda M, Hindu V, Melchinger AE, Boddupalli PM (2020) Genetic dissection of maternal influence on in vivo haploid induction in maize. Crop J 8 (2):287–298. https://doi.org/10.1016/j.cj.2019.09.0082214-5141

1012. Zhong Y, Liu CX, Qi XL, Jiao YY, Wang D, Wang YW, Liu ZK, Chen C, Chen BJ, Tian XL, Li JL, Chen M, Dong X, Xu XW, Li L, Li W, Liu WX, Jin WW, Lai JS, Chen SJ (2019) Mutation of ZmDMP enhances haploid induction in maize. Nat Plants 5 (6):575–580. https://doi.org/10.1038/s41477-019-0443-7

1013. Liu LW, Li W, Liu CX, Chen BJ, Tian XL, Chen C, Li JL, Chen SJ (2018) In vivo haploid induction leads to increased frequency of twin-embryo and abnormal fertilization in maize. BMC Plant Biol 18. https://doi.org/10.1186/s12870-018-1422-2

1014. Khakwani K, Ahsan M, Sadaqat HA, Ahmad R (2018) Development and genetics of maize doubled haploid lines. Maydica 63 (3):1–15

1015. Obert B, Barnabás B (2004) Colchicine induced embryogenesis in maize. Plant Cell Tissue Organ Cult 77(3):283–285. https://doi.org/10.1023/b:ticu.0000018399.60106.33

1016. Ribeiro CB, Pereira FC, Ld NF, Rezende BA, Dias KOG, Braz GT, Ruy MC, Silva MB, Cenzi G, Techio VH (2018) Haploid identification using tropicalized haploid inducer progenies in maize. Crop Breed Appl Biotechnol 18(1):16–23

1017. Wang BB, Zhu L, Zhao BB, Zhao YP, Xie YR, Zheng ZG, Li YY, Sun J, Wang HY (2019) Development of a haploid-inducer mediated genome editing system for accelerating maize breeding. Mol Plant 12 (4):597–602. https://doi.org/10.1016/j.molp.2019.03.006

1018. Tian X, Qin Y, Chen B, Liu C, Wang L, Li X, Dong X, Liu L, Chen S (2018) Hetero-fertilization along with failed egg-sperm cell fusion supports single fertilization involved in in vivo haploid induction in maize. J Exp Bot 69:4689–4701. https://doi.org/10.1093/jxb/ery177

1019. Chase SS (1969) Monoploids and monoploid – derivatives of maize (*Zea mays* L.). Bot Rev 35:117–167

1020. Chase SS (1963) Androgenesis – its use for transfer of maize cytoplasm. J Hered 54 (4):152–158

1021. Samsudeen K, Babu KN, Divakaran M, Ravindran P (2000) Plant regeneration from anther derived callus cultures of ginger (*Zingiber officinale* Rosc.). J Hortic Sci Biotechnol 75(4):447–450

Chapter 4

Analysis of Ploidy in Haploids and Doubled Haploids

Sergio J. Ochatt and Jose M. Seguí-Simarro

Abstract

Determination of the ploidy level is an essential step when trying to produce doubled haploids (DHs) in any species. Each species and method used to produce DHs has its own frequency of DH production, which means that the rest of plants produced stay haploid. Since haploids are of little use for breeding purposes, it is necessary to distinguish them from true DHs. For this, several methodologies are available, including flow cytometry, chromosome counting, chloroplast counting in stomatal guard cells, measurement of stomatal size and length, counting of nucleoli, evaluation of pollen formation and viability, analysis of cell size, and analysis of morphological markers. However, not all of them are equally easy to use, affordable, reliable, reproducible, and resolutive and therefore useful for a particular case. In this chapter, we revise these methods available to assess the ploidy level of plants, discussing their respective advantages and limitations, and provide some troubleshooting tips and hints to help decide which to choose in each case.

Key words Chloroplast counting, Chromosome counting, Cytogenetics, Flow cytometry

1 Introduction

The main interest of developing doubled haploid (DH) technology resides in the fact that they are homozygous homohistonts, that is, in all loci, whereby they are directly exploitable within the context of breeding and lead to a faster acceleration of genome fixation in the progeny of crossings where they have been genitors. Moreover, among novel breeding techniques (NBTs) whose regulation is presently being discussed by the OCDE and controversial in many countries (such as in the EU), DHs still escape the constraints faced by other biotechnology-derived plants, such as GMOs or somatic hybrids between sexually incompatible species. Obtaining haploids is an obvious prerequisite of this breeding approach, and this is done through gynogenesis and/or androgenesis by shifting the development of female gametes (egg cells) or male gamete precursors (microspores or pollen grains), respectively, from a gametophytic to a sporophytic pathway, ultimately permitting the recovery of embryos (*see* Chapter 1 of this volume).

Jose M. Seguí-Simarro (ed.), *Doubled Haploid Technology: Volume 1: General Topics, Alliaceae, Cereals*,
Methods in Molecular Biology, vol. 2287, https://doi.org/10.1007/978-1-0716-1315-3_4,
© Springer Science+Business Media, LLC, part of Springer Nature 2021

Progress in the exploitation of haplodiploidization has been hindered in a number of economically important crops due to their recalcitrance to most tissue culture-based approaches and in particular to plant regeneration from undifferentiated tissues. This was particularly true with grain legumes, and pea (*Pisum sativum* L.), for instance, is probably one of the most recalcitrant legume species studied to date, even though examples of plant regeneration exist from several explants, cell suspensions, callus, and protoplasts [1].

Such difficulty in recovery of intact plants from recalcitrant genotypes is often, if not always, associated with a requirement for a particularly long time in culture required for plant regeneration and coupled with sometimes anecdotal results and/or a reduced reproducibility. It comes therefore as no surprise that producing DHs in such recalcitrant crops has also been difficult despite the urgent need to produce new cultivars while reducing the time needed to do so. This is particularly more appealing today, when approaches for an acceleration of generation cycles exist for several crops, allowing to more rapidly profit of such homozygous DHs obtained in a single generation to be rapidly exploited as genitors.

Triggering divisions in haploid cells and inducing the resulting cells to undergo embryogenesis strongly depends on the species and genotype, but also on the growth conditions of the donor plants, including any pretreatment applied to the flower buds and, most importantly in the particular case of induction of androgenesis, on the developmental stage of microspores and culture media (or media sequence) employed. Stress treatments are known to play a major role in androgenesis in many species and temperature (cold or heat), osmotic stress, microtubule disruption using colchicine or irradiation, and metabolic starvation have been tested in many crops [2–4], while other stress factors like electroporation, centrifugation or sonication are less documented [5–7]. As an example, in protein legumes, with the only exception of one report in chickpea [8], it has been repeatedly shown since 10 years ago [5] that by pyramiding (i.e., superimposing successively) various stress factors prior to culturing isolated pea microspores [7], it was possible to override their recalcitrance, leading to the recovery of haploid and DH plants, albeit at a reduced frequency and for a few genotypes [5, 6].

Once plants are eventually regenerated from any of the in vivo or in vitro methods to produce haploids and DHs, they must be checked for their ploidy level, as spontaneous doubling is not infrequent among various species. Another important aspect that must also be taken into account when inducing haploidization is the impact that artificial in vitro culture conditions may have (and often have for genotypes with an unstable genome or resulting from, more or less, distant hybridizations) on the genome makeup of the regenerated plants and tissues. This is further reinforced with recalcitrant genotypes as above, or whenever a significant amount of stress is required for the shift from the gametophytic to the sporophytic developmental path. Indeed, regenerants of certain species and cultivars tend to present either a nonhomogeneous ploidy level (i.e., they are not homohistonts) or be chimera, whereby their haploid nature is questionable and, following spontaneous or induced chromosome doubling, they may be prone to ending up being subfertile or even sterile, thus annihilating all the efforts undertaken to produce them. On the other hand, it may also happen that prolonged exposures to in vitro culture conditions, and in particular to plant growth regulators, may promote more than one round of chromosome doubling in cells, giving rise to polyploid organisms. Indeed, many researchers have reported an increase in the ploidy level with an increasing age of callus cultures [6, 9]. All these reasons justify the convenience of analyzing the ploidy of regenerated plants obtained by any of the different DH technologies currently available. It is important to underline that, even if many publications entirely omit this step, it is crucial for some DH techniques. For example, when anthers instead of isolated microspores are the tissue source, as they consist of both haploid germ line tissues and diploid, somatic anther wall tissues.

In summary, it is of the outmost importance that plants and tissues derived from DH technologies may be characterized as early as possible and thereafter, in a manner that ensures their true haploid nature prior to such chromosome doubling. This is commonly performed by flow cytometry, chromosome counting, or assessments of stomata size and density, among others. In this chapter, we shall focus on the various tools available to assess the ploidy level of plants, including these widely used methods and other, less known alternatives. We shall discuss the respective pros and cons of each of them and provide some troubleshooting tips and hints to help decide which one to choose.

2   Overview of Techniques for Ploidy Level Determination

At the onset of haploid and doubled-haploid production studies, ploidy level determination of tissues and plants was restricted to karyotypical assessments by counting the number of chromosomes in metaphase plates. This is very time-consuming and laborious, and it requires highly skilled operators and adapted microscopical equipment to be accomplished. Most limiting is also the availability of tissues containing discrete numbers of dividing cells, which may not always be easily available for such (frequently destructive) analyses. All these constraints often contribute to reducing the number of samples that may be practically analyzed at any given time. The advent of flow cytometry as an alternative approach for the determination of the level of ploidy in plant tissues has overturned this trend (Fig. 2). It is important to stress, though, that the microscopical counting of chromosomes generally in root tips stands as the reference and remains the only unambiguous tool to define the chromosome complement of a plant and/or tissue [10]. It should also be evoked that alternative methods to chromosome counting also exist which, although having proven not to be sufficiently reliable sometimes [11, 12], have the advantage that they do not require the presence of actively dividing cells in the tissue samples analyzed. These include, among others, phenotyping in the field, the estimation of leaf stomatal density and size [13–15], measuring the size of pollen grains and cells [16, 17], analyses of isoenzymes, molecular markers, and genomic in situ hybridization (GISH). In any case, since the nuclear DNA content is directly correlated with the ploidy level [18–21], flow cytometry has arisen as a far more reliable and faster methodology for the determination of ploidy level [10].

3   Ploidy Level Determination by Flow Cytometry

In 1986, Sree Ramulu and Dijkhuis [21] were the first to exploit flow cytometry for the study of polysomaty and genetic instability in vitro, with potato tissue cultures, and one year later appeared the first article on the use of flow cytometry for the determination of ploidy level [18]. Following these pioneering studies, a constant and ever-increasing flow of articles on this use of flow cytometry have been published. Interestingly, the most frequently reported use of flow cytometry in the literature has been by far to examine the ploidy level of tissues and/or plants obtained from experiments of either haplodiploidization or chromosome doubling [5, 6, 11, 12, 19, 22, 23], which over the last decade or so are being closely followed by studies assessing the stability and trueness-to-type of plants regenerated in vitro [10, 12, 24].

3.1   Methodology

Although there is a number of methods to prepare the plant materials for the flow cytometry analysis of their nuclear DNA content that have been devised for the different flow cytometer brands and models used (Fig. 1), they all share several common steps, the first one being the need to isolate nuclei from the plant tissues. The most commonly used method for nuclei isolation includes taking a small quantity of the plant tissues, as generally about 1 cm^2 or sometimes even less (i.e., half a leaf or, with composite leaves just one foliole) will largely suffice, and roughly chop it in an appropriate buffer in order to release the nuclei as a suspension. In this respect, whatever the method adopted, the buffer always contains the medium where nuclei will be suspended immediately after being isolated and also includes the stain. In some methods, these

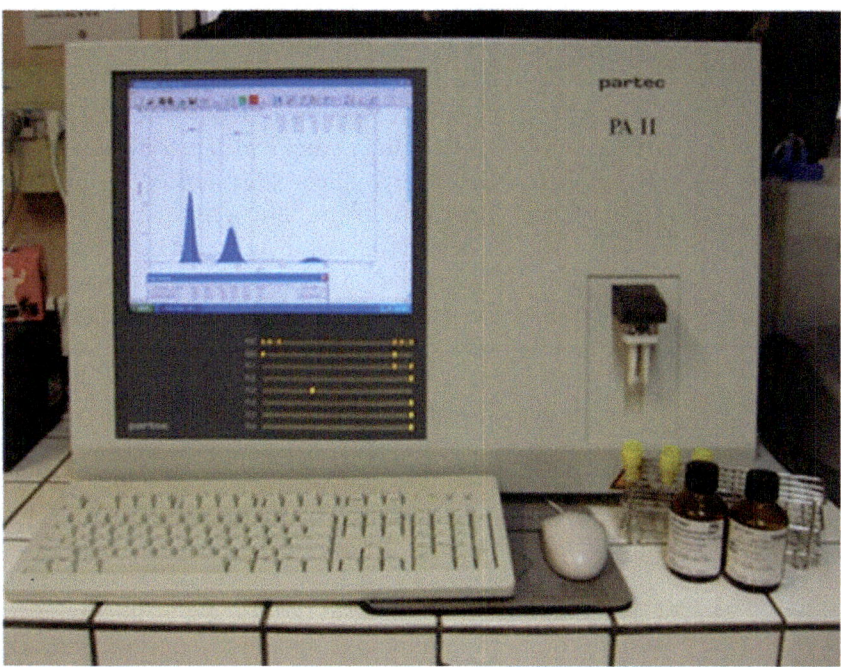

Fig. 1 Example of a flow cytometer. The image shows a Partec ploidy analyzer PA-II

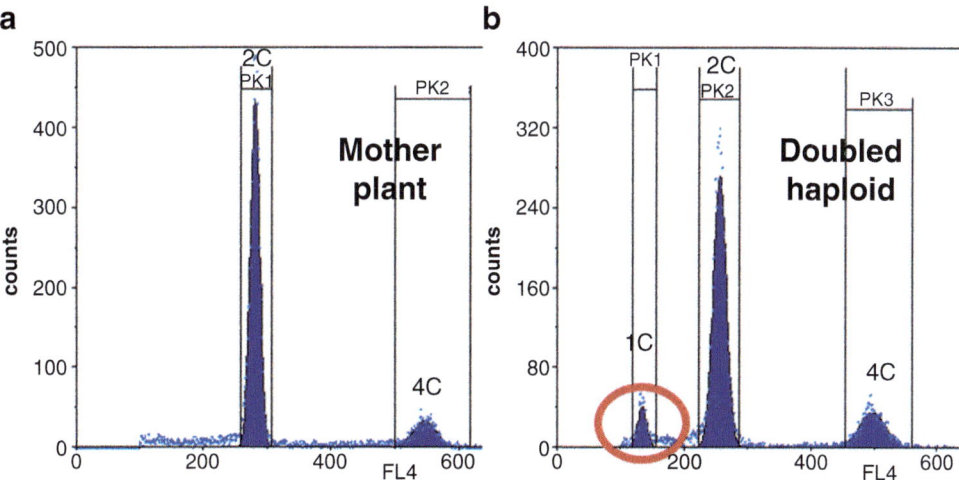

Fig. 2 Flow cytometry profiles for a true-to-type and a DH plant of Caméor pea (*Pisum sativum*). The profiles indicated a rapid spontaneous doubling during early culture stages, with a haploid G0/G1 peak at the 1C DNA level observed (encircled) for the DH derived from anthers

are separate buffers: one for isolation and a second one for staining and can be used either simultaneously or sequentially. There are various brands of ready-made buffers commercially available, but a "home-made" buffer may also be prepared [12] and stored in the fridge for a maximum of one month until use. Either way, it was mostly the pioneering work of Galbraith et al. [23] that paved the way for the widespread use of flow cytometry with plant tissues, where they replaced the enzymes used hitherto for isolating protoplasts whose nuclei were then analyzed, by the simple homogenization achieved by chopping tissues with a razor blade. Thus, not only the production of isolated nuclei became feasible from virtually any type of plant, organ, or tissue, but also attention focused on the analysis of cell nuclei and away from intact cells.

Following the isolation and staining of tissues, practically all methods also include a step of filtration or sieving through a very fine mesh of between 20 and 100 μm (generally 30 μm) pore diameter, generally made of nylon (cheaper) or stainless steel, in order to get rid of most cell material larger than the nuclei to be analyzed. Indeed, this must be done because plant cells are rarely spherical, as opposed to animal ones, and this may introduce errors into the flow cytometry signals, whereas being usually spherical and smaller than the width of the laser or UV beam used to excite them eliminates this problem. It must not be overlooked, however, that this filtration step can generate additional problems, as the suspension resulting from sample chopping not only contains the isolated nuclei but also several organelles as also soluble substances including phenolics, DNase, and RNase, among others, that are released from the cytoplasm and vacuoles. Therefore, in several examples such debris have been eliminated by washing the nuclei through repeated centrifugations and resuspensions or by modifying the composition (fluorescent dye concentration, addition of proteinase K to reduce fading, etc.) and/or pH of the staining buffer in order to block unwanted signals/substances [25].

3.2 Advantages and Limitations

Flow cytometry is as efficient as counting of metaphase chromosomes in root tips, but much cheaper and a lot less time- and labor-consuming. Thus, it is not surprising that this technique has become the standard method to estimate ploidy in in vitro culture protocols, and in particular in protocols for DH production. Indeed, this is the technique used in most full protocols available to produce DHs in different species. As examples, flow cytometry is used to determine the ploidy of regenerants in Chapters 6, 7, 9, 11, 13, 17–19 of volume 1, Chapters 5–8, 10–14, 16–18, 21 of volume 2 and Chapters 4, 5, 10, 11–15, 17–21 of volume 3.

Nevertheless, flow cytometry is complementary to chromosome counting and not meant to entirely replace it, as the ploidy level is calculated. Hence, flow cytometry results should be termed DNA ploidy as opposed to those obtained by karyological determinations, in particular for generative ploidy levels (estimated $2x$, $3x$, $4x$, etc.) but not for somatic endopolyploidy [26]. In any case, it is likely that most progress in techniques for ploidy level determination will come from the generalization of flow cytometry measurements. This is particularly helped by the increasing availability of smaller and more affordable multiparameter instruments which are adapted to measure both fluorescence and light scatter, whereby they allow wide access to a convenient and rapid method for the screening of ploidy levels in living, dried [27, 28] and even frozen [29, 30] samples. In spite of the precision of those analyses that might have also been repeated independently several times, it is likely that when a deviation in chromosome complement is hinted by flow cytometry assessments, this must still be ascertained by conventional chromosome counting in root tips squashes. Several deviations in the relative DNA content per nucleus have been reported and for a rather large number of species. Among these could be cited monocots including banana [31], asparagus [32] or rye grass [33], and hop [34] and the model species *Medicago truncatula* [11, 12, 35] among dicots.

Several other problems have been encountered when the material analyzed is not a nuclei suspension, including for instance a reduced permeability of the cell walls to various stains, the lack of specific cell binding to certain fluorescent probes, the potential occurrence of various autofluorescing pigments and, rather frequently in certain recalcitrant species, the presence in tissues or the release from tissues after chopping of phenolic substances and/or specific secondary metabolites capable of interfering with staining of cells and/or with dye fluorescence [19, 36]. As a result, and having observed that many of these problems for analysis of plant cells were due to their shape and to the existence of a cell wall [12, 37], the evident strategy imagined to solve them was to produce and analyze protoplasts which, being spherical, interfere less with the trajectory of the laminar flow than irregularly shaped objects will do. This, however, implies a preparation of material to be analyzed that becomes more difficult, long, and costly than directly using cells and, in addition, also requires the use of enzyme solutions for tissue digestion where the composition is strongly dependent on the genotype, the tissue source, and its developmental stage. Moreover, isolated protoplasts are fragile and require a very careful handling by experienced operators [38]. Hence the relevance of the pioneering work by Galbraith et al. in 1983 [23].

3.3 Representation of Flow Cytometry Data

Flow cytometry parameters are measured properties of the particles analyzed which are often assimilated to an optical channel, used as unit in the graphical representation of flow cytometry profiles. In turn, data in a one-parameter histogram plot the specific cellular subpopulations which contain a given DNA content, or which are bound to a given number of molecules of the antibody/fluorophore probed [36]. Thus, in a flow cytometry profile, cell content will be assigned to one of several classes or channels which are represented on the *x*-axis, while the number of epifluorescing cells assigned to each channel, named channel content or simply count, will be shown on the *y*-axis. As a result, all cells with about equal quantities of a given cell content, say DNA, will be gathered around the mean to form a peak (Fig. 2) which can be shown as such (brut/raw) or following the statistical analysis of peaks, which are then transformed into Gauss bells at either side of the median of data by the in-built program of the cytometer. In the latter representation of data, it will thus be possible to provide the surface area underneath each peak through the calculation of its integral, and also to include a chi-square analysis of the data represented in the profile. Obviously, for both brut and analyzed peaks the coefficient of variation (%) is provided too.

In a typical DNA histogram for an euploid material, two peaks will be observed (Fig. 2): a first, more prominent peak representing the nuclei in G0/G1 phase (unreplicated, 2C) and a second one (formed at twice the channel value) for the nuclei at G2/M phase (replicated, 4C) of the cell cycle [12, 39]. However, different types of profiles can also appear, as will be discussed below.

Additionally, other graphic representations of data are also possible. These may then be analyzed either automatically by the frequently in-built computer of the flow cytometer, or manually by drawing ranges or numerical fits to identify either the mean intensity of the epifluorescence emitted or the number of cells within a peak. Alternatively, if a correlated two-parameter dot plot is chosen, the quantities of the cell properties (i.e., forward and side scatter intensity) will be assigned to channels on the x- and y-axes, and each cell with a given intensity will be represented as one dot in the dot plot. Accordingly, cells with the same combination of properties (e.g., forward and side scatter intensity) will share the same dot location and subpopulations of such cells with common properties will appear as clusters or clouds, which may subsequently be gated by drawing lines around them in order to analyze the number of cells in each cluster. The most frequent representation of flow cytometry profiles in the literature is the one-parameter histogram plots, but dot plots (e.g., side scatter/FL cytograms) are important in several cases such as for the separation of doublets from endopolyploid nuclei, and of S-phase nuclei from broken 4C nuclei or from cellular fragments [40].

3.4 Stains and Standards

The two most common dyes in the literature are DAPI (4,6-diamidino-2-phenylindole), which is AT-specific and the most frequently used dye to determine ploidy level by flow cytometry, and propidium iodide (PI) which is preferred for the determination of genome size, and usually believed to be non–base pair specific, although it was reported to exhibit a GC-preference [41]. It is important to underline that the absolute DNA content is, in fact, irrelevant for ploidy analysis and, hence, the distribution of nuclear DNA content within a given cell population is obtained by comparing the numbers of objects in the different peaks or clusters. Having said this, it is crucial to include an internal standard as a control, when analyzing the ploidy level of G1 cells by comparing the positions of the major (normally the first) peak in profiles from different genotypes. This is due to the fact that, even in the absence of fluorescence inhibitors, there are various other variables in the isolation and staining procedures which would be difficult to control without an internal standard, including shifts in pH or the quantity of material used, the dye concentration, temperature and time of staining, and measurement conditions. For instance, a very high room temperature in the flow cytometry laboratory may alter the reliability of measurements. Commonly used internal standards are chicken erythrocytes, UV Teflon beads, salmon sperm cells or, more frequently, simply a suspension of nuclei of a genotype as for example the donor plant if comparing populations of in vitro produced regenerants [12], or a model system whose relative DNA content per nucleus for a given tissue source is well documented and consistent among different genotypes within the species. This is the case for pea, rice, tomato, onion, or *Medicago truncatula* [11, 19, 37, 42, 43]. In such cases, though, it is obvious that particular attention should be paid to the choice of species to be used as internal standard so that its peaks do not overlap with those of the sample tissue to be analyzed. Some authors make a distinction between internal standard, when the tissues of the standard have been chopped together with those of the probe, and external if they were chopped separately and mixed at the end immediately prior to the readings, or even analyzed separately and then the histograms compared.

3.5 Flow Cytometry as a Diagnostic Tool Beyond Ploidy Estimation

Due to the mentioned features of this technique, each class of samples will exhibit its exclusive and typical flow cytometry signature. Thus, we can use flow cytometry not only to determine the ploidy level, but also as a diagnostic tool to identify chromosomal abnormalities that might have an impact on the regeneration ability or the future performance of regenerated plants. Thus, a normal euploid material will exhibit profiles with two peaks. For samples having undergone endoreduplication, the profile will include a succession of peaks of decreasing intensity, although a second peak larger than the first (due to a succession of endomitotic DNA replication cycles) is seldom also observed. This phenomenon results in nuclei which constantly increase in size but do not undergo karyokinesis. Moreover, endoreduplication has been shown to block the regeneration competence from protoplast-derived calli in pea [44], and rarely permitted the recovery of normal plants from other tissue sources and, for that matter, nor with other species [9]. For aneuploid tissues, the typical cytometric imprint is a profile with two peaks as in euploid tissue, but where the G1, the G2/M or both peaks is split or "shouldered." When only one of the peaks is split, generally a first shouldered peak is indicative of a missing chromosome while a profile where the

G2/M peak is split would indicate a chromosome in excess; in any such case, the precise chromosome complement of the material in question must be confirmed by root tip or other chromosome counting before drawing any conclusion. It is worthwhile underlying that such aneuploid regenerants are seldom viable and those surviving frequently sterile [12, 22, 35]; nevertheless, aneuploidy remains one of the most frequent chromosomal abnormalities in plants, but also one with large breeding potential, as in cereals [33, 45], fruit [31], horticultural [32] and industrial [34] crops. Another type of profile with only one apparent peak (the G1 nuclei) is that corresponding to senescent tissues where cell division has ceased, as frequently observed when aged plants and tissues are analyzed [11, 12]. Finally, there are also some less frequent situations where a sum of abnormalities is observed in a single piece of undifferentiated tissue, from which no regenerants are recovered [22, 35]. Interestingly, in studies with legumes over the last years [1, 12, 35] any samples deviating from the mother plant genotype in terms of flow cytometry profiles also exhibited deviating isoenzyme and RAPD banding patterns.

4 Chromosome Counting

For many decades now, a large number of useful protocols for chromosome preparations from different cells and for different purposes, including karyotype analysis and chromosome banding, have been published for plants [46]. The most frequently used tissue for chromosome counting are root tip cells from young seedlings, and the chromosome number has been established during mitotic divisions, although it is generally easier and faster to count mitotic chromosomes when they are arrested in metaphase [10, 46, 47]. There are also several studies where other tissues have been used for the determination of the chromosome complement such as the seeds and endosperm, for example in maize [48] and coconut [49], the tapetal cells in maize [50] and microspore mother cells at diakinesis in rice [51]. Having said this, root tips of about 2 cm long remain the most convenient source of mitotic cells. However, if unavailable, young buds and leaves, or callus and even cell suspensions may be used instead.

Logically, if the tissue sources change then the cytological procedures for chromosome preparation and staining applied may have to be modified, as is also the case when the species studied is different. Whatever the species, genotype, or tissue source to be analyzed, there are three main basic operations for the handling of mitotic chromosomes that will remain unchanged, which are (1) material collection and pretreatment, (2) material fixation, and (3) preparation and staining of chromosomes.

A common difficulty is that plant chromosomes can vary in size, and a pretreatment to increase the number of metaphase cells and chromosome condensation may help to resolve it. This is particularly applicable to species with long chromosomes, while in those with short chromosomes, counting can frequently be performed without any pretreatment, especially in late prophases. Most frequently this procedure requires the treatment of the meristematic tissues with colchicine, ice-cold water, 8-hydroxyquinoline, or a solution of saturated a-bromonaphthalene, as reviewed [52]. It should be underlined that the duration of any such pretreatment will vary dependent on the speed of the mitotic cycle which, in turn, depends on the temperature and obviously the species genome size [46].

If such pretreatments are needed, the tissues are thereafter fixed in cold, freshly prepared Carnoy's fixative for 2–4 h at room temperature, and stored at 20 °C until used, or can alternatively be kept at 4 °C in 70% ethanol (for no longer than a month in general) before use. Production of enough well-spread chromosome plates with a good physical separation of the chromosomes is most critical for the handling of mitotic chromosomes, and the two most frequent methods for chromosome preparation are either the staining of the intact tissue source (e.g., the root tips) prior to squashing, or the preparation of the chromosome samples and their subsequent staining [52], which is generally recommended when undertaking experiences with an unknown genotype or species. Examples of the use of this technique in Chinese cabbage, *Saintpaulia ionantha* and marigold can be found in Chapters 9 of volume 2, and Chapters 16 and 18 of volume 3, respectively.

4.1 Chromosome Staining Prior to Squashing

The two most frequently preferred staining methods for chromosome counting prior to squashing are aceto-orcein/acetocarmine or Feulgen stains [52], which stain only the chromosomes while the cytoplasm will remain clear.

4.1.1 Aceto-orcein/ Acetocarmine Stains

Both aceto-orcein and acetocarmine stains (Fig. 3a, b) may be prepared as a stock in hot 45% acetic acid, shaken, taken to the boil, then heated moderately for a further 5 min, allowed to cool down and filtered before use. Interestingly, an old stain solution is more efficient than a freshly prepared one.

For staining, following fixation the material is placed in a drop of stain on a gently warmed slide, and once the stain is dried off, a drop of 45% acetic acid is added prior to placing a cover slip on top and squashing as homogeneously as possible. Importantly, thick or too dense material will not spread, and it is of paramount importance to make a monolayer of cells and nuclei using the correct amount of fluid between the slide and coverslip as needed for squashing. Indeed, excess fluid would move cells to the edges of the coverslip while too little fluid would encourage the formation of

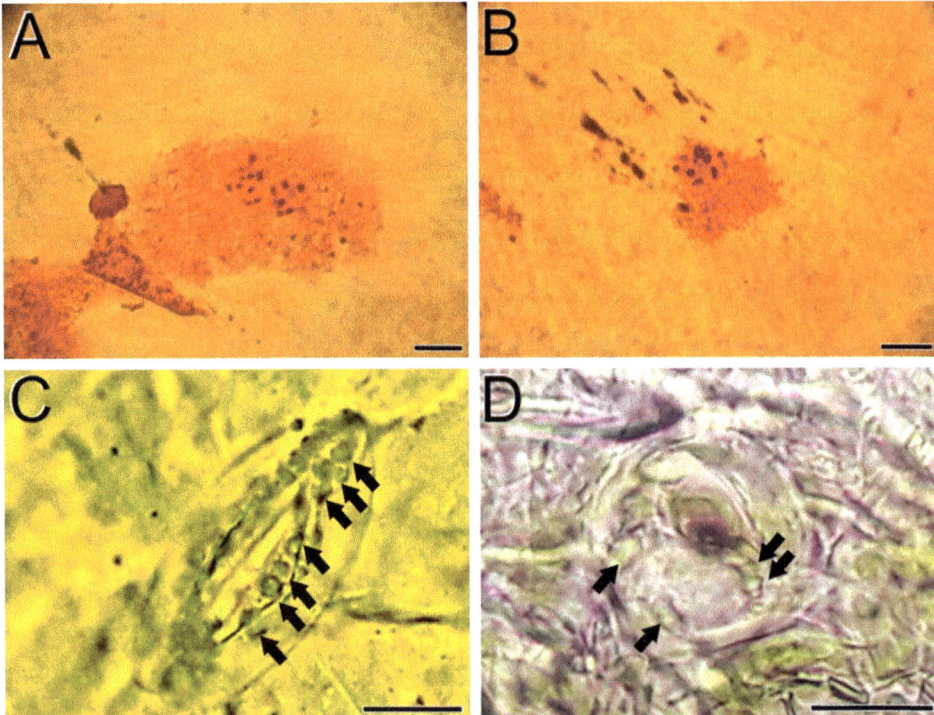

Fig. 3 Determination of ploidy in borage (*Borago officinalis*) by chromosome counting in root tip cells after acetocarmine staining (**a**, **b**) and chloroplast counting in stomatal guard cells (**c**, **d**). (**a**, **c**) Diploid, donor plant cells. (**b**, **d**) haploid cells. Arrowheads point to chloroplasts. Images courtesy of Prof. Mohammad R. Abdollahi. Bars: 10 μm

air bubbles that will render observation troubled. It is also important to apply the right amount of pressure for squashing. Too little pressure may be insufficient to release a sufficient amount of cells to visualize, whereas excessive pressure may break the coverslip.

Sometimes, this procedure may not be enough (as particularly observed for species with very long chromosomes, as in various tropical crops). In such cases, it may be required to make a fine suspension before pressure, to be applied after tissue staining (when using acetic orcein or acetocarmine) and maceration by heating or hydrolysis (when using the Feulgen method), because a simple pressing rarely leads to monolayers, except when analyzing very soft material such as pollen sac contents or enzyme-macerated meristems (a technique to be recommended for simpler and more reliable chromosome counting). Use a soft object not to scratch or break the glass coverslip. Cell separation and squash completion may be eased by gentle throbbing with the tip of a needle onto the mounted and floating coverslip provided this is done with great care to avoid shearing of the material with a lateral movement of the coverslip.

Often, there is also a need to accumulate cells at metaphase stage in order to facilitate chromosome counting, which may be achieved by germinating seeds either on a substrate or in vitro, and then synchronize the cells in metaphase by treating the root tips using a two-step procedure consisting of a treatment with 2 mM hydroxyurea first, followed by amiprophosmethyl solution [46].

4.1.2 The Feulgen Method

The Feulgen method is based on the reaction of Schiff reagent with aldehyde groups in the DNA, that have been previously exposed by an acid hydrolysis. To prepare the Feulgen stain [53], basic fuchsin is dissolved in boiling distilled water, left to cool and filtered. Then, 1 N HCl plus potassium metabisulfite are added, and the solution left to rest in the dark for a minimum of 12 h (or overnight) prior to the addition of activated charcoal. Then, it is subjected to vigorous shaking and a quick vacuum-filtering in order to avoid loss of SO_2 but also before charcoal resuspends, as its concentration and adsorbing power would otherwise be heterogeneous and render staining less efficient. This is so because the activated charcoal acts as an antioxidant by surface absorption (adsorption) of brown oxidation products that derive mostly from the fuchsin dye following the discoloration of the dissolved fuchsin. The resulting solution can then be used immediately or stored in the dark, at 4 °C, until used. Ready-to-use Feulgen stain can also be bought from various suppliers [52], and it can therefore be employed instead of the home-made one, but only provided it does not change color upon storage. For Feulgen staining, the fixed material is first rinsed in distilled water and then hydrolyzed. Hydrolysis can be performed following either of two methods: (1) in 5 N HCl at room temperature for 40 min or (2) in 1 N HCl at 60 °C for 10–12 min and then rinsed in 1 N HCl at room temperature. The excess Schiff reagent is washed off from stained tissues with several quick rinses (<30 s-long each) in SO_2-water (distilled water + HCl + sodium hydrogen sulfite, whereby SO_2 is released). The material is then dried out on paper and may be squashed directly or, for difficult material, following an enzymatic digestion step. This enzymatic digestion is usually performed in 1% (w/v) pectinase for 45–60 min, at room temperature. However, for most recalcitrant tissues, a mixture of cellulase plus pectinase may be used to digest tissues for a longer time (2 h 30 min) and at a higher temperature (29 °C) and in the dark, and coupled with a slight excision of the meristems directly on the slide just after digestion, in order to eliminate vascularized zones that would render the squash too thick or heterogeneous and hence the counting imprecise. The sample is finally washed for a few minutes in the mounting milieu, consisting of 45% acetic acid in deionized water at 4 °C, until tip squashing and chromosome counting. It may be sometimes helpful to air-dry the samples before mounting by instilling 70° ethanol several times between slide and cover, then replacing it by ethanol at 90° and finally 100° until the sample becomes hard and flat. At this point in time the cover is lifted and the preparation air-dried by complete evaporation of ethanol, and then a drop of fluid mounting medium is added, avoiding air bubbles and making sure the preparation is well spread before a new coverslip is placed. Coverslips must obviously be spotless clean, but they must not be cleaned with either alcohol or acid, or the material would then stick to the cover slip and will be lost as the cover slip is removed. There is consensus in the literature that a minimum of five to ten complete metaphase plates should be analyzed [10, 46, 52]. An example of the use of this technique to count chromosomes in haploid and DH plants can be found in Chapter 12 of volume 1.

4.2 Chromosome Staining Following Squashing

There are a number of methods for chromosome staining after squashing. They can be classified as fluorescent or nonfluorescent.

4.2.1 Fluorescence Methods

Today, the fluorescent methods are the most frequently used by most teams and DAPI is undoubtedly the most popular stain. DAPI forms a fluorescent complex by attaching in the minor grove to A-T rich sequences of DNA [10, 12, 54] and, when excited under UV light, the AT-rich sequences of chromosomes fluoresce bright blue. Compared to the Feulgen technique, it has the drawback of requiring a fluorescent microscope. However, DAPI is at present the most popular DNA stain due to many different reasons. First, it is an affordable and easy to use stain. Second, it intercalates into DNA in a manner proportional to DNA length. This means that the fluorescence intensity of the stain will be proportional to the amount of DNA being stained, which allows for quantification of DAPI fluorescence intensity to infer DNA amounts and even genome sizes. In addition, DAPI has the unique feature that, if needed, DAPI-stained samples can be easily destained and subsequently restained with another dye or further used for in situ hybridization. Ready-to-use DAPI solutions are available from several commercial sources and, alternatively, a 0.01% (w/v) DAPI stock solution in water can be prepared and stored at -20 °C, where it can be kept for a very long time until use without any loss of activity. When needed, this stock solution should be diluted to 2–5 μl/ml in McIlvaine or PBS buffer at pH 7.0. Then, a droplet (about 50 μl) of this dilution is added to each slide, left to act for 20–30 min in dark at room temperature, washed off with ultrapure water, mounted, and observed with a photonic microscope under UV light. In this book, there are many examples of the use of DAPI to visualize nuclear DNA in microspores and pollen of different species.

4.2.2 Nonfluorescent Stains

As mentioned, nonfluorescent stains are also available, although its use is being progressively reduced. Some of them have been described in detail and extensively used in the past such as Giemsa [52], which is still occasionally used [55]. Feulgen densitometry [56], that has now become classical for the determination of ploidy level, is in fact a spin off from the original Feulgen method [53]. The densitometric methods involve a staining of fixed tissue preparations on a microscope slide with Feulgen and rely on the fact that the amount of stain bound is stoichiometrically proportional to the amount of DNA present in the sample. However, this methodology should be used with caution because microdensitometric Feulgen measurements of DNA content suffer from several sources of error and, in addition to being time-consuming, they may not be exact [57, 58].

4.2.3 DNA Image Analysis

Against this background, the DNA image analysis-based methods sidestep all these constraints [56, 59], as they are rapid, cost-effective, and user-friendly. This has been proven for the measurement of animal genome sizes, but the number of plant species in which this technique has been used is increasing but still limited so far. Basically, a microscopic field is captured by the microscope CCD camera, and photos are displayed as a series of pixels each with a specific color and intensity. Then, the different intensities of each nuclear pixel correspond to a ready-made point intensity which is converted into an absorbance value by the inbuilt image analysis software of the equipment, so that a color image can be converted into a single linear scale by converting it to grayscale (from 0 for black to 255 for white). The individual pixel values may thus be used for the calculation of the integrated optical density instantaneously and for the whole image, which thereby allows the simultaneous tabulation of the integrated optical densities for all nuclei within the microscope field. As compared, this is equivalent to 500 nuclei measurable in less than 5 min against 50 per hour by densitometry [59].

5   Techniques for Ploidy Level Determination not Requiring Dividing Cells

In contrast to the techniques described above, there are also techniques that permit the determination of ploidy level without requiring an availability of dividing cells. These techniques have been much less used than those described above, partly because they may not be as reliable or reproducible as flow cytometry or chromosome counting, because most of them use leaves, pollen, or entire plants, which implies extended times until plants reach a certain size or even reproductive competence, or because they have been developed and optimized in particular species and their validation in other species is still pending. In any case, these techniques are still being used to evaluate the ploidy of certain species as part of routinely used protocols of DH production. They include chloroplast counting in stomatal guard cells, measurement of stomatal size and length, counting of nucleoli, evaluation of pollen formation and viability, analysis of cell size, and analysis of morphological markers. Examples of some of them can be found in some of the chapters of the three volumes of this book.

5.1 Chloroplast Counting in Stomatal Guard Cells

One of these techniques is based on chloroplast scoring in stomatal guard cells of the epidermis of the abaxial side of leaves. It is known that the number of chloroplasts of these cells is proportional to their ploidy level [13]. Thus, the leaves (preferably young and still unfolding) are fixed in Carnoy's fixative (ethanol–acetic acid 3:1, v/v) and stained with a potassium iodide solution. Thereafter, a minimum of 25 stomata should be counted in each leaf and a minimum of three to five leaves should be examined, comparing the chloroplast number with a diploid reference which obviously should have more chloroplasts (Fig. 3c, d). However, it must be noted that numbers in haploids may not necessarily be the half of those in diploids. Actually they use to be slightly more than half. For example, haploid plants of indica rice are described to have 5–8 chloroplasts, whereas DHs use to have 10–15 (*see* Chapter 20 of volume 1). In watermelon (Chapter 6 of volume 3), haploids have 6 to 7 chloroplasts and diploids have 11 to 12.

5.2 Stomatal Size and Length

Not only chloroplast number of stomatal cells seem to be a good predictor of ploidy, but other parameters of these special type of cells, such as cell size and length, seem to be related to ploidy. To measure them, the procedure is similar to that described above, but a staining with $AgNO_3$ may suffice. Examples of the use of this technique can be found in Chapter 20 of volume 1, Chapters 6–9, and 15 of volume 3. As for chloroplast counts, in this case the sizes of haploid cells are not necessarily half of those of diploids, but slightly more than half. For example, in borage (Chapter 15 of volume 3), stomatal guard cells of haploid plantlets have smaller stomata, averaging 19.8 μm length and 16.2 μm width, whereas those of diploid plants use to have 26.2 μm and 17.5 μm, respectively. In watermelon (Chapter 6 of volume 3), haploids are 17 to 18 μm in length and 10–12 μm in diameter, while diploids are 23 to 24 μm and 18 μm, respectively.

5.3 Counting of Nucleoli

The nucleolus is the nuclear compartment responsible for ribosome biogenesis. They are formed by the nucleolus organizer regions (NORs), present in the secondary constrictions of chromosomes. Only transcriptionally active ribosomal RNA (rRNA) genes give rise to a nucleolus. Thus, it could be assumed that the maximal number of nucleoli per nucleus should correlate with the number of active NORs. In other words, a haploid individual should, in principle, have less nucleoli than a diploid one. Such correlation has been confirmed for several species, including *Spartina pectinata*, *Dactylis glomerata*, *Allium wakegi*, and *A. fistulosum* L. [60–62]. This led to propose this method as an estimator of ploidy level [63]. However, it must be noted that this correlation may not be true in unstable genomes such as those of neopolyploids [62]. It was also shown that in other species, ploidy is not related to nucleolar number but to nucleolar size, at least in tomato fruits [64]. Hence, this estimative method must be used with caution, and ideally, after a confirmation of such a correlation in the species to be applied to.

5.4 Pollen Formation and Viability

It is known that haploids are sterile because they cannot produce pollen or if they do, it is not viable. Therefore, the evaluation of pollen presence in the flowers of regenerants would give an idea about their ploidy status. In turn, the viability of pollen grains, when produced, can also be correlated with genome size. In the genus *Hosta*, data from both traits were analyzed and exploited to confirm the hybrid nature and also the viability of some genotypes [16]. This method, however, would not be recommended, since in addition to the long time one must wait before regenerated plants reach sexual maturity to produce flowers and pollen, the reason why pollen is not present or viable may be other than haploidy. Indeed, gametoclonal variation might have led to some sort of male sterility which would produce false positives. In conclusion, this approach is time-consuming and of limited reliability.

5.5 Analysis of Cell Size

Working with citrus callus, a method was developed to calculate the mitotic index and ploidy level as based on the analysis of cell size [17]. The authors established the formula RD = 0.7937 RM, where RD is the diameter of new daughter cells, and RM that of metaphase cells, which was later validated by morphological, cytological, and fluorescent examination as well as by a protoplast electrofusion test. Finally, the authors then calculated the mitotic index and ploidy level of a cell population according to this formula and concluded that cell size can be used as a morphological marker to calculate both traits, at least for citrus callus. Thus, the range of proven reliability of this method is considerable limited.

5.6 Analysis of Morphological Markers

It is known that haploids are different from diploids in many aspects, and some of them are visible with the naked eye. It seems reasonable to assume that since a haploid cell has half the normal number of chromosomes, its nucleus will be smaller, which will give rise to smaller cells and therefore smaller individuals. This is why size has been considered as a marker to sort haploids and DHs by eye in the field or when no other methods are available. Other, related markers to consider are fruit, leaf or flower size, and in general, organ size. However, this method has some risks, since it is also known that size, as well as color and any other macroscopical, easily identifiable marker, can be highly influenced by plant culture conditions, including nutrition, watering, temperature, light, and pest exposure. In addition, these parameters can be different within among and even within populations. This is more evident when working with nonhomogeneous F_1 hybrids. Thus, the use of these markers must be done with extreme caution, only when working with homogeneous populations, under very well defined and fixed conditions, and when the operator(s) know very well how these traits behave, and how to identify them in the population. It would never be recommended when working with new species or backgrounds, or in not well-defined and controlled environments.

6   Concluding Remarks

Determination of the ploidy level is an essential step when trying to produce DHs in any species. We have revised the main methods available for this, from the fast, widely used, and resolutive flow cytometry to other, less used, and reliable ones. It seems clear that when available, flow cytometry stands out as the method of choice. It is fast, easy to use, reproducible, and resolutive, and has been tested in thousands of species, not only for ploidy determination. Thus, it should be the first option. However, not all laboratories may have this equipment at hand. The purchase of one of these machines (between 20,000 and 30,000 € for the basic equipment) may not be possible for all laboratories. As an alternative, chromosome counting would be advised. It is also easy and reproducible, but very time-consuming. In the time needed to finish a chromosome preparation to be observed, thousands of nuclei can be analyzed in tens of samples with the flow cytometer. In addition, to count chromosomes, dividing cells must be found first, which accounts for the extended times needed to analyze a representative number of cells, which will never be as high as with flow cytometry. Chromosome counting also needs expensive equipment, a microscope in this case. If chromosomes are to be stained with a fluorescent dye, the microscope must have an epifluorescence system. However, microscopes are more accessible than flow cytometers for research groups with low resources. Chloroplast counting in stomatal guard cells would be the third option. It is as time-consuming as chromosome counting, and it also requires a microscope, but there is no need to find dividing cells. Finally, other, less-demanding methods are also available. However, they are not as reliable and reproducible or, at least, they have not been proven useful for as many different species as flow cytometry, chromosome counting, or chloroplast counting. We therefore discourage their use unless there are sufficient evidences of their usefulness in a given material and for certain, fixed conditions.

Acknowledgments

This work was supported by grants AGL2017-88135-R and PID2020-115763RB-I00 to JMSS from Spanish MICINN jointly funded by FEDER.

References

1. Ochatt SJ (2015) Agroecological impact of an *in vitro* biotechnology approach of embryo development and seed filling in legumes. Agron Sustain Dev 35:535–552

2. Lülsdorf MM, Croser JS, Ochatt SO (2011) Androgenesis and doubled haploid production in food legumes. In: Pratap A, Kumar J (eds) Biology and breeding of food legumes. CAB International, Oxford, pp 159–177, 9781845937669

3. Parra-Vega V, Renau-Morata B, Sifres A, Segui-Simarro JM (2013) Stress treatments and in vitro culture conditions influence micro-spore embryogenesis and growth of callus from anther walls of sweet pepper (*Capsicum annuum* L.). Plant Cell Tissue Organ Cult 112:353–360

4. Rivas-Sendra A, Corral-Martínez P, Camacho-Fernández C, Segui-Simarro JM (2015) Improved regeneration of eggplant doubled haploids from microspore-derived calli through organogenesis. Plant Cell Tissue Organ Cult 122:759–765

5. Ochatt S, Pech C, Grewal R, Conreux C, Lülsdorf M, Jacas L (2009) Abiotic stress enhances androgenesis from isolated micro-spores of some legume species (Fabaceae). J Plant Physiol 166:1314–1328

6. Grewal RK, Lülsdorf M, Croser J, Ochatt S, Vandenberg A, Warkentin TD (2009) Doubled haploid production in chickpea (*Cicer arietinum* L.): role of stress treatments. Plant Cell Rep 28:1289–1299

7. Ribalta F, Croser J, Ochatt S (2012) Flow cytometry enables identification of sporophytic eliciting stress treatments in gametic cells. J Plant Physiol 169:104–110. https://doi.org/10.1016/j.jplph.2011.08.013

8. Abdollahi MR, Rashidi S (2018) Production and conversion of haploid embryos in chickpea (*Cicer arietinum* L.) anther cultures using high 2,4-D and silver nitrate containing media. Plant Cell Tissue Organ Cult 133:39–49. https://doi.org/10.1007/s11240-017-1359-4

9. Croser JS, Lülsdorf MM (2004) Progress towards haploid division in chickpea (*Cicer arietinum* L.), field pea (*Pisum sativum* L.)

and lentil (*Lens culinaris* Medik.) using isolated microspore culture. In: Eur grain legume conf, AEP, Dijon, France, pp 189

10. Ochatt SJ, Patat-Ochatt EM, Moessner A (2011) Ploidy level determination within the context of in vitro breeding. Plant Cell Tissue Organ Cult 104:329–341

11. Ochatt SJ (2006) Flow cytometry (ploidy determination, cell cycle analysis, DNA content per nucleus). 13 pp. https://www.noble.org/medicago-handbook/

12. Ochatt SJ (2008) Flow cytometry in plant breeding. Cytometry A 73:581–598

13. Detrez C, Sangwan RS, Sangwan-Norreel BS (1989) Phenotypic and karyotypic status of *Beta vulgaris* plants regenerated from direct organogenesis in petiole culture. Theor Appl Genet 77:462–468

14. Doležel J, Lucretti S, Schubert I (1994) Plant chromosome analysis and sorting by flow cyto-metry. Crit Rev Plant Sci 13:275–309

15. van Duren M, Morpurgo R, Doležel J, Afza R (1996) Induction and verification of autotetra-ploids and diploid banana (*Musa acuminata*) by in vitro techniques. Euphytica 88:25–34

16. Zonneveld BJM, van Iren F (2001) Genome size and pollen viability as taxonomic criteria: application to the genus Hosta. Plant Biol 3:176–185

17. Hao J, You C, Deng X (2002) Cell size as a morphological marker to calculate the mitotic index and ploidy level of citrus callus. Plant Cell Rep 20:1123–1127

18. De Laat AMM, Göhde W, Vogelzakg MJDC (1987) Determination of ploidy of single plants and plant populations by flow cytometry. Plant Breed 99:303–307

19. Doležel J, Greilhuber J, Suda J (2007) Flow cytometry with plants: an overview. In: Doležel J, Greilhuber J, Suda J (eds) Flow cytometry with plant cells. Analysis of genes, chromosomes and genomes. Wiley, Weinheim, pp 41–65

20. Galbraith DW (1984) Flow cytometric analysis of the cell cycle in higher plants. In: Vasil IK (ed) Cell culture and somatic cell genetics of plants. Academic, New York, pp 765–777

21. Sree Ramulu K, Dijkhuis P (1986) Flow cytometric analysis of polysomaty and in vitro genetic inestability in potato. Plant Cell Rep 5:234–237

22. Ochatt SJ, Delaitre C, Lionneton E, Huchette O, Patat-Ochatt EM, Kahane R (2005) One team, PCMV, and one approach, vitro biotechnology, for one aim, the breeding of quality plants with a wide array of species. In: Dris R (ed) Crops growth, quality and biotechnology. WFL, Helsinki, pp 1038–1067

23. Galbraith DW, Harkins KR, Maddox JR, Ayres NM, Sharma DP, Firrozabady E (1983) Rapid flow cytometric analysis of the cell cycle in intact plant tissues. Science 220:1049–1051

24. http://flowerdatabase.blogspot.com. Accessed 12 June 2020

25. Loureiro J, Rodriguez E, Doležel J, Santos C (2006) Comparison of four nuclear isolation buffers for plant DNA flow cytometry. Ann Bot 98:679–689

26. Sùda J, Krahulcovà A, Tràvníbek P, Krahulec F (2006) Ploidy level versus DNA ploidy level: an appeal for consistent terminology. Taxon 55:447–450

27. Suda J, Tràvníbek P (2006) Reliable DNA ploidy determination in dehydrated tissues of vascular plants by DAPI flow cytometry—new prospects for plant research. Cytometry A 69:273–280

28. Roberts AV (2007) The use of bead beating to prepare suspensions of nuclei for flow cytometry from fresh leaves, herbarium leaves, petals and pollen. Cytometry A 71:1039–1044

29. Chiatante D, Brusa P, Levi M, Sgorbati S, Sparvoli E (1990) A simple protocol to purify fresh nuclei from milligram amounts of meristematic pea root tissue for biochemical and flow cytometry applications. Physiol Plant 78:501–506

30. Hopping ME (1993) Preparation and preservation of nuclei from plant tissues for quantitative DNA analysis by flow cytometry. New Zealand J Bot 31:391–401

31. Roux N, Toloza A, Radecki Z, Zapata-Arias FJ, Doležel J (2003) Rapid detection of aneuploidy in Musa using flow cytometry. Plant Cell Rep 21:483–490

32. Ozaki Y, Narikiyo K, Fujita C, Okubo H (2004) Ploidy variation of progenies from intra- and inter-ploidy crosses with regard to trisomic production in asparagus (Asparagus officinalis L.). Sexual Plant Reprod 17:157–164

33. Barker RE, Kilgore JA, Cook RL, Garay AE, Warnke SE (2001) Use of flow cytometry to determine ploidy level of ryegrass. Seed Sci Technol 29:493–502

34. Sěsek P, Sustar-Vozlic J, Bohanec B (2000) Determination of aneuploids in hop (*Humulus lupulus* L.) using flow cytometry. Pflügers Archiv 439(Suppl):R16–R18

35. Elmaghrabi A, Ochatt S (2006) Isoenzymes and flow cytometry for the assessment off true-to-typeness of calluses and cell suspensions of barrel medic prior to regeneration. Plant Cell Tissue Organ Cult 85:31–43

36. Robinson JP (2006) Introduction to flow cytometry. Flow cytometry talks. Purdue Univ Cytometry Laboratory, West Lafayette, IN. http://www.cyto.purdue.edu/flowcyt/educate/pptslide.htm

37. Galbraith DW, Bartoš J, Doležel J (2005) Flow cytometry and cell sorting in plant biotechnology. In: Sklar LA (ed) Flow cytometry for biotechnology. Oxford University Press, Oxford, pp 291–322

38. Harkins KR, Galbraith DW (1984) Flow sorting and culture of plant protoplasts. Physiol Plant 60:43–52

39. Greilhuber J, Doležel J (2009) 2C or not 2C: a closer look at cell nuclei and their DNA content. Chromosoma 118:391–400

40. Greilhuber J, Doležel J, Leitch IJ, Loureiro J, Sùda J (2010) Genome size. J Bot 2010:Article ID 946138, 4 pages. https://doi.org/10.1155/2010/946138

41. Vinogradov AE (1994) Measurement by flow cytometry of genomic AT/GC ratio and genome size. Cytometry 16:34–40

42. Bennett MD, Leitch IJ (2005) Nuclear DNA amounts in angiosperms: progress, problems and prospects. Ann Bot 95:45–90

43. Zonneveld BJM, Leitch IJ, Bennett MD (2005) First nuclear DNA amounts in more than 300 angiosperms. Ann Bot 96:229–244

44. Ochatt SJ, Mousset-Déclas C, Rancillac M (2000) Fertile pea plants regenerate from protoplasts when calluses have not undergone endoreduplication. Plant Sci 156:177–183

45. Lee JH, Arumuganathan K, Kaeppler SM, Parl S-W, Kim K-Y, Chung Y-S, Kim D-H, Fukui K (2002) Variability of chromosomal DNA contents in maize (*Zea mays* L.) inbred and hybrid lines. Planta 215:666–671

46. Fukui K, Nakayama S (1996) Plant chromosomes: laboratory methods. CRC Press, Tokyo

47. Sharma AK, Sharma A (1999) Plant chromosomes. Harwood, Amsterdam

48. Jones DF (1941) Natural and induced changes in chromosome structure in maize endosperm. Proc Natl Acad Sci U S A 27:431–435

49. Abraham A, Mathew PM (1963) Cytology of coconut endosperm. Ann Bot 27:505–512

50. Chiavarino AM, Rosato M, Manzanero S, Jiménez G, González Sánchez M et al (2000) Chromosome nondisjunction and instabilities in tapetal cells are affected by B chromosomes in maize. Genetics 155:889–897

51. Dane F, Meric C (2005) Cytological and embryological studies of anther in rice (Oryza sativa) cv. 'Rocca'. Acta Bot Hung 47:257–272

52. Maluszynska J (2003) Cytogenetic tests for ploidy level analyses—chromosome counting. In: Maluszynski M, Kasha KJ, Forster BP, Szarejko I (eds) IAEA. Kluwer, Dordrecht, pp 391–395

53. Feulgen R, Rossenbeck H (1924) Mikroskopisch-chemischer Nachweis einer Nukleinsaüre von Typus der Thymonukleinsaüre und die darauf beruhende selektive Färbung von Zellkernen in mikroskopischen Präparaten. Z Physiol Chem 135:203–248

54. Kapuscinski J (1995) DAPI: a DNA-specific fluorescent probe. Biotech Histochem 70:220–233

55. Ribeiro CB, Pereira FC, Nóbrega Filho L, Rezende BA, Dias KOG, Braz GT, Ruy MC, Silva MB, Cenzi G, Techio VH (2018) Haploid identification using tropicalized haploid inducer progenies in maize. Crop Breed Appl Biotechnol 18(1):16–23

56. Hardie DC, Gregory TR, Hebert PDN (2002) From pixels to picograms: a beginners' guide to genome quantification by Feulgen image analysis densitometry. J Histochem Cytochem 50:735–749

57. Bedi KS, Goldstein DJ (1976) Apparent anomalies in nuclear Feulgen-DNA contents: role of systematic microdensitometric errors. J Cell Biol 71:68–88

58. Bertino B, Knape WA, Pytlinska M, Strauss K, Hammou JC (1994) A comparative study of DNA content as measured by flow cytometry and image analysis in 1864 specimens. Anal Cell Pathol 6:377–394

59. Vilhar B, Greilhuber J, Koce JD, Temsch EM, Dermastia M (2001) Plant genome size measurement with DNA image cytometry. Ann Bot 87:719–728

60. Vilhar B, Vidic T, Jogan N, Dermastia M (2002) Genome size and the nucleolar number as estimators of ploidy level in Dactylis glomerata in the Slovenian Alps. Plant Syst Evol 234 (1):1–13. https://doi.org/10.1007/s00606-002-0186-0

61. Adaniya S, Ardian (1994) A new method for selecting cytochimeras by the maximum number of nucleoli per cell in Allium wakegi Araki and A. fistulosum L. Euphytica 79(1):5–12. https://doi.org/10.1007/bf00023570

62. Kim S, Lee D, Rayburn AL (2015) Analysis of active nucleolus organizing regions in polyploid prairie cordgrass (Spartina pectinata Link) by silver staining. Cytologia 80 (2):249–258

63. Chawla HS (2004) Plant biotechnology: laboratory manual for plant biotechnology. Oxford & IBH Publishing, New Delhi

64. Bourdon M, Pirrello J, Cheniclet C, Coriton O, Bourge M, Brown S, Moïse A, Peypelut M, Rouyère V, Renaudin J-P, Chevalier C, Frangne N (2012) Evidence for karyoplasmic homeostasis during endoreduplication and a ploidy-dependent increase in gene transcription during tomato fruit growth. Development 139(20):3817–3826. https://doi.org/10.1242/dev.084053

Chapter 5

Methods for Chromosome Doubling

Mehran E. Shariatpanahi, Mohsen Niazian, and Behzad Ahmadi

Abstract

The completely homozygous genetic background of doubled haploids (DHs) has many applications in breeding programs and research studies. Haploid induction and chromosome doubling of induced haploids are the two main steps of doubled haploid creation. Both steps have their own complexities. Chromosome doubling of induced haploids may happen spontaneously, although usually at a low rate. Therefore, artificial/induced chromosome doubling of haploid cells/plantlets is necessary to produce DHs at an acceptable level. The most common method is using some mitotic spindle poisons that target the organization of the microtubule system. Colchicine is a well-known and widely used antimitotic. However, there are substances alternative to colchicine in terms of efficiency, toxicity, safety, and genetic stability, which can be applied in in vitro and in vivo pathways. Both pathways have their own advantages and disadvantages. However, in vitro-induced chromosome doubling has been much preferred in recent years, maybe because of the dual effect of antimitotic agents (haploid induction and chromosome doubling) in just one step, and the reduced generation of chimeras. Plant genotype, the developmental stage of initial haploids, and type–concentration–duration of application of antimitotic agents, are top influential parameters on chromosome doubling efficiency. In this review, we highlight different aspects related to antimitotic agents and to plant parameters for successful chromosome doubling and high DH yield.

Key words Antimitotic agents, Chromosome doubling, Haploid, Microtubules

1 Introduction

Plant materials with the completely homozygous genetic background are very valuable and applicable resources for genetics and breeding studies [1]. In breeding programs, the most difficult aspect of hybrid production is the creation of inbred parental lines, as it takes several generations of selfing/inbreeding [2, 3]. The most convenient alternative to conventional selfing techniques to create inbred lines is doubled haploid (DH) production [4]. The development of completely homozygous lines is the major advantage of DH production [5]. In addition to breeding purposes, the completely homozygous background of DHs is useful for genetic studies such as QTL

Jose M. Seguí-Simarro (ed.), *Doubled Haploid Technology: Volume 1: General Topics, Alliaceae, Cereals*,
Methods in Molecular Biology, vol. 2287, https://doi.org/10.1007/978-1-0716-1315-3_5,
© Springer Science+Business Media, LLC, part of Springer Nature 2021

mapping, marker assisted selection, *Agrobacterium*-mediated gene transformation, mutation and selection, and reverse breeding, among others [6].

Agrobacterium-mediated gene transformation of androgenic microspore cultures led to primary transgenic (T_0) plants in barley containing only a single copy of the transferred DNA [7] and thus avoiding hemizygosity following artificial chromosome doubling. In addition, isolated microspores are the proper targets for gene transformation through short peptide nanocarriers [8] and cell-penetrating peptides [9]. One of the important barriers in genetic studies, such as genetic stability and multienvironment studies using conventional F_2 and backcross populations, is the unrepeatability of the results in different years and locations. When a completely homozygous DH population is used for these studies, the effect of location and year can be easily studied on different agronomical characteristics without the interference of genotype-associated differences [10]. Using a DH population to create linkage maps is more economical and faster, as compared to conventional mapping populations in QTL analysis. The completely homozygous nature of DHs is also valuable for marker-assisted selection [11] and genome-wide association studies [12], allowing for a reliable and reproducible selection based on completely homozygous genotypes. Selection inside an artificial genetic diversity, created through mutation induction, can help to find a reliable selection of desired mutants, and this fact makes isolated microspore cultures, for example, an ideal method for mutation studies [13]. Breeders have utilized induced mutation and in vitro selection of desired mutants to improve valuable agronomic characteristics such as yield, quality, and resistance to disease and pests in various economical crops [6]. Selection of plants more tolerant to biotic and abiotic stresses is another application of DHs. Screening the homozygous genome of DHs is more reproducible and reliable in order to find plant materials with disease resistance [14]. One of the applied plant breeding strategies using DHs is reverse breeding, in which the knockdown of a key gene responsible for crossover leads to nonrecombinant chromosomes and gamete precursors. Upon DH generation from these nonrecombinant gamete precursors (e.g., microspores), genetically complementary lines are created. The cross of the screened complementary lines can lead to the identification of the proper complementary lines to be used as parents for reconstitution of the initial heterozygotes [15]. Nonrecombinant chromosomes should be present in all loci; therefore, a completely homozygous background and subsequently a DH induction system is necessary for the reverse breeding method.

DH production is a two-step procedure including (1) haploid induction and (2) DH production through chromosome doubling of induced haploids [16]. Haploid refers to a sporophytic plant with a gametophytic chromosome number [17]. The absence of one set of chromosomes leads to sterility of haploids [18]. Therefore, chromosome doubling of haploids is a vital prerequisite for restoring fertility and achieving the DHs [19]. Chromosome doubling can occur either spontaneously or artificially by targeting the organization of the microtubular system, using some antimitotic agents and/or inducing stresses [16, 19]. Induced haploids often exhibit very low rate of spontaneous genome doubling. Therefore, an efficient induced chromosome doubling protocol is usually required for a successful DH production pipeline [4, 20]. To date, different mechanisms have been reported for spontaneous chromosome doubling [4, 16] which are discussed in the following sections.

The metaphase inhibitors are the largest group of antimitotic agents used to prevent chromosome separation by binding to α- and β-tubulin dimers of microtubules [21]. Induced chromosome doubling of haploids can be carried out via in vivo and/or in vitro methods and various factors are involved in both pathways. Antimitotic agents and initial plant materials have their own influential parameters which should be taken into account for efficient induced chromosome doubling (Fig. 1).

Efficiency and toxicity of antimitotic agents applied and genetic stability of DHs obtained are critical points in the selection of proper doubling agent in the desired plant genotype(s). The production of large number of DHs with high level of genetic stability

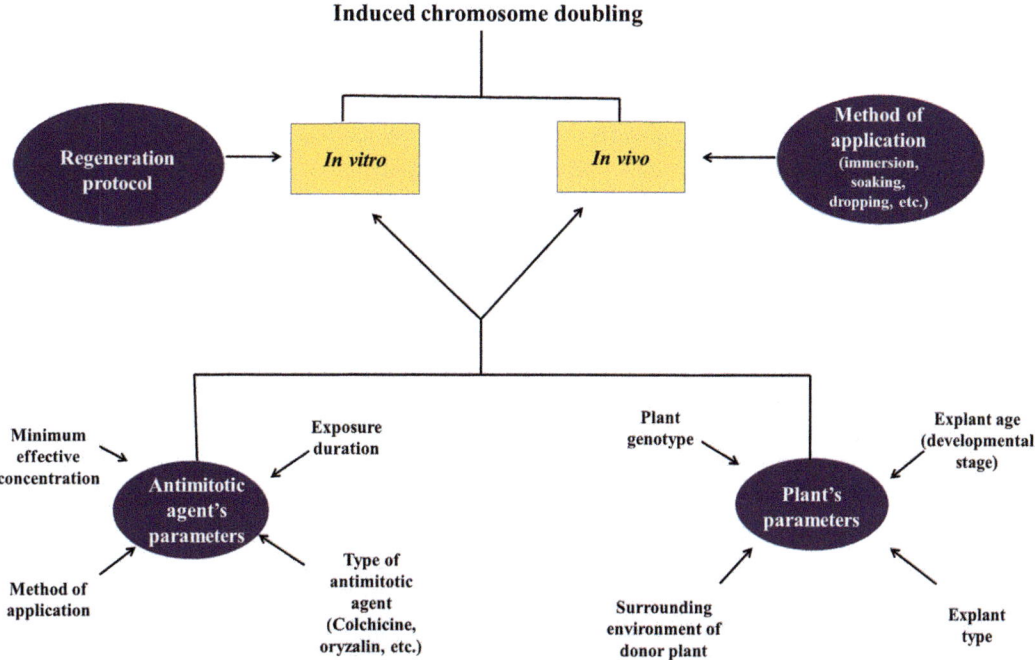

Fig. 1 The multivariable nature of induced chromosome doubling in haploids to produce doubled haploids

is critical to develop an efficient DH breeding program. The interaction effect of antimitotic agent and plant genotype makes induced chromosome doubling a complex biological process with a multivariable nature. Finding the best combination of these parameters is essential for successful chromosome doubling of haploids and subsequently obtaining DHs.

2 Importance of Chromosome Doubling in Haploid Systems

Both plant characteristics and morphological attributes are regulated by genetic background, environmental cues, and their interaction. Like other diploid organisms, the nuclear genome of land plants consists of two sets of parental homologous chromosomes (maternal and paternal), which are paired together to form diploid somatic cells and then haploid gametes (male/female), after the reductional homologous chromosome segregation (meiosis). The absence of one set of the homologous chromosomes can change the morphology (smaller size) and disrupt meiosis, failing to produce "n" gametes, and subsequently becoming sterile.

Haploids are gametophyte-originated plants containing only half of the chromosome number of a zygote. In other words, it is a sporophytic plant with a gametophytic chromosome number [17]. Haploid plants are sterile and morphologically weaker than their diploid relatives [18]. Therefore, duplicating the chromosome number of haploids to recover both sets of chromosomes is crucial for their survival and complete fertility [16]. Access to the pure homozygous genome of DHs and exploiting their subsequent applications heavily depends on an efficient genome doubling procedure as the second step of DH production [5]. This step is a major concern for the utilization of induced haploid in plant breeding programs and genetic studies [16]. Chromosome doubling in haploid regenerants may occur without the application of additional treatments (commonly known as *spontaneous* genome doubling) or artificially induced, using antimitotic agents (artificial genome doubling). There are several factors affecting both spontaneous and induced chromosome doubling pathways, which will subsequently affect the efficiency of DH production.

3 *Spontaneous* Genome Doubling

As already discussed, the DH production includes two major steps: haploid induction and chromosome doubling of the regenerants. Theoretically, microspores or egg cells contain only half of the chromosomes of the parental somatic tissues. Therefore, plantlets derived from microspores or egg cells should be haploid [22]. However, in some cases, derived embryos can *spontaneously* double their chromosomes during the very early stages of development giving rise to DHs ($2n$), triploids ($3n$), tetraploids ($4n$), or even higher ploidy levels [23]. The ease of undergoing this phenomenon, referred to as "spontaneous haploid genome duplication," is an inherited trait tendency in a wide range of species, allowing for the generation and rapid fixation of genetic variants in a homozygous state [16, 24] with drastically lower cost of DH production [4, 22, 25]. Cytogenetic assay of regenerated plantlets following isolated microspore culture in *H. vulgare* indicated that 65–76% spontaneous haploid genome duplication of the regenerated population were completely fertile [26]. Depending on accession, 17.5–63.6% of spontaneous doubling rate has already been reported in *Daucus carota* L. [27]. Exploiting this phenomenon for DH generation allows to avoid the use of artificial genome doubling treatments and, at the same time, increases the efficiency of DH production [28].

The exact mechanism of spontaneous chromosome doubling still remains obscure. The high frequency of completely fertile plants indicates that chromosome doubling must occur very early, most likely at the time of embryogenesis induction [29]. Many factors such as genotype, developmental stage of microspores at the time of culture, type of stresses applied, use of plant hormones, type of explant, and culture conditions greatly affect the efficiency of genome duplication [4, 22]. Spontaneous doubling rates of about 70–90% have been reported in *H. vulgare* L., 25–70% in *T. aestivum* L., 70% in *T. turgidum* L., 50–60% in *Oryza sativa* L., 50–90% in *Secale cereale* L. [30, 31], 16–71% in *B. napus* L. [32], 43–88% in *B. oleracea* [22, 33], and 20% in *Zea mays* L. [34], pointing out that the rate of this phenomenon varies among crop species and genotypes. Compelling evidence indicates that stresses applied for diverting undifferentiated gametes toward embryogenic state such as heat, cold, and starvation also affect the ratio of spontaneous doubling possibly by changing the structure and density of microfilament and microtubule [16, 35, 36]. In cereals, addition of mannitol or 2-hydroxynicotinic acid in the pretreatment medium has provided consistently high spontaneous genome doubling frequencies [26, 29, 37]. Moreover, mannitol pretreatment when accompanied with cold or heat treatment could further increase the rate of spontaneous doubling [20]. Since the majority of triggering stresses exhibit antimicrotubular properties, it is possible that they can cause both embryo induction and chromosome doubling [16]. As seen, many of the factors used to induce haploid development may be inducing chromosome doubling too. This is why the concept of "spontaneity," although widely extended and used, should be treated with precaution, always keeping in mind that "spontaneous" means, in this context, occurring without the application of additional treatments.

The efficiency of spontaneous doubling has also proved to be affected by the developmental stage of cultured gametes. About 54–66% of spontaneous doubling rate was observed in cultured microspores of *T. aestivum* L. at late uninucleate to early binucleate stage whereas this rate drastically decreased to 33% when younger microspores, at mid to late uninucleate stage, were selected [38]. Therefore, in order to obtain high frequencies of completely fertile DH plants, the stress treatments should be carried out at the early to late uninucleate microspores and its application at later stages could give rise more triploids or even higher ploidy levels [16, 20]. Duration of tissue culture can also change the ploidy level of the microspore-derived regenerants. In *Z. mays* L., for instance, haploid calli did not change their ploidy level for 97 days but by 466 days, there were up to 50% diploid or higher ploidy levels, showing that spontaneous doubling occurred throughout successive calli subcultures. Similar results were also observed in *B. oleracea* L. where after one or more years of tissue culture, the number of chromosomes of the haploid plantlets was doubled spontaneously, the majority of which turned into DHs or mixoploid plants [22]. Methods of haploid induction such as androgenesis (anther culture, isolated microspore culture, or shed microspore culture), gynogenesis, irradiated pollen-induced parthenogenesis, and intra and interspecific hybridization, may have a considerable impact on spontaneous genome duplication. A striking example is observed in isolated microspore cultures of *B. rapa* L., which yielded up to a threefold higher rate of spontaneous doubling than anther culture [39].

As already indicated, the mechanisms underlying the spontaneous doubling and regeneration of fertile plants from haploid gametophytes are rather ambiguous and have yet not to be fully understood. It is noteworthy that some of the fertile plants could have arisen from unreduced gametes generated via meiotic restitution [40]. Reports of seed set in haploid plants are scarce. However, production of unreduced gametes in haploids of *T. durum* L. [41] and *T. aestivum* L. [42] have led to viable seed production. In *Brassicas*, however, molecular assays using codominant simple sequence repeat (SSR) markers revealed that fully fertile, homozygote diploid individuals are derived from spontaneously doubled haploid gametophyte precursors, rather than from unreduced gametes [43].

It has been postulated that genome duplication of haploid cells may occur via three to four different mechanisms, that is, endoreplication, nuclear fusion, C-mitosis, and endomitosis [4, 16].

Endoreplication, also known as endoreduplication or endocycle, is the most likely mechanism causing chromosome doubling at the outset of androgenic switch prior to first embryogenic mitosis [4]. Endoreplication is the specialized cell cycle in which one or several rounds of nuclear DNA synthesis occurs, mitosis is bypassed, and chromatid separation and cytokinesis are disrupted. Generally, endoreduplication increases the number of chromatids of a chromosome without changing total chromosome number, leading to formation of giant chromosomes known as "polytene" chromosomes within the hypertrophying nuclei [44]. This phenomenon is the most common mode of cell polyploidization and estimated to be occurred in over 90% of angiosperms [44]. Endoreduplication has proved to be regulated by genetic, environmental, and developmental cues [4]. Exogenous application of plant growth regulators such as 2,4-D to the in vitro cultures has also exhibited endoreduplication-enhancing effects [45, 46].

Nuclear fusion is another proposed mechanism responsible for doubling total chromosome number. Theoretically, this phenomenon can take place at three different stages: (A) after pollen mother cell meiosis followed by incomplete cytokinesis allowing for daughter cells to stay physically attached together, fuse their nuclei, giving rise to a single, larger nucleus with twice as many the chromosome number as the original nucleus leading to generate diploid but non-DH individuals, (B) in the induced microspores after normal karyogenesis followed by disrupted cytokinesis that allows daughter nuclei to coalesce. Mannitol pretreatment in microspore culture of cereals blocks cell wall formation during the first cell divisions, resulting in coenocytic cells in which the nuclei are able to fuse and thereby leading to high rate of spontaneous doubling rate in this species. Nuclear fusion at this stage has also been reported in *Z. mays* L., in which the majority of cells at the beginning of microspore development contained a 1C level of DNA with the haploid number ($n = 10$) of chromosomes. However, 5–7 days following initial culture, several rounds of nuclear multiplication occurred in the embryo-like domain via karyokinesis followed by nuclear fusion and genome duplication [47], (C) in the induced binucleate pollen grains that vegetative and generative nuclei fuse together within the same cytoplasm [4, 16, 48]. Live-cell imaging of microspore embryogenesis in *H. vulgare* L. revealed that nuclear fusion is the most common process that attributes to spontaneous genome duplication especially in the mid to late uninucleate microspores [25]. Nuclear fusion can happen throughout the whole process of embryogenesis due to the failure of cell wall formation, which explains the mixoploidy often observed within the regenerated individuals [25]. However, the degree of spontaneous doubling rate via nuclear fusion is considered to be species-specific and also considered to be dependent upon the extent of gene action encoding for proteins that may promote nuclear envelope fusion [16].

In some cases, both endoreduplication and nuclear fusion may occur simultaneously. If one or both nuclei undergo endoreduplication prior to nuclear fusion, triploid or higher ploidy levels might be observed [29]. It has been hypothesized that nuclear fusion followed by endoreplication during the formation of mature andro-genetic structures is responsible for increasing the polyploidy level of microspore-derived embryos in *Z. mays* L. [47].

Endomitosis is another postulated mechanism with regard to polyploidy in cultured microspores [4, 49]. Like normal mitosis, endomitosis takes place inside the nuclear membrane, but it differs essentially by the absence of a mitotic spindle, resulting in the duplication of chromosome number without accompanying nuclear division [50, 51]. Though endomitosis is less common in plants compared with animal taxa, it was found to occur in anther tapetal cells, aiding to provide nutrients and regulatory molecules to the forming pollen grain [52, 53]. However, reports on its occurrence throughout microspore embryogenesis are rather scarce [4].

4   Induced Genome Doubling

Specifically induced (artificial) chromosome doubling is important to efficiently produce DH plants when in some species the fre-quency of spontaneous doubling is usually low [20].

A proper cell cycle progression is required for mitotic division of cells into two new, identical daughter cells. DNA replication and chromosome segregation takes place normally during the S and M phases, respectively. Microtubular dynamics is critical for segrega-tion of sister chromatids at metaphase [54]. The correct bipolar attachment of sister kinetochores of chromosomes to the microtubule-based mitotic spindle apparatus is also considered to be essential for accurate chromosome segregation. Any disruption in the correct formation of the mitotic spindle and kinetochore–microtubule attachments can inhibit the metaphase-to-anaphase transition, and thus lead to the mitotic arrest [55, 56]. Two well-known microtubule-binding agents, including microtubule-stabilizing and destabilizing agents, suppress the dynamics of microtubules rather than changing their net polymer mass [57]. Different binding sites have been reported for microtubule-binding agents, including (1) taxane site on β-tubulin within the lumen of the microtubules, (2) the *Vinca* domain close to the GTP binding site on β-tubulin, and (3) the colchicine binding site at the interface between the α- and β-tubulin dimers [58].

As already mentioned, the exact mechanisms underlying the spontaneous doubling, endomitosis, endoreduplication, C-mitosis, and nuclear fusion mechanism, are not fully understood. For exam-ple, a delay in cell wall formation and subsequent nuclear fusion may be induced by either embryogenesis-inducing agents like man-nitol or antimicrotubule agents (blocking the microtubules of the phragmoplast, essential for plant cell cytokinesis), while endomito-sis (blocking of divisions) may be induced by antimicrotubule agents [16]. It has been asserted that the mechanism behind induced chromosome doubling using substances that promote mitotic arrest (antimicrotubular agents) is blocking the formation of microtubules by binding to their specific target sites and preventing formation of tubulin dimers (C-mitosis) [20, 59].

5   Antimicrotubular Agents

There are some critical parameters such as type of antimitotic agent, optimum concentration, duration of exposure, and its method of application, that should be taken into account to induce chromosome doubling efficiently [60].

Different kinds of mitotic inhibitors, such as colchicine, oryzalin, amiprophosmethyl, pronamide, and trifluralin have been applied for chromosome doubling studies in different species [21, 61, 62]. However, colchicine is the oldest and most widely used antimitotic agent [63]. Colchicine is a well-known mitosis-arresting alkaloid that blocks the formation of microtubules by binding to β-tubulin and prevents formation of tubulin dimers [59]. In spite of being routinely used for chromosome doubling, colchicine has exhibited low affinity to plant tubulins when compared to vertebrates [64]. High concentrations of colchicine are generally applied for artificial genome doubling in plant cells [4, 65]. Therefore, a compromise should be taken between its proper concentration on the one hand and its detrimental effects on living cells on the other. There are other less toxic, alternative antimitotic agents, including oryzalin, trifluralin, and flufenacet, a chemical mixture of amiprophosmethyl + pronamide + dimethyl sulfoxide, and nitrous oxide (N_2O), a gas that can be used instead of colchicine [59]. Unlike colchicine, which is applicable for chromosome doubling in many plant species, other antimitotic agents may not be useful in a wide range of species. For example, N_2O has been widely used in chromosome doubling of monocots, mostly cereals; however, it has efficiency lower than colchicine in dicots [4]. Compared to colchicine, amiprophosmethyl (APM) and oryzalin have exhibited no affinity to animal tubulin binding sites, thus they are considered to be far safer to work with [60, 61]. Moreover, their effective concentrations are 50- to 250-fold lower than colchicine [65]. The best type, concentration, and exposure duration of antimitotic agents are the variables that have been studied in the chromosome doubling procedure in DH studies [61, 66–68].

It is clear that an optimum concentration of antimitotic agents is necessary for a successful induced chromosome doubling. It is also clear that concentrations higher than the effective one have lethal effects on treated explants. Therefore, finding the minimal effective concentration is critical. Different plant species have different responses to applied concentrations of a specific type of antimitotic agent. There is a significant interaction between plant species/genotype and the concentration of applied antimitotic agents. The negative effect of higher concentrations on the survival rate, which leads to higher mortality, has been reported in different induced chromosome doubling studies [63, 69–71]. In artificial polyploidy induction, it has been reported that physiological disturbances may occur by the inverse relationship between the concentration of antimitotic agents applied and the survival rate of treated explants [72]. There is also a negative relationship between the concentration of antimitotic agents and the duration of treatment. Desired results can be obtained when high concentrations of antimitotic agents combined with short treatment durations or conversely, with low concentrations together with longer durations [73, 74]. Environmental factors such as culture temperature and medium composition are effective to establish an efficient in vitro chromosome doubling [75]. The ambient temperature and heat shock following antimitotic treatment can also increase the efficiency of induced chromosome doubling [76]. For example, the highest percent of polyploidy induction in *Andrographis paniculata* was reported when the seeds were first soaked in colchicine solution at 4 °C for 40 min followed by exposing to fresh colchicine solution at 40 °C for 20 min [77].

The genotype is the other main factor affecting chromosome doubling efficiency [20], as different genotypes of plant species show different responses to concentrations of an antimitotic agent applied. Understanding the molecular mechanisms underlying the chromosome doubling process and response to antimitotic agents is of great importance. Sun et al. [78] studied the expression of microRNAs in response to colchicine treatment in *H. vulgare* L. via high-throughput sequencing and reported a microRNA-target regulation network under colchicine treatment that involves actin, cell cycle regulation, cell wall synthesis, and the regulation of oxidative stress, having a crucial role in response to colchicine treatment in barley.

Although it has been reported that antimitotic agents do not significantly affect the rates of variation, the genetic stability and the effect of antimitotic agents on the somaclonal variation of the DHs obtained, should be considered after chromosome doubling [4].

In summary, chromosome doubling is a procedure that depends upon the method of haploid induction. Therefore, different strategies should be considered, according to the different androgenic and gynogenic pathways used [16].

6 In Vivo Induced Chromosome Doubling

Induced chromosome doubling can be carried out either via in vitro or in vivo procedures. In comparison with in vitro method, a higher concentration of antimicrotubular agents is required for chromosome doubling through in vivo methods. Higher concentrations of antimicrotubular agents result in higher rates of mortality, mixoploids, and chimeric plants as well, which are considered to be disadvantages of in vivo chromosome doubling [20].

To date, different methods of in vivo chromosome doubling have been described. These methods are mainly immersion, soaking, dropping, and injection [19, 21]. Tissue wounding is the oldest method of in vivo chromosome doubling [76]. The capping and immersion are the two main methods of in vivo chromosome doubling using different antimicrotubular agents [20]. The concentration of applied chemicals and duration of application strongly depends on the methods adopted. However, the abovementioned negative correlation between the dosage of applied antimitotic agents and duration of treatment should be taken into account. High concentration for shorter time, or low concentration for longer time [73, 74] may be a choice and its results can be assessed for each plant species. Different plant parts, including seeds, roots, and scutellar nodes, can be exposed to chromosome doubling agents. The developmental stage of primary haploid materials is very important for in vivo chromosome doubling. In barley, the percentage of produced DHs in treated plants at the four-tiller stage (90%) [79] was more than those treated in 2–3 leaf stage (56%) [80]. The active growing phase of plants, with the maximum cell division rate, is ideal to maximize the uptake of the antimitotic agent [16]. It is also noteworthy that the efficiency of in vivo doubling strongly increases when DMSO (dimethyl sulfoxide) and wetting agents (such as Tween 20) are applied to enhance the penetration of microtubule inhibiting agents. The application of gibberellic acid (GA), after treatment of antimitotic agents, has also proved to be beneficial for recovery of treated materials and reducing their mortality [16].

Most of the chromosome doubling protocols currently developed have utilized chemical agents during the early stages of androgenesis or gynogenesis (in vitro) [20]. Some recently published studies have adopted in vivo chromosome doubling methods, in different plant species. Immersion of different types of explants, such as roots, nodal segments, crowns, and leaves, in colchicine solutions is the most frequently applied procedure. Scarce results with the use of other doubling agents such as trifluralin and oryzalin have been documented so far (Table 1).

Table 1
Examples of currently applied in vivo chromosome doubling induction in different doubled haploidy studies

Plant species	Applied method (s)	Applied antimitotic agent(s)	Chromosome doubling efficiency (%)	Reference
Broccoli (*Brassica oleracea* var. italica L.)	Immersion of roots	Colchicine (0.05%)	>50	[22]
Cabbage (*Brassica oleracea* var. capitata L.)	Immersion of roots	Colchicine (0.2–0.4%)	>50	[22]
Chrysanthemum (*Chrysanthemum morifolium*)	Immersion of nodal segments	Colchicine (500 mg/ L)	11.25	[90]
Maize (*Zea mays*)	Injection to shoot apical meristem	Colchicine (0.125%)	–	[91]
Oat (*Avena sativa* L.)	Immersion of roots	Colchicine (0.1%)	85	[92]
	Immersion of roots	Colchicine	–	[93]
	Immersion of roots	Colchicine (0.1%)	1.64	[94]
Oil palm (*Elaeis guineensis*)	Immersion of roots	Colchicine	96	[95]
Triticale (× *Triticosecale* Wittmack)	Immersion of roots and crowns	Colchicine (0.06%)	6–98	[96]
	Immersion of leaves and roots	Colchicine (500 mg/ L)	81.5–83.5	[97]
Wheat (*Triticum aestivum* L.)	Injection to uppermost internodes	Colchicine (100-10,000 ppm)	75	[98]
	Tiller injection	Colchicine (0.1–0.5%)	–	[99]
	Dropping	Trifluralin	75	[100]

7   In Vitro Induced Chromosome Doubling

Antimicrotubular agents can be applied to induce genome doubling at the onset of in vitro culture, at the single haploid cell level, providing a great opportunity of screening recessive mutants in the first generation, avoiding chimeras of different ploidy levels, and allowing for rapid exploitation of true breeding lines in the DH production pipeline [16, 81]. In addition, this technique drastically reduces the labor-intensive steps and time-consuming chemical treatment of established haploid plants [38]. Moreover, a relatively fewer amount of antimitotic agent is required for this method [19]. Chemical agents can also be applied to the in vitro regenerated embryos. However, one drawback to this method is that ploidy level of the regenerants is not determined prior to treatment and thus spontaneous DHs would be treated simultaneously along with haploid embryos. Above all, varying degrees of chimerism might accompany this procedure [82]. Therefore, several extra months can be added to the regeneration time to completely recover DH lines [83].

The most important superiority of in vitro chromosome doubling to the in vivo method is the higher doubling rates and lower occurrence of chimeras [70]. There have been several approaches for in vitro induced chromosome doubling to develop DHs in many crops or plant species. However, isolated microspore culture and anther culture are the dominant approaches. A list of publications and protocols related to genome doubling of haploid plants for DH production in different species using in vitro approach is presented in Table 2.

8   Methods to Verify Chromosome Doubling

Ploidy level confirmation through the direct or indirect methods, such as chromosome counting, morphological and anatomical indicators, and flow cytometry is another important step following treating the regenerants. Rapid and early-stage diagnosis of ploidy level is a prerequisite to establish an efficient chromosome doubling methodology [84]. Classical markers, such as stomatal morphometric data (stomatal density per unit area, the number and size of stomata), density of chloroplasts per stomatal guard cell, guard cell size, and pollen size are rapid and simple indirect methods to distinguish haploids from diploid plants at early growth stages [19, 82]. The most reliable and unambiguous method to determine ploidy level is chromosome counting directly under light microscopes [85]. In order to obtain clear images, tissues with actively dividing cells (mitotic cells) are needed, and the three main steps of a chromosome counting procedure, including explant preparation and pretreatment, cell fixation and preparation, and chromosome staining, should be well optimized for the desired plant species. Chromosome counting also requires highly skilled operators; therefore, it is a complicated, laborious, and time-consuming method to determine the ploidy level, especially in crops with relatively small chromosomes [86, 87]. Flow cytometry, which benefits from a direct correlation between nuclear DNA content and ploidy level, is a highly reliable method. However, it has been recently reported that the comparison of DNA content of standardized leaf-punch samples is not an applicable method to

Table 2
Published in vitro chromosome doubling protocols in different plant species

Plant Species	Method	Explant	Agent	Concentration	Duration	Efficiency, %	References
Allium cepa L.	Gynogenesis	Plantlets	Colchicine	250 mg/l	2 days	100	[101]
A. cepa L.	Gynogenesis	Plantlets	Colchicine	300–400 mg/l	2 days	16–19	[102]
Avena sativa L.	IMC	Plantlets	Colchicine	1000–2000 mg/l	4 h	93.6–97.9	[103]
Beta vulgaris L.	Gynogenesis	Buds	Colchicine	50–100 mg/l	1–5 days	17.6–70	[104]
Brassica juncea L.	IMC	Plantlets	Colchicine	3400 mg/l	45–180 min	70–100	[105]
Brassica napus L.	IMC	Embryos	Colchicine	250–500 mg/l	24–36 h	64–66	[82]
B. napus L.	IMC	Microspores	Colchicine	50 mg/l	14 days	30.8–89.5	[32]
B. napus L.	IMC	Microspores	Colchicine	10 mg/l	3 days	69.6	[106]
Brassica oleracea L.	IMC	Plantlets	Colchicine	150 mg/l	36 h	73.3	[107]
Coffea arabica L.	IMC	Microspores	Colchicine	100 mg/l	2 days	95	[108]
Cucumis melo L.	Parthenogenesis	Shoot tips	Colchicine	250–500 mg/l	3–6 h	6	[109]
Cucumis sativus L.	Parthenogenesis	Nodes	Colchicine	750 mg/l	18 h	57.1	[60]
Nicotiana tabacum L.	AC	Anthers	Colchicine	200 mg/l	3 days	60	[110]
N. tabacum L.	AC	Plantlets	Colchicine	200 mg/l	7–48 h	33.3–42.3	[110]
Triticum aestivum L.	IMC	Anthers	Colchicine	100 mg/l	–	4–9	[111]
T. aestivum L.	IMC	Microspores	Colchicine	50–150 mg/l	3 days	78–86	[111]
T. aestivum L.	AC	Anthers	Colchicine	300 mg/l	2 days	118	[38]
T. aestivum L.	IMC	Microspores	Colchicine	300 mg/l	2 days	75	[38]
Triticosecale Wittmack L.	IMC	Microspores	Colchicine	400–1200 mg/l	1–3 days	10.8–12.5	[112]
Zea mays L.	AC	Anthers	Colchicine	40 mg/l	3 days	18.1–20	[113]
Z. mays L.	CWHI	Plantlets	Colchicine	600 mg/l	1 days	20.6	[114]
A. cepa L.	Gynogenesis	Embryos	APM	15 mg/l	1–3 days	32.6–34.8	[115]

Species	Method	Explant	Agent	Concentration	Duration	%	Ref.
A. cepa L.	Gynogenesis	Plantlets	APM	15–60 mg/l	1–2 days	34–40	[101]
A. cepa L.	Gynogenesis	Embryos	APM	7.5–15 mg/l	1–3 days	10–11	[116]
A. cepa L.	Gynogenesis	Embryos	APM	30–45 mg/l	2 days	9–13	[102]
B. vulgaris L.	Gynogenesis	Ovule	APM	45 mg/l	5 h	56–64	[117]
B. napus L.	IMC	Microspores	APM	1.0 mg/l	3 days	33.4	[106]
Quercus suber L.	AC	Embryos	APM	3.0 mg/l	1–3 days	38–40	[67]
T. aestivum L.	IMC	Microspores	APM	3.0 mg/l	2 days	74.0	[118]
Z. mays L.	AC	Calli	APM	1.5–6.0 mg/l	3 days	51.1–67	[119]
Z. mays L.	CWHI	Plantlets	APM	6.0 mg/l	1 days	38.2	[114]
A. cepa L.	Gynogenesis	Embryos	Oryzalin	15 mg/l	1–3 days	26.5–47.1	[115]
A. cepa L.	Gynogenesis	Embryos	Oryzalin	30–45 mg/l	2 days	5–9	[102]
C. sativus L.	Parthenogenesis	Nodes	Oryzalin	50 mg/l	18 h	92.3	[60]
Q. suber L.	AC	Embryos	Oryzalin	3.0 mg/l	2 days	46.0	[67]
Z. mays L.	AC	Calli	Oryzalin	1.75–3.5 mg/l	2–3 days	57.1–77.8	[119]
Z. mays L.	CWHI	Plantlets	Oryzalin	7.5 mg/l	1 days	19.4	[114]
A. cepa L.	Gynogenesis	Embryos	Trifluralin	15 mg/l	1–3 days	21.7–47.1	[115]
C. sativus L.	Parthenogenesis	Nodes	Trifluralin	50 mg/l	18 h	83.3	[60]
T. aestivum L.	AC	Anthers	Trifluralin	0.3–1.0 mg/l	2 days	51–53	[1]
T. aestivum L.	IMC	Microspores	Trifluralin	3.0 mg/l	2 days	65.0	[118]
Z. mays L.	AC	Calli	Trifluralin	1.5 mg/l	3 days	100	[119]
B. napus L.	IMC	Microspores	Pronamide	0.75 mg/l	3 days	52.0	[106]
Z. mays L.	AC	Calli	Pronamide	0.25–2.5 mg/l	2–3 days	52.4–73.5	[119]
T. aestivum L.	AC	Anthers	n-Butanol	0.1–0.2% (v/v)	5 h	38–40	[120]

AC anther culture, IMC isolated microspore culture, CWHI crossing with haploid inducer line

recognize putative DHs as there is equivalent DNA content between haploid and diploid samples [18]. The size of epidermal and mesophyll cell types is reduced in haploids. However, the denser packing of haploid cells (more cells per same unit area) offset the reduced cell expansion and absence of missed genome. Therefore, it leads to equivalent DNA content per unit leaf area for haploids and their diploid counterparts [18]. Haploids keep normal cellular patterning in the epidermis and mesophyll, but with more cells into the same amount of space. Therefore, epidermal cell size could also be used as a selection criterion for recognizing haploids, diploids, and DHs [18]. For a more comprehensive review of the methods available to analyze ploidy levels of haploids and DHs, *see* of this volume Chapter 4.

Image-based modeling can also be used to determine the probability of a desired ploidy level in chromosome engineering studies based on the use of imaging techniques, for correct records of leaf morphology, cell size content, of each ploidy group, and subsequent modeling of captured images (classification modeling) using machine learning approaches. It could be a relatively more precise, fast, and cost-effective distinguishing method. A coupled image processing, artificial neural network modeling technique has already been used for in vitro somatic embryogenesis [88] and androgenesis-based haploid induction [89].

9   Conclusion

The production of large number of DHs with high level of genetic stability, after haploid induction, is critical to develop an efficient DH breeding program. Application of an efficient chromosome doubling protocol can increase the number of produced DHs, especially in plant species with low rate of spontaneous genome doubling. The use of metaphase-inhibiting antimitotic agents is the most efficient approach to induce chromosome doubling in different plant species. Artificial chromosome doubling can be conducted in in vitro and in vivo pathways.

Different parameters can affect both in vivo and in vitro methods. Adjusting the proper antimitotic agent parameters (e.g., type, concentration, and duration of application) and plant material (e.g., genotype, type of explant and developmental stage of primary haploid materials) are critical in developing an efficient chromosome doubling protocols. The in vitro pathway is much faster and simpler than in vivo methods, as application of fewer quantity of doubling agent during early stages of embryogenesis can stimulate both haploid induction and chromosome doubling simultaneously, and reduce the number of chimeric regenerants. In addition, one other drawback to the in vivo procedures is regeneration of plants with varying degrees of chimerism. Furthermore, higher rate of mortality and the labor-intensive and time-consuming steps of chemical treatment of established haploid plants are other disadvantages accompanying the in vivo procedures. Therefore, the in vitro chromosome doubling has become popular and increasingly utilized in the recent years, especially in androgenesis—isolated microspore culture and anther culture—studies.

Although colchicine has been widely used for artificial chromosome doubling of haploids in both in vitro and in vivo methods, there are other alternatives that should be considered according to their efficiency and less toxicity in future studies. Overall, the induced chromosome doubling is a rather complex and multifactorial biological process due to the interactive effect of antimicrotubular chemicals and plant parameters. Nowadays, some machine learning methods have been developed, based on artificial neural networks, for solving complex computational problems and better interpretation of nonlinear relationships of involved factors to find the best combination of antimitotic agents and plant parameters for higher rate of chromosome doubling in the DH production pipeline.

References

1. Broughton S, Castello M, Liu L, Killen J, Hepworth A, O'Leary R (2020) The effect of caffeine and trifluralin on chromosome doubling in wheat anther culture. Plan Theory 9(105):1–14

2. Germanà MA (2011) Gametic embryogenesis and haploid technology as valuable support to plant breeding. Plant Cell Rep 30 (5):839–857

3. Seguí-Simarro JM (2016) Androgenesis in Solanaceae. In: Germana M, Lambardi M (eds) In Vitro embryogenesis in higher plants. Methods in molecular biology, vol 1359. Humana, New York, NY

4. Segui-Simarro JM, Nuez F (2008) Pathways to doubled haploidy: chromosome doubling during androgenesis. Cytogenet Genom Res 120:358–369

5. Ren J, Wu P, Trampe B, Tian X, Lübberstedt T, Chen S (2017) Novel technologies in doubled haploid line development. Plant Biotechnol J 15(11):1361–1370

6. Shariatpanahi ME, Ahmadi B (2016) Isolated microspore culture and its applications in plant breeding and genetics. In: Anis M, Ahmad N (eds) Plant tissue culture: propagation, conservation and crop improvement. Springer, Singapore

7. Kumlehn J, Serazetdinova L, Hensel G, Becker D, Loerz H (2006) Genetic transformation of barley (*Hordeum vulgare* L.) via infection of androgenetic pollen cultures with agrobacterium tumefaciens. Plant Biotechnol J 4(2):251–261

8. Eudes F, Shim YS, Jiang F (2014) Engineering the haploid genome of microspores. Biocatal Agric Biotechnol 3(1):20–23

9. Chugh A, Amundsen E, Eudes F (2009) Translocation of cell-penetrating peptides and delivery of their cargoes in triticale microspores. Plant Cell Rep 28(5):801–810

10. Collard BCY, Jahufer MZZ, Brouwer JB, Pang ECK (2005) An introduction to markers, quantitative trait loci (QTL) mapping and marker-assisted selection for crop improvement: the basic concepts. Euphytica 142(1–2):169–196

11. Collard BC, Mackill DJ (2008) Marker-assisted selection: an approach for precision plant breeding in the twenty-first century. Philos Trans R Soc Lond B Biol Sci 363 (1491):557–572

12. Sanchez DL, Liu S, Ibrahim R, Blanco M, Lübberstedt T (2018) Genome-wide association studies of doubled haploid exotic introgression lines for root system architecture traits in maize (*Zea mays* L.). Plant Sci 268:30–38

13. Seyis F, Aydin E, Çatal Mİ (2014) Haploids in the improvement of crucifers. Türk Tarım Doğa Bilimleri 7(7):1419–1424

14. Dong YQ, Zhao WX, Li XH, Liu XC, Gao NN, Huang JH, Wang WY, Xu XL, Tang ZH (2016) Androgenesis, gynogenesis, and parthenogenesis haploids in cucurbit species. Plant Cell Rep 35(10):1991–2019

15. Dirks R, Van Dun K, De Snoo CB, Van Den Berg M, Lelivelt CL, Voermans W, Woudenberg L, De Wit JP, Reinink K, Schut JW, Van Der Zeeuw E (2009) Reverse breeding: a novel breeding approach based on engineered meiosis. Plant Biotechnol J 7 (9):837–845

16. Kasha KJ (2005) Chromosome doubling and recovery of doubled haploid plants. In: Palmer CE, Keller WA, Kasha KJ (eds) Haploids in crop improvement II, vol 56. Springer-Verlag, Berlin, pp 123–152

17. Kasha KJ, Maluszynski M (2003) Production of doubled haploids in crop plants. An introduction. In: Maluszynski M, Kasha KJ, Forster BP, Szarejko I (eds) Doubled haploid production in crop plants. Springer, Dordrecht

18. Santeramo D, Howell J, Ji Y, Yu W, Liu W, Kelliher T (2020) DNA content equivalence in haploid and diploid maize leaves. Planta 251(1):30

19. Ahmadi B, Ebrahimzadeh H (2020) In vitro androgenesis: spontaneous vs. artificial genome doubling and characterization of regenerants. Plant Cell Rep 39(3):299–316

20. Castillo AM, Cistué L, Vallés MP, Soriano M (2009) Chromosome doubling in monocots. In: Touraev A, Forster BP, Jain SM (eds) Advances in haploid production in higher plants. Springer, Dordrecht

21. Dhooghe E, Van Laere K, Eeckhaut T, Leus L, Van Huylenbroeck J (2011) Mitotic chromosome doubling of plant tissues in vitro. Plant Cell Tissue Organ Cult 104 (3):359–373

22. Yuan S, Su Y, Liu Y, Li Z, Fang Z, Yang L, Zhuang M, Zhang Y, Lv H, Sun P (2015) Chromosome doubling of microspore-derived plants from cabbage (Brassica oleracea var. capitata L.) and broccoli (Brassica oleracea var. italica L.). Front Plant Sci 6:1118

23. Palmer CE, Keller WA, Arnison PG (1996) Utilization of Brassica haploids. In: Jain SM, Sopory SK, Veilleux RE (eds) In vitro haploid production in higher plants. Kluwer Academic Publishers, Dordrecht, pp 173–192

24. Hofinger BJ, Huynh OA, Jankowicz-Cieslak J, Müller A, Otto I, Kumlehn J, Till BJ (2013) Validation of doubled haploid plants by enzymatic mismatch cleavage. Plant Methods 9 (43):1–10

25. Daghma DES, Hensel G, Rutten T, Melzer M, Kumlehn J (2014) Cellular dynamics during early barley pollen embryogenesis revealed by time-lapse imaging. Front Plant Sci 5(675):1–14

26. Kahrizi D, Mohammadi R (2009) Study of androgenesis and spontaneous chromosome doubling in barley (Hordeum vulgare L.) genotypes using isolated microspore culture. Acta Agron Hung 57(2):155–164

27. Li JR, Zhuang FY, Ou CG, Hu H, Zhao ZW, Mao JH (2013) Microspore embryogenesis and production of haploid and doubled haploid plants in carrot (Daucus carota L.). Plant Cell Tissue Organ Cult 112:275–287

28. Kleiber D, Prigge V, Melchinger AE, Burkard F, San Vicente F, Palomino G, Gordillo GA (2012) Haploid fertility in temperate and tropical maize germplasm. Crop Sci 52:623–630

29. Kasha KJ, Hu TC, Oro R, Simion E, Shim YS (2001) Nuclear fusion leads to chromosome doubling during mannitol pretreatment of barley (Hordeum vulgare L.) microspores. J Exp Bot 52(359):1227–1238

30. Maluszynska J (2003) Cytogenetic tests for ploidy level analyseschromosome counting. In: Maluszynski M, Kasha KJ, Forster BP, Szarejko I (eds) Doubled haploid production in crop plants: a manual. Kluwer Academic Publishers, Dordrecht, pp 391–395

31. Cistue L, Castan MS, Castillo A, Valles MP, Sanz JM, Echavari B (2006) Production of doubled haploids in durum wheat (Triticum turgidum L.) through isolated microspore culture. Plant Cell Rep 25(4):275–264

32. Weber S, Nker F, Friedt W (2005) Improved doubled haploid production protocol for Brassica napus using microspore colchicine treatment in vitro and ploidy determination by flow cytometry. Plant Breed 124:511–513

33. da Silva Dias JC (2003) Protocol for broccoli microspore culture. In: Maluszynski M, Kasha KJ, Forster BP, Szarejko I (eds) Doubled haploid production in crop plants: a manual. Kluwer Academic, Dordrecht, pp 195–204

34. Martin B, Widholm JM (1996) Ploidy of small individual embryo-like structures from maize anther cultures treated with chromosome doubling agents and calli derived from them. Plant Cell Rep 15:781–785

35. Gervais C, Newcomb W, Simmonds DH (2000) Rearrangement of the actin filament and microtubule cytoskeleton during induction of microspore embryogenesis in Brassica napus L. cv. Topas. Protoplasma 213 (3–4):194–202

36. Dubas E, Custers J, Kieft H, Wędzony M, van Lammeren AAM (2011) Microtubule configurations and nuclear DNA synthesis during initiation of suspensor-bearing embryos from *Brassica napus* cv Topas microspores. Plant Cell Rep 30(11):2105–2116

37. Li H, Devaux P (2003) High frequency regeneration of barley doubled haploid plants from isolated microspore culture. Plant Sci 164(3):379–386

38. Soriano M, Cistue L, Valles MP, Castillo AM (2007) Effects of colchicine on anther and microspore culture of bread wheat (*Triticum aestivum* L.). Plant Cell Tissue Organ Cult 91 (3):225–234

39. Sato S, Katoh N, Iwai S, Hagimori M (2005) Frequency of spontaneous polyploidization of embryos regenerated from cultured anthers or microspores of *Brassica rap* a var. *pekinensis* L. and *B. oleracea* var. *capitata* L. Breed Sci 55:99–102

40. Hu T, Kasha KJ (1996) Performance of isolated microspore-derived doubled haploids of wheat (*Triticum aestivum* L.). Can J Plant Sci 77(4):549–555

41. Jauhar PP, Dogramacı-Altuntepe M, Peterson TS, Almouslem AB (2000) Seedset on synthetic haploids of durum wheat: cytological and molecular investigations. Crop Sci 40:1742–1749

42. Jauhar PP (2007) Meiotic restitution in wheat polyhaploids and amphihaploids: a potent force in evolution of the Triticeae. J Hered 98:188–193

43. Takahira J, Cousin A, Nelson MN, Cowling WA (2011) Improvement in efficiency of microspore culture to produce doubled haploid canola (*Brassica napus* L.) by flow cytometry. Plant Cell Tissue Organ Cult 104 (1):51–59

44. Joubés J, Chevalier C (2000) Endoreduplication in higher plants. Plant Mol Biol 43:735–745

45. Weber J, Georgiev V, Pavlov A, Bley T (2008) Flow cytometric investigations of diploid and tetraploid plants and in vitro cultures of *Datura stramonium* and *Hyoscyamus niger*. Cytometry A 73:931–939

46. Perera PIP, Ordoñez CA, Lopez-Lavalle LAB, Dedicova B (2014) A milestone in the doubled haploid pathway of cassava. Protoplasma 251:233–246

47. Testillano PS, Georgiev S, Mogensen HL, Coronado MJ, Dumas C, Risueno MC, Matthys-Rochon E (2004) Spontaneous chromosome doubling results from nuclear fusion during *in vitro* maize induced microspore embryogenesis. Chromosoma 112(7):342–349

48. Shim YS, Kasha KJ, Simion E, Letarte J (2006) The relationship between induction of embryogenesis and chromosome doubling in microspore cultures. Protoplasma 228 (1–3):79–86

49. Acharya BC, Ramji MV (1977) Experimental androgenesis in plants – a review. Proc Indian Acad Sci 86(6):337–360

50. Sarto GE, Stubblefield PA, Therman E (1982) Endomitosis in human trophoblast. Human Genet 62(3):228–232

51. Lee HO, Davidson JM, Duronio RJ (2009) Endoreplication: polyploidy with purpose. Gen Dev 23:2461–2477

52. Malallah GA, Afzal M, Attia TA, Abraham D (1996) Tapetal cell nuclear characteristics of some Kuwaiti plants. Cytologia 61:259–267

53. Sharma SK, Kumaria S, Tandon P, Rao SR (2012) Endomitosis in tapetal cells of some *Cymbidiums* (Orchidaceae). Nucleus 55:21–25

54. Hüsemann LC, Reese A, Radine C, Piekorz RP, Budach W, Sohn D, Jänicke RU (2020) The microtubule targeting agents eribulin and paclitaxel activate similar signaling pathways and induce cell death predominantly in a caspase-independent manner. Cell Cycle 19 (4):464–478

55. Silva P, Barbosa J, Nascimento AV, Faria J, Reis R, Bousbaa H (2011) Monitoring the fidelity of mitotic chromosome segregation by the spindle assembly checkpoint. Cell Prolif 44(5):391–400

56. Dube D, Tiwari P, Kaur P (2016) The hunt for antimitotic agents: an overview of structure-based design strategies. Expert Opin Drug Discov 11(6):579–597

57. Schmidt M, Bastians H (2007) Mitotic drug targets and the development of novel antimitotic anticancer drugs. Drug Resist Up 10 (4–5):162–181

58. Downing KH (2000) Structural basis for the interaction of tubulin with proteins and drugs that affect microtubule dynamics. Annu Rev Cell Dev Biol 16:89–111

59. Chaikam V, Molenaar W, Melchinger AE, Boddupalli PM (2019) Doubled haploid technology for line development in maize: technical advances and prospects. Theor Appl Genet 132:3227–3243

60. Ebrahimzadeh H, Soltanloo H, Shariatpanahi ME, Eskandari A, Ramezanpour SS (2018) Improved chromosome doubling of parthenogenetic haploid plants of cucumber (*Cucumis sativus* L.) using colchicine, trifluralin,

and oryzalin. Plant Cell Tissue Organ Cult 135(3):407–417

61. Grosso V, Farina A, Giorgi D, Nardi L, Diretto G, Lucretti S (2018) A high-throughput flow cytometry system for early screening of *in vitro* made polyploids in *Dendrobium* hybrids. Plant Cell Tissue Organ Cult 132(1):57–70

62. Sood S, Dwivedi S (2015) Doubled haploid platform: an accelerated breeding approach for crop improvement. In: Bahadur B, Venkat RM, Sahijram L, Krishnamurthy K (eds) Plant biology and biotechnology, volume II: plant genomicsand biotechnology. Springer, New Delhi, pp 89–111

63. Noori SAS, Norouzi M, Karimzadeh G, Shirkool K, Niazian M (2017) Effect of colchicine-induced polyploidy on morphological characteristics and essential oil composition of ajowan (*Trachyspermum ammi* L.). Plant Cell Tissue Organ Cult 130 (3):543–551

64. Sivakumar G, Alba K, Phillips GC (2017) Biorhizome: a biosynthetic platform for colchicine biomanufacturing. Front Plant Sci 8:1137

65. Eng WH, Ho WS (2019) Polyploidization using colchicine in horticultural plants: a review. Sci Hortic 246:604–617

66. Melchinger AE, Molenaar WS, Mirdita V, Schipprack W (2016) Colchicine alternatives for chromosome doubling in maize haploids for doubled-haploid production. Crop Sci 56 (2):559–569

67. Pintos B, Manzanera JA, Bueno MA (2007) Antimitotic agents increase the production of doubled-haploid embryos from cork oak anther culture. J Plant Physiol 164 (12):1595–1604

68. Pazuki A, Aflaki F, Gürel S, Ergül A, Gürel E (2018) Production of doubled haploids in sugar beet (*Beta vulgaris*): an efficient method by a multivariate experiment. Plant Cell Tissue Org Cult 132(1):85–97

69. Aqafarini A, Lotfi M, Norouzi M, Karimzadeh G (2019) Induction of tetraploidy in garden cress: morphological and cytological changes. Plant Cell Tissue Organ Cult 137 (3):627–635

70. Fu L, Zhu Y, Li M, Wang C, Sun H (2019) Autopolyploid induction via somatic embryogenesis in *Lilium distichum* Nakai and *Lilium cernuum* Komar. Plant Cell Tissue Organ Cult 139:237–248

71. Khalili S, Niazian M, Arab M, Norouzi M (2019) *In vitro* chromosome doubling of

African daisy, *Gerbera jamesonii* bolus cv. Mini red. Nucleus 63:59–65

72. Sabzehzari M, Hoveidamanesh S, Modarresi M, Mohammadi V (2019) Morphological, anatomical, physiological, and cytological studies in diploid and tetraploid plants of *Plantago psyllium*. Plant Cell Tissue Organ Cult 139(1):131–137

73. Ye YM, Tong J, Shi XP, Yuan W, Li GR (2010) Morphological and cytological studies of diploid and colchicine-induced tetraploid lines of crape myrtle (*Lagerstroemia indica* L.). Sci Hortic 124(1):95–101

74. He M, Gao W, Gao Y, Liu Y, Yang X, Jiao H, Zhou Y (2016) Polyploidy induced by colchicine in *Dendranthema indicum* var. *Aromaticum*, a scented chrysanthemum. Eur J Hortic Sci 81(4):219–226

75. Xu C, Zhang Y, Huang Z, Yao P, Li Y, Kang X (2018) Impact of the leaf cut callus development stages of *Populus* on the tetraploid production rate by colchicine treatment. J Plant Growth Regul 37(2):635–644

76. Rao PS, Suprasanna P (1996) Methods to double haploid chromosome numbers. In: Jain SM, Sopory SK, Veilleux RE (eds) In vitro haploid production in higher plants. Current plant science and biotechnology in agriculture, vol 23. Springer, Dordrecht

77. Qi-Qing LI, Zhang J, Ji-Hua LIU, Bo-Yang YU (2018) Morphological and chemical studies of artificial *Andrographis paniculata* polyploids. Chin J Nat Med 16(2):81–89

78. Sun FY, Liu L, Yu Y, Ruan XM, Wang CY, Hu QW, Wu DX, Sun G (2020) MicroRNA-mediated responses to colchicine treatment in barley. Planta 251(2):1–14

79. Jensen CJ (1975) Barley monoploids and doubled monoploids: technique and experience. In: H. Gaul Proceedings of the 3rd international barley genetics symposium., 316–345. Verlag Karl Thiemeg, Munich

80. Subrahmanyam NC, Kasha KJ (1975) Chromosome doubling of barley haploids by nitrous oxide and colchicine treatments. Can J Genet Cytol 17(4):573–583

81. Premvaranon P, Vearasilp S, Thanapornpoonpong S, Karladee D, Gorinstein S (2011) *In vitro* studies to produce double haploid in Indica hybrid rice. Biologia 66(6):1074–1081

82. Mohammadi PP, Moini A, Ebrahimi A, Javidfar F (2012) Doubled haploid plants following colchicine treatment of microspore-derived embryos of oilseed rape (*Brassica napus* L.). Plant Cell Tissue Organ Cult 108:251–256

83. Yemets AI, Blume YB (2008) Progress in plant polyploidization based on antimicrotubular drugs. Open Hortic J 1(1):15–20

84. Sood S, Dhawan R, Singh K, Bains NS (2003) Development of novel chromosome doubling strategies for wheat × maize system of wheat haploid production. Plant Breed 122 (6):493–496

85. Pan-pan H, Wei-xu L, Hui-hui L (2018) *In vitro* induction and identification of autotetraploid of *Bletilla striata* (Thunb.) Reichb. f. by colchicine treatment. Plant Cell Tissue Organ Cult 132(3):425–432

86. Ochatt SJ, Patat-Ochatt EM, Moessner A (2011) Ploidy level determination within the context of *in vitro* breeding. Plant Cell Tissue Organ Cult 104(3):329–341

87. Dwivedi SL, Britt AB, Tripathi L, Sharma S, Upadhyaya HD, Ortiz R (2015) Haploids: constraints and opportunities in plant breeding. Biotechnol Adv 33(6):812–829

88. Niazian M, Sadat-Noori SA, Abdipour M, Tohidfar M, Mortazavian SMM (2018) Image processing and artificial neural network-based models to measure and predict physical properties of embryogenic callus and number of somatic embryos in ajowan (*Trachyspermum ammi* (L.) Sprague). In Vitro Cell Dev Biol Plant 54(1):54–68

89. Niazian M, Shariatpanahi ME, Abdipour M, Oroojloo M (2019) Modeling callus induction and regeneration in an anther culture of tomato (*Lycopersicon esculentum* L.) using image processing and artificial neural network method. Protoplasma 256(5):1317–1332

90. Wang H, Dong B, Jiang J, Fang W, Guan Z, Liao Y, Chen S, Chen F (2014) Characterization of *in vitro* haploid and doubled haploid *Chrysanthemum morifolium* plants via unfertilized ovule culture for phenotypical traits and DNA methylation pattern. Front Plant Sci 5:738

91. Vanous K, Vanous A, Frei UK, Lübberstedt T (2017) Generation of maize (*Zea mays*) doubled haploids via traditional methods. Curr Protoc Plant Biol 2(2):147–157

92. Warchoł M, Skrzypek E, Nowakowska A, Marcińska I, Czyczyło-Mysza I, Dziurka K, Juzoń K, Cyganek K (2016) The effect of auxin and genotype on the production of *Avena sativa* L. doubled haploid lines. Plant Growth Regul 78(2):155–165

93. Marcińska I, Nowakowska A, Skrzypek E, Czyczyło-Mysza I (2013) Production of double haploids in oat (*Avena sativa* L.) by pollination with maize (*Zea mays* L.). Open Life Sci 8(3):306–313

94. Warchoł M, Czyczyło-Mysza I, Marcińska I, Dziurka K, Noga A, Skrzypek E (2018) The effect of genotype, media composition, pH and sugar concentrations on oat (*Avena sativa* L.) doubled haploid production through oat× maize crosses. Acta Physiol Plant 40(5):93

95. Dunwell JM, Wilkinson MJ, Nelson S, Wening S, Sitorus AC, Mienanti D, Alfiko Y, Croxford AE, Ford CS, Forster BP, Caligari PD (2010) Production of haploids and doubled haploids in oil palm. BMC Plant Biol 10 (1):218

96. Arzani A, Darvey NL (2001) The effect of colchicine on triticale anther-derived plants: microspore pre-treatment and haploid-plant treatment using a hydroponic recovery system. Euphytica 122(2):235–241

97. Ślusarkiewicz-Jarzina A, Pudelska H, Woźna J, Pniewski T (2017) Improved production of doubled haploids of winter and spring triticale hybrids via combination of colchicine treatments on anthers and regenerated plants. J Appl Genet 58(3):287–295

98. Tayeng T, Chaudhary HK, Kishore N (2012) Enhancing doubled haploid production efficiency in wheat (*Triticum aestivum* L. em. Thell) by *in vivo* colchicine manipulations in *Imperata cylindrica*-mediated chromosome elimination approach. Plant Breed 131 (5):574–578

99. Srivastava P, Gill RS, Sharma A, Kumar S, Raghupati N, Mahal GS, Bains NS (2012) Colchicine administered as post pollination tiller injection is deleterious for doubled haploid production in durum wheat x maize crosses. Crop Improv 39(1):60–64

100. Srivastava P, Bains NS (2018) Accelerated wheat breeding: doubled haploids and rapid generation advance. In: Gosal S, Wani S (eds) Biotechnologies of crop improvement, volume 1. Springer, Cham

101. Foschi ML, Martinez LE, Ponce MT, Galmarini CR, Bohanec B (2013) Effect of colchicine and amiprophos-methyl on the production of in vitro doubled haploid onion plants and correlation assessment between ploidy level and stomatal size. Rev FCA UNCUYO 45(2):155–164

102. Alan AR, Lim W, Mutschler MA, Earle ED (2007) Complementary strategies for ploidy manipulations in gynogenic onion (*Allium cepa* L.). Plant Sci 173:25–31

103. Ferrie AMR, Irmen KI, Beattie AD, Rossnagel BG (2014) Isolated microspore culture of oat (*Avena sativa* L.) for the production of doubled haploids: effect of pre-culture and

post-culture conditions. Plant Cell Tissue Organ Cult 116(1):89–96

104. Bossoutrot D, Hosemans D (1985) Gynogenesis in *Beta vulgaris* L.: from *in vitro* culture of unpollinated ovules to the production of doubled haploid plants in soil. Plant Cell Rep 4:300–303

105. Lionneton E, Beuret W, Delaitre V, Ochatt S, Rancillac M (2001) Improved microspore culture and doubled-haploid plant regeneration in the brown condiment mustard (*Brassica juncea*). Plant Cell Rep 20:126–130

106. Klutschewski S (2012) Methodical improvements in microspore culture of *Brassica napus* L. Dissertation. University of Gottingen, Gottingen

107. Bhatia R, Dey SS, Sood S, Sharma K, Parkash C, Kumar R (2017) Efficient microspore embryogenesis in cauliflower (Brassica oleracea var. botrytis L.) for development of plants with different ploidy level and their use in breeding programme. Sci Hortic 216:83–92

108. Herrera JC, Moreno LG, Acuna JR, De Pena M, Osorio D (2002) Colchicine-induced microspore embryogenesis in coffee. Plant Cell Tissue Organ Cult 71(1):89–92

109. Lotfi M, Alan AR, Henning MJ, Jahn MM, Earle ED (2003) Production of haploid and doubled haploid plants of melon (*Cucumis melo* L.) for use in breeding for multiple virus resistance. Plant Cell Rep 21:1121–1128

110. Burun B, Emiroglu U (2008) A comparative study on colchicine application methods in obtaining doubled haploids of tobacco (*Nicotiana tabacum* L.). Turk J Biol 32 (2):105–111

111. Islam SMS (2010) The effect of colchicine pretreatment on isolated microspore culture of wheat (*Triticum aestivum* L.). Austr J Crop Sci 4(9):660–665

112. Würschum T, Tucker MR, Reif JC, Maurer HP (2012) Improved efficiency of doubled haploid generation in hexaploid triticale by in vitro chromosome doubling. BMC Plant Biol 12(109):1–7

113. Obert B, Barnabas B (2004) Colchicine induced embryogenesis in maize. Plant Cell Tissue Organ Cult 77(3):283–285

114. Ren X, Ci J, Cui X, Yang W (2018) Doubling effect of anti-microtubule herbicides on the maize haploid. Emirates J Food Agric 30 (10):903–908

115. Grzebelus E, Adamus A (2004) Effect of anti-mitotic agents on development and genome doubling of gynogenic onion (*Allium cepa* L.) embryos. Plant Sci 167:569–574

116. Fayos O, Valles MP, Graces-Claver A, Moller C, Castillo AM (2015) Doubled haploid production from Spanish onion (*Allium cepa* L.) germplasm: embryogenesis induction, plant regeneration and chromosome doubling. Front Plant Sci 6:384

117. Hansen AL, Gertz A, Joersbo M, Andersen SB (2000) Chromosome doubling *in vitro* with amiprophos-methyl in *Beta vulgaris* ovule culture. Acta Agric Scand Sect B Soil Plant Sci 50:89–95

118. Hansen NJP, Andersen SB (1998) Efficient production of doubled haploid wheat plants by *in vitro* treatment of microspores with trifluralin or APM. Plant Breed 117:401–405

119. Wan Y, Duncan DR, Rayburn AL, Petolino JF, Widholm JM (1991) The use of anti-microtubule herbicides for the production of doubled haploid plants from anther-derived maize callus. Theor Appl Genet 81 (2):205–211

120. Soriano M, Cistué L, Castillo AM (2008) Enhanced induction of microspore embryogenesis after n-butanol treatment in wheat (*Triticum aestivum* L.) anther culture. Plant Cell Rep 27(5):805–811

Part II

Doubled Haploids in Alliaceae

Chapter 6

Doubled Haploid Onion (*Allium cepa* L.) Production Via In Vitro Gynogenesis

Ali Ramazan Alan

Abstract

Onion (*A. cepa* L.) is an outcrossing biennial species with a very large genome. Development of genetically uniform (inbred) lines highly desired by onion breeders is a difficult process due to high level of heterozygosity. Inbred onion development may take up to five generations (~10 years) by classical selfing technique. Onion shows severe inbreeding depression, which further complicates production of lines with stabilized important agronomic traits. When applied successfully, haploidization technology can be useful in the development of fully homozygous onion lines in 2 years. Although production of haploid and doubled haploid (DH) onions via gynogenesis was reported more than three decades ago, successful production and utilization of DHs in onion breeding is still far behind of expectations of breeders. The main obstacles in front of the success include high variation in the response of donor materials to gynogenesis induction and difficulties faced in the process of obtaining DHs from haploid plants. We use a DH production procedure enabling us to develop DH plants from a wide range of onion donor materials. This procedure is based on production of haploid plants via single step culture of unopened flower buds, detection of haploid plants among gynogenic regenerants, and converting these plants to fecund DHs using a combination of ploidy manipulation techniques. The bulbs of DHs are produced in about 1 year after the initiation of induction cultures and selfed seeds are produced from fecund DH plants when they flower in the second year.

Key words *Allium cepa* L., Breeding, Chromosome doubling, Doubled haploid, Flow cytometry, Gynogenesis, Onion

1 Introduction

Onion (*A. cepa* L.) is one of the most important crop species grown all around the world. It is a diploid ($2n = 2x = 16$) species with a 2C nuclear DNA amount of ~32 pg (~1.57×10^4 Mbp/1C) deduced by flow cytometry. Onion has been cultivated for more than 5000 years [1]. It is thought that cultivation of onion took place in the mountainous region between Turkmenistan and northern Iran. South West Asia is considered as the primary domestication and diversity area, while the Mediterranean region with large diversity is thought to be the secondary center [2, 3]. According to

Jose M. Seguí-Simarro (ed.), *Doubled Haploid Technology: Volume 1: General Topics, Alliaceae, Cereals*,
Methods in Molecular Biology, vol. 2287, https://doi.org/10.1007/978-1-0716-1315-3_6,
© Springer Science+Business Media, LLC, part of Springer Nature 2021

Hanelt [2], today's onions have resulted from selection for biennial plants with large bulbs during the course of domestication.

Today, onion is the second most produced vegetable after tomato. According to FAOSTAT [4], annual onion production in the world is ~98 million tons. About half of world onion production comes from China and India. Onion is highly valued worldwide as vegetable, spice, and medicine. The bulb and leaves can be consumed fresh as salad or it can be used as a processed product such as dried flakes, rings, and powder. Onion is a rich of source of health-promoting compounds [5, 6]. It contains high amount of quercetin, the main dietary flavonol [7, 8].

Onion can be propagated by seeds, seedlings, sets, and bulbs. Seeds are preferred for large scale production systems. While open pollinated (OP) standard varieties are still used in onion production worldwide, production with hybrid varieties is more common in Western Europe and North America. There are onion genotypes specifically bred for bulb production in different photoperiods and temperatures, storage, dry matter, taste, bulb shape, skin color, and so on. Onion genotypes are conventionally classified according to their day-length response as long day- (>14 h), intermediate- (12–13 h), and short- (10–11 h) day types.

Breeding companies are interested in producing new high yielding F1 hybrid onion varieties with high uniformity and quality in agronomical traits. Majority of F1 hybrid onion varieties used in developed countries are bred to produce high-quality bulbs under long-day conditions. There is also strong interest in developing early maturing and high yielding F1 hybrid onion varieties from intermediate- and short-day onion types for the production in warmer regions of the world. Conventional F1 hybrids are produced by crossing inbred lines. Development of an inbred onion line may take up to five generations of selfing (~10 years) to stabilize agronomically important traits. Inbreeding depression is a commonly encountered problem during the inbred production process. It often leads to rapid loss of plant vigor and fecundity. Therefore, conventionally developed onion inbreds retain certain amount of heterozygosity.

Genetically homozygous onion lines can be produced only by the production of doubled haploid (DH) plants. Onion was reported to be recalcitrant to androgenesis induction [9]. The first groups of haploid onion plants were obtained by in vitro gynogenesis induction three decades ago [10–12]. Researchers used whole immature flower buds, ovaries and ovules as explants from different onion genotypes and cultured them on various gynogenesis induction media. They detected plants with haploid, diploid, mixoploid, tetraploid, and other ploidy levels among the gynogenic regenerants. Dore and Marie [13] suggested the use of irradiation-inactivated pollen in crosses with male sterile plants (in situ gynogenesis) as an alternative strategy for production of gynogenic onion plants. Cytological evaluations confirmed the haploid origin of some of the plants developed from the seeds obtained through in situ gynogenesis. In vitro gynogenesis induction is the major technique for haploid plant production in onion. In general, researchers used one or two step culture of explants on MS [14], B5 [15], and BDS [16] media with various modifications. Researchers compared the possible influence of different explant types (whole immature flower buds and extracted ovules and ovaries) on different gynogenesis induction media [10–12, 17–26]. Response to gynogenesis induction, regeneration of gynogenic plants, and rates of spontaneous diploid plants are strongly influenced by the donor onion genotype [18, 22, 23].

The low gynogenesis response of the majority of onion genotypes and the difficulties faced during conversion of haploids to fecund DH plants are the two main issues restricting utilization of the gynogenesis technique in onion breeding programs. In onion, the haploid state is very persistent through the life cycle of the gynogenic plants. Although some of the gynogenic onion regenerants are shown to be diploids, researchers should be cautious about these plants since they may have been originated from unreduced egg cells or somatic tissues. These plants must be checked by molecular techniques to determine their true origin. Although not very common, spontaneous chromosome doubling can occur in early developmental stages converting haploid plants to DHs or mixoploids for haploid and diploid cells. These spontaneous DHs and some of mixoploids can produce normal egg and pollen gametes and can set seeds [27]. The conventional way of converting haploid onion plants to diploids is treating explants (newly emerged plantlets, whole plants, or basal portions of plants) with antimitotic agents in vitro [27–30]. Unfortunately, this method is not well optimized for onion and it often leads to loss of valuable gynogenic plants due to the toxic effects of antimitotic agents. Alan et al. [27] developed three strategies for maximizing the frequency of gynogenic DH onion plants obtained. These strategies included induced chromosome doubling by treatment with antimitotic agents, spontaneous chromosome doubling by somatic shoot regeneration from cultured flower buds, and ploidy reduction by second-cycle gynogenesis.

Very few publications report the fecundity levels of DH onion lines. A detailed study published by Alan et al. [23] showed that there may be substantial differences in fecundity levels of gynogenic onion lines. They reported the recovery of viable selfed seeds obtained from many diploid as well as mixoploid and tetraploid gynogenic plants. The plants grown from the seeds of a DH line are genetically identical to each other. They grow uniformly and show similar responses to environmental conditions under greenhouse and field conditions.

Onion researchers have strong interest in the utilization of DH onion lines in studying genetic basis of important traits and in the improvement of new varieties. Recently, Kosha et al. [31] declared the promotion of the use of a DH onion line 'CUDH2107', developed from long-day onion improvement program at Cornell University (Ithaca, NY, USA) by Alan et al. [23], to simplify the discovery and analysis of genes underlying important onion traits. This DH line is also used in continuing de novo assembly of onion genome [32]. The study reported by Hyde et al. [33] showed that the DH lines developed from long-day onion improvement program by Alan et al. at Cornell University [22, 23, 27] can replace conventional inbred pollen donor lines in the production of new F1 hybrid onion varieties. F1 hybrids developed using DH lines as pollen donors showed higher levels of heterosis and better uniformity of bulbs than the conventional F1 hybrids.

The protocol presented here can be utilized in all onion types for the production of DH plants. Gynogenic onion plants are produced by a single step induction culture of whole immature flower buds and DH plants are recovered by a combination of several ploidy manipulation strategies.

2 Materials

2.1 Plant Material

This protocol can be used to produce DH onion plants from long-, short-, and intermediate-day genotypes.

2.2 Equipment

1. Autoclave.
2. Calipers.
3. Desktop centrifuge.
4. Digital timer.
5. Dissecting microscope with digital camera.
6. Flow cytometer (Beckman Coulter Cell Lab QUANTA).
7. Fume hood.
8. Greenhouse set for 25 °C during the day and 15 °C at night.
9. Growth chamber at 17 °C with a 16/8 h photoperiod.
10. High precision digital balance.
11. Laboratory ice maker.
12. Laminar flow hood.
13. Light microscope with digital camera.
14. Magnetic hotplate stirrer and mixing fish.
15. pH meter.
16. Refrigerator with freezer.

17. Sterilizator.

18. Tissue culture room adjusted for constant 25 °C with a 16/8 h photoperiod.

19. Ultrapure water purification system.

20. Vortex.

2.3 Laboratory Materials

1. Aluminum foil.

2. Autoclavable glass bottles.

3. Baby food jars.

4. Beakers.

5. Eppendorf tubes 1.5 mL.

6. Falcon tubes 15 mL.

7. Falcon tubes 50 mL.

8. Filter papers.

9. Flow Check beads (Beckman-Coulter).

10. Forceps.

11. Glass culture tubes 50 mL.

12. Graduated cylinder.

13. Magenta boxes.

14. Microscope slides and coverslips.

15. Nonpyrogenic syringe-R filters 0.20μm.

16. Nylon mesh with 60μm pores.

17. Parafilm.

18. Permanent marker.

19. Razor blades.

20. Scalpels.

21. Sharp blunt surgical scissors.

22. Syringes 30 mL.

23. Spatulas.

24. Stainless steel tea strainers.

25. Sterile filter paper.

26. Sterile Pasteur or automatic pipettes.

27. Sterile petri plates 60×15 mm (diameter × height).

28. Sterile petri plates 90×15 mm (diameter × height).

29. Sterile pipette tips 100μL.

30. Sterile pipette tips 1000μL.

31. Sterile six-well cell culture plates.

32. Test tube racks.

33. Trays.

34. Vi-CELL 4 mL sample vials (for use in flow cytometry analysis).

35. Weight boats.

2.4 Greenhouse/Growth Chamber Materials

1. Autoclavable bags.

2. Labels.

3. Peat and perlite (2:1) mixture.

4. Plant fertilizer NPK 20 20 20 + TE.

5. Plastic agricultural trays—104 cavities.

6. Plastic agricultural trays—24 cavities.

7. Plastic labels 12 cm.

8. Plastic pots 0.5 L.

9. Plastic pots 7 L.

10. Plastic sandwich bags (20 × 30 cm).

11. Plastic tags.

12. Rubber bands.

2.5 Solutions

1. 70% ethanol (v/v).

2. 0.4 g/L colchicine: in a 50 mL Falcon tube, dissolve 20 mg colchicine in 0.1 mL dimethyl sulfoxide (DMSO) by vortexing and fill the tube up to 50 mL with autoclaved liquid MSO. Filter-sterilize the colchicine solution through 0.20μm non-pyrogenic syringe-R filter.

3. Alexander stain (as modified by Peterson et al. [34]: to prepare 50 mL stain, add the following constituents in order and store the stain in a light-protected bottle. Add 5 mL 95% ethanol, 0.5 mL malachite green (1% solution in 95% ethanol, 25 mL double distilled (dd) water, 12.5 mL glycerol, 2.5 mL acid fuchsin (1% solution in dd water), 0.25 mL Orange G (1% solution in dd water), 2 mL glacial acetic acid, and ~2 mL dd water to a total of 50 mL.

4. Autoclaved dd water.

5. Disinfection solution: 20% commercial bleach (1.12% sodium hypochlorite) and 0.1% Tween 20.

6. Fluorescent dye: dissolve 100 mg propidium iodide (PI) in 10 mL dd water in a 15 mL Falcon tube to prepare a 10 mg/mL PI stock solution. Cover the PI stock with aluminum foil. Keep PI stock in the refrigerator (4 °C) until use.

7. Sheath fluid for the flow cytometry system.

8. Nuclei Isolation Buffer (NIB): 15 mM HEPES, 1 mM Na_2EDTA, 80 mM KCl, 20 mM NaCl, 300 mM sucrose,

0.2% (v/v) Triton X-100, 0.5 mM spermine, 1% (w/v) poly-vinylpyrrolidone (PVP40). Adjust pH to 7.5. Fill 15 mL or 50 mL tubes with NIB and keep them in the freezer (−20 °C) until use.

9. RNase (sold at 4 mg/mL concentration).

10. Shutdown solution for nuclear packing efficiency (NPE) analysis in flow cytometry.

11. Wetting agent: Tween 20 (1 mL per 1 L disinfection solution).

2.6   Culture Media

1. Gynogenesis induction medium: BDS (Table 1) supplemented with 2 mg/L 2,4-dichlorophenoxyacetic acid (2,4-D) and 2 mg/L 6-Benzylaminopurine (BAP). Adjust pH to 6. Add 7.4 g/L agar for microbiological use. Autoclave media at 121 °C for 20 min, let it cool down until you can hold with your hands (~60 °C), and then pour ~25 mL medium into each 90 × 15 mm sterile petri plate and let them solidify in the sterile hood.

2. Elongation medium (EM): BDS medium with half strength major and minor salts and 30 g/L sucrose (Table 1). Prepare as described for the gynogenesis induction medium. Pour ~25 mL EM into 90 × 15 mm plates and 15 mL into culture tubes and let them solidify in sterile hood.

3. Shooting medium (SM): BDS medium supplemented with 2 mg/L BAP and 30 g/L sucrose (Table 1). Prepare as described for the gynogenesis induction medium. Pour ~25 mL SM into 90 × 15 mm plates and let them solidify in sterile hood.

4. MSO: Liquid MS medium with 30 g/L sucrose for induced chromosome doubling (Table 1). Prepare as described for the induction medium. Keep in the refrigerator until use.

3   Methods

3.1   Donor Plant Growth Conditions

1. Use healthy and fully developed dried bulbs of the donor plant materials and vernalize them at 4 °C in a cold room (*see* **Note 1**).

2. After vernalization, place the bulbs in plastic agricultural trays with 24 cavities or 0.5 L pots filled with peat and perlite (2:1 v/v) for sprouting them.

3. Keep the bulbs in a growth chamber adjusted for 20 °C/12 h light and 17 °C/12 h dark until sufficient level of shoot development is obtained.

Table 1
Composition of the media used in whole flower bud culture–based gynogenesis induction and regeneration of gynogenic onion plants. BDS, EM, and GM are modified from Dunstan and Short [16]. MSO medium is based on Murashige and Skoog [14]

	BDS (mg/L)	EM (mg/L)	SM (mg/L)	MSO (mg/L)
Macroelements				
$CaCl_2 \cdot 2H_2O$	1500.00	750.00	1500.00	4400.00
KNO_3	25300.00	12650.00	25300.00	19000.00
$NH_4(NO_3)$	3202.00	1601.00	3202.00	16500.00
$MgSO_4 \cdot 7H_2O$	2470.00	1235	2470.00	3700.00
KH_2PO_4	–	–	–	1700.00
$(NH_4)_2SO_4$	1340.00	670.00	1340.00	–
$NaH_2PO_4 \cdot H_2O$	1690.00	845.00	1690.00	–
$NH_4H_2PO_4$	2300.00	1150	2300.00	–
Microelements				
$CoCl_2 \cdot 6H_2O$	2.50	1.25	2.50	0.025
$CuSO_4 \cdot 5H_2O$	3.90	1.95	3.90	0.025
H_3BO_3	300.00	150	300.00	620.00
KI	75.00	37.50	75.00	83.00
$MnSO_4 \cdot H_2O$	925.00	462.50	925.00	1690.00
$Na_2MoO_4 \cdot 2H_2O$	25.00	12.50	25.00	25.00
$ZnSO_4 \cdot 7H_2O$	200.00	100	200.00	860.00
Iron source				
$EDTANa_2 \cdot 2H_2O$	37.25	37.25	37.25	37.25
$FeSO_4 \cdot 7H_2O$	27.85	27.85	27.85	27.85
Vitamins				
Myoinositol	500.00	500.00	500.00	100.00
Nicotinic acid	1.00	1.00	1.00	0.5
Pyridoxine hydrochloride	1.00	1.00	1.00	0.5
Thiamine hydrochloride	10.00	10.00	10.00	0.40
Amino acids				
Glycine	–	–	–	2.00
L-Proline	200.00	200.00	200.00	–
Plant growth regulators				
2,4-Dichlorophenoxyacetic acid (2,4-D)	2.00	–	–	–

(continued)

Table 1
(continued)

	BDS (mg/L)	EM (mg/L)	SM (mg/L)	MSO (mg/L)
6-Benzylaminopurine (BAP)	2.00	–	2.00	–
Carbohydrate				
Sucrose	100,000	30,000	30,000	30,000
Solidifying agent				
Agar (for microbiology)	7400	7400	7400	7800
pH	6.00	6.0	6.00	5.8

4. Transfer the sprouted bulbs to 7 L pots and place them in the greenhouse and control the conditions to keep the plants at 25 °C/16 h light and 17 °C/8 h dark.

5. Grow the donor onion plants under the optimal conditions necessary for vegetative and generative development (*see* **Note 2**).

6. Observe the donor onion genotypes for growth and formation of flower scapes (*see* **Note 3**).

7. Stake the plants to keep leaves and flower scapes straight.

8. Watch the donor plants closely when the umbel capsule start to break (*see* **Note 4**).

3.2 Establishment of Gynogenesis Induction Cultures and Obtaining Gynogenic Plants

1. Collect the scapes when ~20% of the flowers in the umbels are at anthesis stage (*see* **Note 5**).

2. Label the scapes to indicate their donor origin, place them in a beaker filled with sterile water, and bring them to the lab.

3. Perform a pollen viability test with the pollen collected from the anthers of several flowers of each donor plant at the anthesis stage (*see* **Note 6**).

4. Observe the flower buds of the umbels carefully and choose those at the right stage for gynogenesis induction (*see* **Note 7**).

5. Excise the preferred unopened flower buds without pedicels using sterile scissors (Fig. 1a).

6. For surface disinfection, place the buds into metal tea filters and dip them into a Magenta box filled with 70% ethanol for 30 s and then into another Magenta box filled with disinfection solution for 30 min.

7. In the laminar flow hood, pour off the disinfection solution and then wash the buds three times for 5 min each with autoclaved dd water.

Fig. 1 In vitro gynogenesis induction in onion. (**a**) An immature flower bud suitable for gynogenesis work. (**b**) An anthesis-stage flower in induction culture. (**c**) A flower bud with an enlarged ovary about 2 weeks after culture initiation. (**d**) A responsive flower bud. (**e**) A gynogenic onion plantlet. (**f**) A gynogenic onion in vitro plant

8. After disinfection, spread the buds on a dry sterile filter paper to eliminate the excessive water from bud surface.

9. Using sterile forceps, culture flower buds by dipping their bottom halves into the medium.

10. Place 30 buds in each plate, label with marker, seal with Parafilm, and transfer to the tissue culture room (*see* **Note 8**).

11. Observe the cultures on a weekly basis. Cultured flower buds generally reach anthesis within several days and reach the maximum size within 2 weeks (Fig. 1b, c).

12. Check the cultures for microbial contamination (*see* **Note 9**).

13. Gynogenic onion plantlets are detected as early as 8 weeks after culture (Fig. 1d).

14. Transfer responsive buds to plates with EM (Table 1) using sterile forceps, record the transfer date, seal with Parafilm, and place them back in the tissue culture room (Fig. 1e; *see* **Note 10**).

15. Transfer well-formed plantlets to culture tubes with EM and place them back to the same culture room (Fig. 1f; *see* **Note 11**).

16. When gynogenic plants reach the 3–4 leaf stage with established root systems, determine their ploidy.

3.3 Determination of Ploidy of Androgenic Regenerants with Flow Cytometry Analysis

1. Collect equal amounts (50 mg from each plant) of fresh leaf tissues from onion and *Hordeum vulgare* (used as internal control) plants, put them in 60 × 15 mm Petri plates placed in an ice box, and coprocess for the isolation of nuclei. Diploid donor onion cells contain ~32 pg DNA/2C. *H. vulgare* cells contain 10.1 pg DNA/2C [35].

2. Thaw two tubes of NIB on the lab bench and place them in an ice box. Use regular NIB for isolation of nuclei from tissues by chopping. Prepare NIB with 25μg/mL RNase to dissolve nuclei and eliminate RNA from the samples.

3. Add 1.5 mL of ice-cold NIB into the sample plate and gently chop the leaves using a razor blade on ice.

4. Filter the finely chopped tissue though a nylon mesh with 60μm pore size into an Eppendorf tube placed in ice.

5. Centrifuge the samples at 8000 × *g* for 5 s.

6. Take the samples out of centrifuge and pour off the supernatant carefully.

7. Place the tubes with nuclei pellets upside down on a paper towel for several min to drain the remaining supernatant.

8. Resuspend the nuclei pellet in 300μL NIB with 25μg/mL RNase and vortex the samples.

9. Add 10μL of the propidium iodide solution into each sample and vortex the samples.

10. Keep the tubes of nuclei sample in ice until the start of the flow cytometry analysis.

11. Start the flow cytometry instrument and check it for linearity with fluorescent Flow Check beads.

12. Adjust the amplification to position the 2C peak of diploid onion nuclei approximately at channel 400 and analyze a minimum of 3000 nuclei from each sample to determine DNA amounts and ploidy levels.

13. The nuclear DNA content (2C) of onion is calculated according to the formula adapted from Dolezel and Bartos [36]:

$$\text{Gynogenic onion 2C DNA (pg)} = \left(\frac{\text{onion sample 2C peak mean}}{H.vulgare\ \text{2C peak mean}} \right)$$
$$\times \text{2C reference nuclear DNA (pg)}$$

14. Gynogenic regenerants with half of the nuclear DNA content of diploid onion plants (~16 pg DNA/2C) are classified as haploids. Regenerants with nuclear DNA contents identical to donor onion plants (~32 pg DNA/2C) are considered as DHs.

15. Haploids must be converted to diploids via chromosome doubling procedures.

3.4 Chromosome Doubling

Use this technique to convert haploid gynogenic plants into DH plants.

3.4.1 Induced Chromosome Doubling Through Treatment of Whole Basal Explants with Antimitotic Agents

1. At the day of chromosome doubling work, prepare the chromosome doubling solution by adding required amount of the antimitotic agent to MSO (400 mg/L colchicine in MSO).

2. In a sterile hood, take haploid onion plants out of their tubes (Fig. 2a) using sterile forceps and place on a sterile paper towel.

3. Remove their roots and leaves leaving basal sections (0.5–1 cm) with sterile scalpels (Fig. 2b).

4. Place the basal explants in 6-well cell culture plates filled with chromosome doubling solution (Fig. 2c).

5. Cover the treated explants with aluminum foil to keep them in darkness and leave them in the sterile hood for 48 h.

6. At the end of the treatment, pour the doubling solution into a safe container and wash the treated explants three times with autoclaved dd water (10 min each wash).

7. Place the basal explants on sterile filter paper to eliminate the excessive water from explant surface (Fig. 2d).

8. Transfer treated basal explants into Magenta boxes or baby food jars. Dip the roots of explants into the medium and keep them in straight position. Label and seal the containers with Parafilm strips and place them in the culture room to promote explant growth (Fig. 2e, f).

9. Observe the treated explants and note those continuing growth.

10. Clean the death tissue surrounding the regenerating roots and shoots (*see* **Note 12**).

11. When plants reach the 3–4 leaf stage, transfer them to in vivo conditions as explained below.

3.4.2 Spontaneous Chromosome Doubling in Somatic Regenerants from Cultured Flower Buds of Haploid Plants

Use this technique to obtain DH plants from flowering haploid plants.

1. Follow the immature flower bud culture procedure explained above.

Fig. 2 Induced chromosome doubling treatment. (**a**) A well-developed gynogenic onion plant in culture tube. (**b**) Preparation of a basal explant. (**c**) Treatment of basal explants with antimitotic agent. (**d**) A basal explant after the chromosome doubling treatment. (**e**) Shoot development from a basal explant. (**f**) A well-developed plant after chromosome doubling treatment

2. Establish cultures with immature flower buds from the umbels of haploid plants (remained haploid after the chromosome doubling treatment).

3. Culture the buds in plates containing BDS medium to induce somatic shoots.

4. Transfer flowers to plates with shooting medium (Fig. 3a).

5. Excise somatic shoots from the buds and transfer them to Magenta boxes/Baby food jars containing EM medium (Fig. 3b–d).

6. When plants reach the 3–4 leaf stage, transfer them to in vivo conditions as explained below.

Fig. 3 Regeneration of somatic shoots from the flower buds of haploid plants. (**a**) Flower buds in shooting medium. (**b**) Regeneration of shoots from a basal section of a cultured bud. (**c**) Multiple shoots emerging from a cultured bud. (**d**) Fully developed somatic plants developed from the flower buds of a haploid plant

3.4.3 Recovery of Diploid Plants from Tetraploid Gynogenic Plants by Second Cycle Gynogenesis

Use this technique to obtain DH plants from flowering tetraploid plants.

1. Establish cultures with immature flower buds from flowering gynogenic tetraploid plants that were obtained after chromosome doubling treatment.

2. Follow the immature flower bud culture procedure explained above.

3. Culture the buds in plates containing BDS medium for gynogenic plant production.

4. Place the gynogenic regenerants in culture tubes to allow them grow.

5. When plants reach the 3–4 leaf stage, determine their ploidy levels and transfer them to in vivo conditions.

3.5 Acclimatization to Ex Vitro Conditions

1. Remove the well-rooted plants from their containers and wash off the remnants of culture medium by washing the roots with running tap water.

2. Transplant the onion plants to 0.5 L pots filled with autoclaved peat and perlite mix (2:1 v/v), irrigate with sterile water and cover the pots with plastic sandwich bags and fix with a rubber band.

3. Place the potted plants on trays and transfer to a growth chamber set for constant 17 °C with a 18 h photoperiod.

4. After about 10 days, start making small openings on the plastic cover to help acclimatization of the plants.

5. Remove the plastic cover completely by the end of the third week, when plants are acclimatized.

6. Keep the plants in the chamber for another 2 weeks before moving them to a greenhouse.

3.6 Production of Seeds and Checking for Plant Uniformity in DH Lines

1. Grow the gynogenic onion plants in a greenhouse under regular conditions suitable for onion bulb production.

2. Do not allow the plants to lodge. This can be prevented by staking the plants.

3. Eventually, plants will come to a stage when bulb formation should be induced by gradually increasing the day length (about 3 months after transferring the plants to greenhouse).

4. Keep a good care of onion plants during their growth in the greenhouse.

5. Follow the growth of plants in the greenhouse until maturity (*see* **Note 13**).

6. Pull out the mature bulbs with drying leaves out of pots and let dry.

7. Cut the dried leaves, place the dried bulbs in paper bag along with the labels, and store them in the cold storage facility for vernalization.

8. Resprout bulbs in the following fall and grow the plants for seed production.

9. Check the viability of pollen produced by gynogenic onion plants.

10. Self the flowers using a paint brush.

11. Check for seed set (Fig. 4a, b).

12. Collect umbels with seeds and let them dry.

13. Clean, count, and store the seeds (Fig. 4c).

14. Produce seedlings of DH lines in the following fall (Fig. 4d), transfer the seedlings to 7 L pots (Fig. 4e) and grow in the greenhouse (or transplant to the field if there are sufficient amounts of seeds) until obtaining mature bulbs (Fig. 4f).

15. Characterize the bulbs of DH lines for the traits of interest.

4   Notes

1. The vernalization period depends on the genotype(s) used. Long-day genotypes require up to 5 months of vernalization, whereas shorter periods are required for short-day and intermediate-day genotypes.

Fig. 4 Fecund DH onion plants and evaluation of their offspring. (**a**) A greenhouse-grown DH plant producing seeds. (**b**) A close picture of an umbel with maturing seeds. (**c**) Cleaned seeds produced from a DH plant. (**d**) Seedlings of a DH line. (**e**) Onion plants grown from the seeds of a DH onion line. (**f**) Some bulbed plants of a DH onion line

2. Growth conditions of the donor plants can influence the gynogenic plantlet yield in onion. In general, bulbs of donor onion genotypes are produced under field conditions. Bulbs selected for DH plant production are sprouted in growth chamber and greenhouse with control. Fertilize the plants with liquid fertilizers (NPK 20 20 20) and micronutrients. Plants with scapes must be protected against temperatures below 5 °C and above 30 °C.

3. Record the morphological appearances of the donor plants at the time of umbel collection.

4. Bud size is a subjective parameter, since same size flower buds of the same umbel may have different phenological stages. Researchers are advised to modify bud selection procedure to suit the genotypes used or changes in donor growth conditions. I suggest the use of buds several days prior to anthesis stage.

5. The optimal stage of the umbel suitable for gynogenesis work is when it contains 40–50% of flower buds at the right stage for induction.

6. Place a drop of Alexander staining solution on a microscope slide. Remove anthers from several flowers of an umbel using forceps, squeeze them in staining solution to release the pollen and cover it with a coverslip. After several hours of incubation at room temperature, inspect the pollen under the light microscope. If pollen is stained red or pink, they are viable. If pollen is green. They are nonviable. To obtain selfed seeds from DH plants, they must be fully fertile. Cytoplasmic male sterile (CMS) plants are not generally preferred in DH onion production, although they can provide gynogenic plants.

7. To determine the right stage buds, use a dissecting microscope to take measures of the flower buds and image them. Follow the buds cultured in gynogenesis induction cultures. Choose the buds reaching anthesis within 3 days in culture.

8. Excise flower buds carefully. Damaged buds do not respond to gynogenesis induction.

9. One of the major problems in whole flower bud culture is microbial contamination. This may occur as a result of microbes residing inside the bud or carried by thrips (*Thrips tabaci*). Antibiotics can be used against microbial contaminations in cultures when there is no other choice. Supplementation of induction medium with 300 mg/L timentin may help reducing the development of bacterial growth in and around the flower bud. Do not use antibiotics in culture since it may interfere with the development of gynogenic embryos. Some of the contaminations occurring in tissue culture can be prevented by keeping the flower bud donor plants clean from diseases and pests in the greenhouse.

10. The majority of plantlets emerge between the third and sixth months after culture initiation. Although very rare, emergence of plantlets may continue up to a year. Normal regenerants emerge from responsive flower buds as if they were precociously germinating (viviparous) seedlings. As they emerge, they come with single greenish coleoptiles and roots. Abnormal regenerants may have many different morphologies, all of which fail to develop to plants.

11. Emerging embryos must be transferred to EM as soon as they are detected, as extended maintenance on gynogenesis induction medium causes abnormal root development and senescence.

12. Antimitotic agents are cytotoxic chemicals that may cause tissue death on treated explants. Dying tissues must be cleaned away when they inhibit growth of regenerating roots and shoots.

13. When a plant (bulb) comes to maturity, it stops producing new leaves, the pseudostem gives way to lodging, leaves cannot stand straight and start drying. The bulbs have to be harvested, dried, and stored in a cold room.

Acknowledgments

This research was supported by grants from The Scientific and Technological Research Council of Turkey (TUBITAK-TOVAG, Project Nos. 1100095 and 1170386). Special thanks to members of PAU BIYOM (Pamukkale University, Denizli, Turkey) and Earle Lab (Cornell University, Ithaca, NY, USA).

References

1. Brewster JL (2008) Onions and other vegetable alliums, 2nd edn. CAB International, Wallingford, UK

2. Hanelt P (1990) Taxonomy, evolution and history. In: Rabinowitch HD, Brewster JL (eds) Onions and allied crops. botany, physiology, and genetics, vol I. CRC Press, Boca Raton, Florida, pp 1–26

3. Fritsch RM, Friesen N (2002) Evolution, domestication and taxonomy. In: Rabinowitch HD, Currah L (eds) Allium crop science: recent advances. CABI Publishing, New York, pp 5–30

4. FAO. (2017). FAOSTAT. Food and Agriculture Organization of the United Nations FAO Statistical Database. http://faostat.fao.org/

5. Bijl JR (1994) *Allium*-flowering onions. Herbertia 50:88–94

6. Kamenetsky R, Frisch RM (2002) Ornamental Alliums. In: Rabinowitch HD, Brewster JL (eds) Allium crop science: recent advances. CABI Publishing, New York, pp 459–491

7. Huber LS, Hoffmann-Ribani R, Rodriguez-Amaya DB (2009) Quantitative variation in Brazilian vegetable sources of flavonols and flavones. Food Chem 113:1278–1282

8. Zill-E-Huma, Vian MA, Maingonnat JF, Chemat F (2009) Clean recovery of antioxidant flavonoids from onions: Optimising solvent free microwave extraction method. J Chromatogr A 1216:7700–7707

9. Keller ERJ, Korzun L (1996) Haploidy in onion (*Allium cepa* L.) and other allium species. In: Jain SM, Sopory SK, Veilleux RE (eds) In vitro haploid production in higher plants, vol 3. Kluwer, Dordrecht, pp 51–71

10. Muren RC (1989) Haploid plant induction from unpollinated ovaries in onion. Hortic Sci 24:833–834

11. Keller J (1990) Culture of unpollinated ovules, ovaries, and fower buds in some species of the genus allium and haploid induction via gynogenesis in onion (*Allium cepa* L.). Euphytica 47:241–247

12. Campion B, Alloni C (1990) Induction of haploid plants in onion (*Allium cepa* L.) by in vitro culture of unpollinated ovules. Plant Cell Tissue Organ Cult 20:1–6

13. Dore C, Marie F (1993) Production of gynogenetic plants of onion (*Allium cepa* L.) after crossing with irradiated pollen. Plant Breed 111:142–147

14. Murashige T, Skoog F (1962) A revised medium for rapid growth and bio assays with tobacco tissue cultures. Physiol Plant 15:473–497

15. Gamborg OL, Miller RA, Ojima K (1968) Nutrient requirements of suspension cultures of soybean root cells. Exp Cell Res 50:151–158

16. Dunstan DI, Short KC (1977) Improved growth of tissue cultures of onion, *Allium cepa*. Physiol Plant 41:70–72

17. Campion B, Azzimonti MT, Vicini E, Schiavi M, Falavigna A (1992) Advances in haploid plant induction in onion (*Allium cepa* L.) through *in vitro* gynogenesis. Plant Sci 86:97–104

18. Bohanec B, Jakse M, Ihan A, Javornik B (1995) Studies of gynogenesis in onion (*Allium cepa* L.) induction procedures and genetic analysis of regenerants. Plant Sci 104:215–224

19. Geoffriau E, Kahane R, Rancillac M (1997) Variation of gynogenesis ability in onion (*Allium cepa* L.). Euphytica 94:37–44

20. Martinez LE, Aguero CB, Lopez ME, Galmarini CR (2000) Improvement of *in vitro* gynogenesis induction in onion (*Allium cepa* L.) using polyamines. Plant Sci 156:221–226

21. Michalik B, Adamus A, Nowak E (2000) Gynogenesis in polish onion cultivars. J Plant Physiol 156:211–216

22. Alan AR, Mutschler MA, Brants A, Cobb E, Earle ED (2003) Production of gynogenic plants from hybrids of *Allium cepa* L. and *A. roylei* Stearn. Plant Sci 165:1201–1211

23. Alan AR, Brants A, Cobb E, Goldschmied PA, Mutschler MA, Earle ED (2004) Fecund gynogenic lines from onion (*Allium cepa* L.) breeding materials. Plant Sci 167:1055–1066

24. Celebi-Toprak F, Kaska A, Alan V, Aykut A, Alan AR (2015) Evaluation of doubled haploid onion (*Allium cepa* L.) lines developed from Turkish landraces. In: ISEA 2015|7th International Symposium on Edible Alliaceae 21st–25th May 2015. Nigde, Turkey, p 81

25. Fayos O, Valles MP, Garces-Claver A, Mallor C, Castillo AM (2015) Doubled haploid production from Spanish onion (*Allium cepa* L.) germplasm: embryogenesis induction, plant regeneration and chromosome doubling. Front Plant Sci 6:384. https://doi.org/10.3389/fpls.2015.00384

26. Khar A, Kumar A, Islam S, Kumar A, Agarwal A (2018) Genotypic response towards haploid induction in short day tropical Indian onion (*Allium cepa* L.). Ind J Agric Sci 88:709–713

27. Alan AR, Lim W, Mutschler MA, Earle ED (2007) Complementary strategies for ploidy manipulations in gynogenic onion (*Allium cepa* L.). Plant Sci 173:25–31

28. Geoffriau E, Kahane R, Bellamy C, Rancillac M (1997) Ploidy stability and *in vitro* chromosome doubling in gynogenic clones of onion (*Allium cepa* L.). Plant Sci 122:201–208

29. Jakse M, Havey MJ, Bohanec B (2003) Chromosome doubling procedures of onion (*Allium cepa* L.) gynogenic embryos. Plant Cell Rep 21:905–910

30. Grzebelus E, Adamus A (2004) Efect of antimitotic agents on development and genome doubling of gynogenic onion (*Allium cepa* L.) embryos. Plant Sci 167:569–574

31. Khosa JS, Lee R, Brauning S, Lord J, Pither-Joyce M, McCallum J, Macknight RC (2016) Doubled haploid 'CUDH2107' as a reference for bulb onion (*Allium cepa* L.) research: development of a transcriptome catalogue and identification of transcripts associated with male fertility. PLoS One 11:e166568

32. Finkers R, Van Workum W, Van Kaauwen M, Huits H, Jungerius A, Vosman B, Scholten OE (2015) A *de novo* assembly for the 16 GB *Allium cepa* genome, tears of joy. San Diego: Plant & Animal Genome XXIII; 2015. https://pag.confex.com/pag/xxiii/webprogram/Paper17794.html

33. Hyde PT, Earle ED, Mutschler MA (2012) Doubled haploid onion (*Allium cepa* L.) lines and their impact on hybrid performance. Hortic Sci 47:1690–1695

34. Peterson R, Slovin JP, Chen C (2010) A simplified method for differential staining of aborted and non-aborted pollen grains. Int J Plant Biol 1:e13

35. Arumuganathan K, Earle ED (1991) Nuclear DNA content of some important plant species. Plant Mol Biol Report 9(3):211–215

36. Dolezel J, Bartos J (2005) Plant DNA flow cytometry and estimation of genome size. Ann Bot 95:99–110

Chapter 7

In Vitro Gynogenesis in Leek (*Allium ampeloprasum* L.)

Fevziye Celebi-Toprak and Ali Ramazan Alan

Abstract

Leek (*A. ampeloprasum* L.) is an economically important vegetable crop from *Alliaceae* family. It is a non–bulb forming biennial species grown for its pseudostem and leaves. Leek is a tetraploid with one of the largest genomes known among cultivated plant species. It has enormous economic importance all around the world for many purposes such as vegetable, medicinal herb, and food seasoning. Production and consumption of leek is in rise all around the world and breeders are trying to develop new F1 hybrid varieties with desired agronomical traits. Although self-compatible, leek shows high tendency toward outcrossing and display severe inbreeding depression when selfed with its own pollen. Therefore, inbred development through classical breeding techniques is very difficult in this crop. Traditional leek genotypes are highly heterozygous, open pollinated varieties. There is a high demand for F1 hybrid varieties with resistance to biotic and abiotic stresses and high-quality plants. Our group is trying to incorporate gynogenesis-based doubled haploid technology to leek improvement programs. Over the years, many experiments were carried out to determine the gynogenic potential of donor leek genotypes of different genetic backgrounds in different induction media. Here, we report a protocol allowing production of green gynogenic leek plants via single step culture of unopened flower buds. Ploidy levels of gynogenic regenerants are determined by flow cytometry analysis. A majority of the gynogenic leek regenerants produced survived well in vivo.

Key words *Allium ampeloprasum L.*, Breeding, Diploid, Gynogenesis, Leek, Ploidy

1 Introduction

Leek (*Allium ampeloprasum* L.) is one of the economically important crop species from *Alliaceae*. Commercial leek types are generally tetraploid ($2n = 4x = 32$) [1]. The nuclear DNA content of tetraploid leek is 59.74 ± 2.61 pg DNA/2C. It has 1C genome size of 29,213 Mbp [2]. This biennial species has a high tendency toward outcrossing [3]. It is a day length insensitive (day-neutral) crop, which allows for its production in different parts of the world [4].

Leek is one of the oldest crop species cultivated and consumed by humans since ancient times [5]. Wild forms of *A. ampeloprasum* are found in the Mediterranean area [6, 7]. Leek is thought to

Jose M. Seguí-Simarro (ed.), *Doubled Haploid Technology: Volume 1: General Topics, Alliaceae, Cereals*,
Methods in Molecular Biology, vol. 2287, https://doi.org/10.1007/978-1-0716-1315-3_7,
© Springer Science+Business Media, LLC, part of Springer Nature 2021

originate from the eastern Mediterranean region [8, 9]. Today, cultivated leek spread all over the world. According to FAOSTAT [10] world leek production is ~2 million tons. Indonesia is the world's leading leek-producing country. About one fourth of world leek production comes from this Southeast Asian country. Turkey and Belgium are leading leek production in Europe with about 200 tons of leek production each. Leek is mainly marketed as a fresh vegetable, while 10% of marketed leek is processed. It is a nutritious vegetable. Leek pseudostems and leaves are consumed as cooked vegetable, fresh salad, medicinal herb, and food seasoning. Leek is also valued for its antimicrobial, anticancer, cardioprotective, cholesterol-lowering, and antioxidant activities [11–18].

Leek is a seed-propagated crop. Due to its long cultivation period, leek seedlings are used as transplants in large-scale field productions. Most of the varieties used in leek production worldwide are open pollinated (OP) standard varieties, whereas production with hybrid varieties is becoming more common in Western Europe and North America. Leek genotypes can be divided into four groups based on season of maturity as summer, autumn, autumn/winter and winter types. Summer and autumn leek types are generally produced in areas with mild winter climates, whereas winter types are preferred in the regions with hard winters. Summer and winter leek types have distinct features. Summer leek types generally produce light green leaves and white pseudostems with soft texture. They may produce pseudostems reaching up to 80 cm. However, their lives are short and they are sensitive to cold (below -8 °C). Winter leek types produce large leaves with darker green color. Their pseudostems are shorter and thicker than summer types and they can endure cold temperatures as low as -18 °C.

Breeding new leek varieties through classical methods is a very difficult and time-consuming process because of its biennial nature, polyploidy, inbreeding depression, tendency for outcrossing, and high level of heterozygosity [3, 19–22]. Severe inbreeding depression is the major obstacle for the development of high yielding leek varieties with a uniform crop [3, 23]. Therefore, OP leek varieties have been developed with mass and/or family selections. However, OP leek varieties often suffer from low yield, poor quality, and nonuniform crop. The first commercial leek F1 hybrid varieties were developed and used in Europe in early 1990s. F1 hybrid leek varieties generally provide higher yield and crop uniformity than OP varieties [20, 22, 24, 25]. Commercial hybrid leek cultivars are produced mainly by private seeds companies [26, 27]. Desired agronomical traits in leek varieties include winter hardiness, non-bulbing, longer and softer pseudostems, resistance to pests and diseases, resistance to senescence, green leaf color, time of maturity, high crop yield, and crop uniformity at harvest [28]. The majority of these desirable traits are governed by multiple genes and breeding new varieties fixed for these agronomical traits require novel approaches such as haploid embryogenesis. Gynogenesis induction is the major technique used in the production of diploid plants from tetraploid donor leek lines. Gynogenic leek plants were obtained by culturing ovules, ovaries, and immature flower bud explants in gynogenesis induction media [2, 29–33]. Response to gynogenesis induction, regeneration of gynogenic plants and polyploidy levels of regenerants are influenced by the donor genotype [2, 33]. Several research groups suggested that production of diploid leek lines may allow selection against recessive alleles with undesired effects [33]. Accumulation of favorable alleles in diploid plants may lead to production of new lines with superior agronomical traits. Gynogenesis can help accelerate the inbred development processes in leek breeding programs. Partially or fully homozygous leek lines obtained via gynogenesis can provide significant levels of heterosis in hybrid production.

We suggest the use of fecund gynogenic leek lines as male parents to improve crop uniformity in F1 hybrid varieties. Detailed experiments were carried out to optimize the production of gynogenic leek lines from different genetic backgrounds. We were able to obtain green diploid leek plants from the majority of donor leek lines and grow them in the greenhouse for characterization. Dihaploid leek plants generally show reduced growth and fail to produce viable selfed seeds. On the other hand, tetraploid plants originating from dihaploid gynogenic lines via spontaneous doubling seem to provide viable selfed seeds.

The protocol presented here allows for the production of green gynogenic leek plants via single step culture of unopened flower buds from all leek donor genotypes. Ploidy levels of gynogenic regenerants are determined with flow cytometry analysis. A majority of gynogenic leek regenerants produced by this method survive after their transference to in vivo conditions and their growth in the greenhouse for characterization.

2 Materials

2.1 Plant Material

This protocol can be used to produce gynogenic leek plants from all commercial leek varieties we have tested.

2.2 Equipment

1. Autoclave.
2. Calipers.
3. Desktop centrifuge.
4. Digital timer.
5. Dissecting microscope with digital camera.
6. Flow cytometer (Beckman Coulter Cell Lab QUANTA).
7. Greenhouse set for 25 °C during the day and 15 °C at night.

8. Growth chamber at 17 °C with a 16/8 h photoperiod.

9. High- precision digital balance.

10. Laboratory ice maker.

11. Laminar flow hood.

12. Light microscope with digital camera.

13. Magnetic hot plate stirrer and mixing fish.

14. pH meter.

15. Refrigerator with freezer.

16. Sterilizator.

17. Tissue culture room adjusted for constant 25 °C with a 16/8 h photoperiod.

18. Ultrapure water purification system.

19. Vortex.

2.3 Laboratory Materials

1. Aluminum foil.

2. Autoclavable glass bottles.

3. Beakers.

4. Eppendorf tubes 1.5 mL.

5. Falcon tubes 15 mL.

6. Falcon tubes 50 mL.

7. Filter papers.

8. Flow Check beads (Beckman-Coulter).

9. Forceps.

10. Glass culture tubes 50 mL.

11. Graduated cylinder.

12. Magenta boxes.

13. Microscope slides and coverslips.

14. Nylon mesh with 160μm pore size.

15. Parafilm.

16. Permanent marker.

17. Razor blades.

18. Scalpels.

19. Sharp blunt surgical scissors.

20. Spatulas.

21. Stainless steel tea strainers.

22. Sterile filter paper.

23. Sterile Pasteur or automatic pipettes.

24. Sterile petri plates 60 × 15 mm (diameter × height).

25. Sterile petri plates 90 × 15 mm (diameter × height).

26. Sterile pipette tips 100μL.

27. Sterile pipette tips 1000μL.

28. Test tube racks.

29. Trays.

30. Vi-CELL 4 mL sample vials (for use in flow cytometry analysis).

31. Weight boats.

2.4 Greenhouse/ Growth Chamber Materials

1. Autoclavable bags.

2. Labels.

3. Peat and perlite mixture (2:1 v/v).

4. Plant fertilizer NPK 20 20 20 + TE.

5. Plastic agricultural trays—104 cavities.

6. Plastic agricultural trays—24 cavities.

7. Plastic labels 12 cm.

8. Plastic pots 0.5 L.

9. Plastic pots 7 L.

10. Plastic sandwich bags (20 × 30 cm).

11. Plastic tags.

12. Rubber bands.

2.5 Solutions

1. 70% ethanol (v/v).

2. Alexander stain, as modified by Peterson et al. [34]: to prepare 50 mL stain, add the following constituents in order and store the stain in a light-protected bottle. Add 5 mL 95% ethanol, 0.5 mL malachite green (1% solution in 95% ethanol, 25 mL double distilled (dd) water, 12.5 mL glycerol, 2.5 mL acid fuchsin (1% solution in dd water), 0.25 mL Orange G (1% solution in dd water), 2 mL glacial acetic acid, and ~2 mL dd water to a total of 50 mL.

3. Autoclaved dd water.

4. Disinfection solution: 20% commercial bleach (1.12% sodium hypochlorite) and 0.1% (v/v) Tween 20 (wetting agent).

5. Propidium iodide (PI) stock solution: dissolve 100 mg PI (fluorescent dye) in 10 mL dd water in a 15 mL Falcon tube to prepare a 10 mg/mL PI stock solution. Cover the PI stock with aluminum foil. Keep PI stock in the refrigerator (4 °C) until use.

6. Sheath fluid for the flow cytometry system.

7. Nuclei Isolation Buffer (NIB): 15 mM HEPES, 1 mM Na$_2$EDTA, 80 mM KCl, 20 mM NaCl, 300 mM sucrose, 0.2% (v/v) Triton X-100, 0.5 mM spermine, 1% (w/v) polyvinylpyrrolidone (PVP40). Adjust pH to 7.5. Fill 15 mL or 50 mL tubes with NIB and keep them in the freezer (−20 °C) until use.

8. RNase (sold at 4 mg/mL concentration).

9. Shutdown solution for nuclear packing efficiency (NPE) analysis in flow cytometry.

2.6 Culture Media

1. Gynogenesis induction medium: BDS (Table 1) supplemented with 2 mg/L 2,4-dichlorophenoxyacetic acid (2,4-D) and 2 mg/L 6-benzylaminopurine (BAP). Adjust pH to 6 Add 7.4 g/L agar for microbiological use. Autoclave media at 121 °C for 20 min, let it cool down until it can be hold with the hands (~60 °C), and then pour ~25 mL medium into each 90 × 15 mm sterile petri plate and let them solidify in the sterile hood.

2. Elongation medium (EM): BDS medium with half strength major and minor salts and 30 g/L sucrose (Table 1). Prepare as described for the gynogenesis induction medium. Pour ~25 mL EM into 90 × 15 mm plates and 15 mL into culture tubes and let them solidify in sterile hood.

3   Methods

3.1 Donor Plant Growth Conditions

1. In the first year, grow the plants of donor leek genotypes by transplanting the seedling in late spring.

2. Harvest the healthy and fully developed plants in late fall. Clean the plants by washing their roots and dying leaves.

3. Place the cleaned leek plants in plastic bags and vernalize them at 4 °C in a cold room (*see* **Note 1**).

4. After vernalization, transplant the leek plants to 7 L pots filled with peat and perlite mixture, place them in the greenhouse and control the conditions to keep the plants at 25 °C/16 h light and 17 °C/8 h dark.

5. Grow the donor leek plants under the optimal conditions necessary for vegetative and generative development (*see* **Note 2**).

6. Observe the plants of donor leek genotypes for growth and formation of flower scapes (*see* **Note 3**).

7. If necessary, stake the plants to keep leaves and flower scapes straight.

8. Watch the donor plants closely when the umbel capsules start to break (*see* **Note 4**).

Table 1
Composition of the media used in whole flower bud culture–based gynogenesis induction and regeneration of leek plants. BDS and EM are modified from Dunstan and Short [16]

	BDS (mg/L)	EM (mg/L)
Macro elements		
$CaCl_2 \cdot 2H_2O$	1500.00	750.00
KNO_3	25300.00	12650.00
$NH_4(NO_3)$	3202.00	1601.00
$MgSO_4 \cdot 7H_2O$	2470.00	1235
KH_2PO_4	–	–
$(NH_4)_2SO_4$	1340.00	670.00
$NaH_2PO_4 \cdot H_2O$	1690.00	845.00
$NH_4H_2PO_4$	2300.00	1150
Micro elements		
$CoCl_2 \cdot 6H_2O$	2.50	1.25
$CuSO_4 \cdot 5H_2O$	3.90	1.95
H_3BO_3	300.00	150
KI	75.00	37.50
$MnSO_4 \cdot H_2O$	925.00	462.50
$Na_2MoO_4 \cdot 2H_2O$	25.00	12.50
$ZnSO_4 \cdot 7H_2O$	200.00	100
Iron source		
$EDTANa_2 \cdot 2H_2O$	37.25	37.25
$FeSO_4 \cdot 7H_2O$	27.85	27.85
Vitamins		
Myoinositol	500.00	500.00
Nicotinic acid	1.00	1.00
Pyridoxine hydrochloride	1.00	1.00
Thiamine hydrochloride	10.00	10.00
Amino acids		
Glycine	–	–
L-proline	200.00	200.00
Plant growth regulators		
2,4-Dichlorophenoxyacetic acid (2,4-D)	2.00	–
6-Benzylaminopurine (BAP)	2.00	–

(continued)

Table 1
(continued)

	BDS (mg/L)	EM (mg/L)
Carbohydrate		
Sucrose	100,000	30,000
Solidifying agent		
Agar (for microbiology)	7400	7400
pH	6.00	6.00

3.2 Establishment of Gynogenesis Induction Cultures and Obtaining Gynogenic Plants

1. Collect the scapes when 10–15% of the flowers in the umbels are at anthesis stage (Fig. 1a; *see* **Note 5**).

2. Label the scapes to track their donor origin, place them in a beaker filled with sterile water, and bring them to the lab.

3. Perform a pollen viability test with the pollen collected from the anthers of several flowers of each donor plant at the anthesis stage (*see* **Note 6**).

4. Observe the flower buds of the umbels carefully and choose those at the right stage for gynogenesis induction (*see* **Note 7**).

5. Excise the preferred unopened flower buds without pedicels using sterile scissors (Fig. 1b).

6. For surface disinfection, place the flower buds into metal tea filters and dip them into a Magenta box filled with 70% ethanol for 30 s and then into another Magenta box filled with disinfection solution for 45 min.

7. In the laminar flow hood, pour off the disinfection solution and then wash the flower buds three times for 5 min each with autoclaved dd water.

8. After disinfection, spread the buds on a dry sterile filter paper to eliminate the excessive water from flower bud surface.

9. Using sterile forceps, culture flower buds in plates with induction medium by dipping their bottom halves into the medium.

10. Place 30 flower buds in each plate, label with marker, seal with Parafilm, and transfer to the tissue culture room (*see* **Note 8**).

11. Observe the cultures on a weekly basis. Cultured flower buds generally reach anthesis within a week and reach the maximum size within 2–3 weeks (Fig. 1c, d).

12. Check the cultures for microbial contamination (*see* **Note 9**).

13. Gynogenic leek plantlets start to emerge from inside the buds about 10 weeks after culture initiation (Fig. 1e).

Fig. 1 Production of gynogenic leek plants. (**a**) A leek umbel with a large number of immature flower buds suitable for in vitro gynogenesis induction. (**b**) An immature leek flower bud used for gynogenesis induction. (**c**) An anthesis-stage leek flower in induction culture. (**d**) A leek flower bud with an enlarged ovary about 2 weeks after culture initiation. (**e**) A responsive leek flower bud with a gynogenic plantlet

14. Transfer responsive buds to plates with elongation medium using sterile forceps, record the transfer date, seal with Parafilm, and place them back in the tissue culture room.

15. Observe the responsive buds to determine the gynogenic regenerants converted to plants (Fig. 2a; *see* **Note 10**).

16. Transfer well-formed plants to culture tubes with elongation medium and place them back to the same culture room.

17. When gynogenic plants reach the 3–4 leaf stage with established root systems (Fig. 2b), determine their ploidy (*see* **Note 11**).

3.3 Determination of Ploidy of Androgenic Regenerants with Flow Cytometry Analysis

1. Collect fresh leaf tissues from leek (~50 mg) and *Hordeum vulgare* (~15 mg) plants, put them in 60 × 15 mm Petri plates placed in an ice box, and coprocess for the isolation of nuclei. Donor leek cells contain ~60 pg DNA/2C [2]. *H. vulgare* cells contain 10.1 pg DNA/2C [35].

2. Thaw two tubes of NIB on the lab bench and place them in an ice box. Use regular NIB for isolation of nuclei from tissues by chopping. Prepare NIB with 25μg/mL RNase to dissolve nuclei and eliminate RNA from the samples.

Fig. 2 Gynogenic leek plants. (**a**) Emergence of gynogenic plantlets from cultured flower buds about three and a half months in induction plate. (**b**) Gynogenic leek plantlets in culture tubes. (**c**) Gynogenic leek plants acclimatized to in vivo conditions in a growth chamber

3. Add 1.5 mL of ice-cold NIB into the sample plate and gently chop the leaves using a razor blade on ice.

4. Filter the finely chopped tissue though a nylon mesh with 160μm pore size into an Eppendorf tube placed in ice.

5. Centrifuge the samples at $8000 \times g$ for 5 s.

6. Take the samples out of centrifuge and pour off the supernatant carefully.

7. Place the tubes with nuclei pellets upside down on a paper towel for several min to drain the remaining supernatant.

8. Resuspend the nuclei pellet in 300μL NIB with 25μg/mL RNase and vortex the samples.

9. Add 10μL of the propidium iodide solution into each sample and vortex the samples.

10. Keep the tubes of nuclei sample in ice until the start of the flow cytometry analysis.

11. Start the flow cytometry instrument and check it for linearity with fluorescent flow check beads.

12. Adjust the amplification to position the 2C peak of diploid leek nuclei approximately at channel 400 and analyze a minimum of 3000 nuclei from each sample to determine DNA amounts and ploidy levels.

13. The nuclear DNA content (2C) of leek is calculated according to the formula adapted from Dolezel and Bartos [36]:

$$\text{Gynogenic leek 2C DNA (pg)} = \left(\frac{\text{leek sample 2C peak mean}}{H.\textit{vulgare} \text{ 2C peak mean}} \right)$$
$$\times \text{2C reference nuclear DNA (pg)}$$

14. Gynogenic regenerants with half of the nuclear DNA content of tetraploid leek plants (~30 pg DNA/2C) are classified as dihaploids. Regenerants with nuclear DNA contents identical to donor leek plants (~60 pg DNA/2C) are considered as tetraploids.

3.4 Acclimatization to Ex Vitro Conditions

1. Remove the well-rooted plants from their containers and wash off the remnants of culture medium by carefully washing the roots with running tap water.

2. Transplant the leek plants to 0.5 L pots filled with autoclaved peat and perlite mix, irrigate with sterile water and cover the pots with plastic sandwich bags and fix with a rubber band.

3. Place the potted plants on trays and transfer to a growth chamber set for constant 17 °C with a 18 h photoperiod.

4. After about 10 days, start making small openings on the plastic cover to help acclimatization of the plants.

5. Remove the plastic cover completely by the end of the third week, when plants are acclimatized.

6. Keep the plants in the chamber for another 2 weeks before moving them to a greenhouse (Fig. 2c).

7. Transfer the gynogenic plants to a greenhouse to grow them.

8. Grow the gynogenic leek plants in a climate-controlled greenhouse to obtain flowers and determine their fecundity levels as suggested for DH onions [37].

4 Notes

1. Six weeks of vernalization is sufficient for summer and fall type leek genotypes if plants are grown under Mediterranean climate. Plants harvested after the fall frosts generally do not require vernalization. Flowering of leek plants is a slow process.

2. Growth conditions of the donor plants may influence the gynogenic plantlet yield in leek. In general, plants of donor genotypes are produced under field conditions in the first year. In the second year, donor leek plants are grown in the greenhouse for their umbels. Fertilize the plants with liquid fertilizers (NPK 20 20 20) and micronutrients. Keep the plants free from pests and pathogens. Spray them with necessary chemicals to make sure they are healthy. Plants with scapes must be protected against temperatures below 5 °C and above 30 °C. Gametes start to die off if flowering plants are exposed to high temperatures for extended periods of time.

3. Record the morphological appearances of the donor plants at the time of umbel collection. Make sure the scapes and umbels are not diseased or damaged.

4. Bud size is a subjective parameter, since same sized flower buds of the same umbel may be at different phenological stages. We use immature buds >3 mm in length for gynogenic plant production. Researchers are advised to modify the bud selection procedure to suit the genotypes used or changes in donor growth conditions. We suggest the use of buds two to 5 days prior to anthesis stage.

5. The optimal stage of the umbel suitable for gynogenesis work is when it contains 30–40% of flower buds at the right stage for induction. If the plant material is highly valuable, harvest only buds at the right stage for culture and let the smaller buds to grow and use them later.

6. Place a drop of Alexander staining solution on a microscope slide. Remove anthers from several flowers of an umbel using forceps, squeeze them in staining solution to release the pollen and cover it with a coverslip. After several hours of incubation at room temperature, inspect the pollen under the light microscope. If pollen is stained red or pink, it is viable. If pollen is green, it is nonviable. To obtain selfed seeds from gynogenic plants, they must be fully fertile. Generally, male sterile plants are not preferred for gynogenic leek production, although they can produce gynogenic plants.

7. To determine the right stage buds, use a dissecting microscope to take measures of the flower buds and image them. Follow the buds cultured in gynogenesis induction cultures. Choose the buds reaching anthesis within 3 days in culture.

8. Excise flower buds carefully. Damaged buds do not respond to gynogenesis induction.

9. One of the major problems in whole flower bud culture is microbial contamination. This may occur as a result of microbes residing inside the bud or carried by thrips (*Thrips tabaci*). Antibiotics can be used against microbial

contaminations in cultures when there is no other choice. Supplementation of induction medium with timentin (300 mg/L) may help reducing the development of bacterial growth in and around the flower bud. Try not to use antibiotics in embryo cultures since it may interfere with the development of gynogenic embryos. Some of the contaminations occurring in tissue culture can be prevented by keeping the flower bud donor plants clean from diseases and pests in the greenhouse.

10. The majority of plantlets emerge between the third and fifth months after culture initiation. Although very rare, emergence of plantlets may continue up to a year. Normal regenerants emerge from responsive flower buds as if they were precociously germinating (viviparous) seedlings. As they emerge, they come with single greenish coleoptiles and roots. Abnormal regenerants may have many different morphologies, all of which fail to develop to plants.

11. Emerging embryos must be transferred to elongation medium plates as soon as they are detected, as extended maintenance on gynogenesis induction medium may cause abnormal root development and senescence.

Acknowledgments

This research was supported by grant from The Scientific and Technological Research Council of Turkey (TUBITAK-TOVAG, Project No. 113O232). Special thanks to members of PAU BIYOM (Pamukkale University, Denizli, Turkey).

References

1. van der Meer QP, Hanelt P (1990) Leek (*Allium ampeloprasum* var. *porrum*). In: Brewster JL, Rabinowitch HD (eds) Onions and allied crops, Vol. III. Biochemistry, food science, and minor crops. CRC Press, Boca Raton, FL, pp 179–196

2. Alan AR, Celebi-Toprak F, Kaska A (2016) Production and evaluation of gynogenic leek (*Allium ampeloprasum* L.) plants. Plant Cell Tissue Organ Cult 125:249–259

3. Berninger E, Buret P (1967) Etude des deficients chlorophylliens chez deux especes cultivees du genre *Allium*: l'oignon *A. cepa* L. et le *A. porrum*. Ann Amel Plant 17:175–194

4. De Clercq H, Van Bockstaele E (2002) Leek: advances in agronomy and breeding. In: Rabinowitch HD, Currah L (eds) Allium crop science: recent advances. CABI Publishing, Wallingford, UK, pp 431–438

5. Block E (2010) Garlic and other *Alliums*: the lore and the science. Royal Society of Chemistry, Cambridge

6. De Wilde-Duyfjes B (1976) A revision of the genus *Allium* L. *(Liliaceae)* in Africa. Meded Landbouwhogeschool Wageningen, Netherlands, pp 76–311

7. Stearn WT (1978) European species of *Allium* and allied genera of Alliaceae: a synonymic enumeration. Annal Musei Goulandris 4:83–198

8. Jones HA, Mann LK (1963) Onion and their allies: botany, cultivation and utilization. Leonard Hill, London

9. Vavilov NI (1926) Studies on the origins of cultivated plants. Bull Appl Bot Plant Breed 16:1–245

10. FAO (2019) Food and Agriculture Organization of the United Nations. http://www.fao.

org/faostat/en/#data/QC.   Accessed
10 August 2019

11. Alfonso C, Ernesto F, Virginia L, Silvana M
(1998) Porric acids a−C new antifungal diben-
zofurans from the bulbs of *Allium porrum*
L. Eur J Org Chem 4:661–663

12. Ayumi U, Jun O, Hitomi K et al (2009)
Mechanisms of sulfide components expression
and structural determination of substrate pre-
cursor in jumbo leek (*Allium ampeloprasum*
L.). Nippon Shokuhin Kagaku Kogaku Kaishi
56(5):280–285

13. Ben Arfa A, Najjaa H, Yahia B et al (2015)
Antioxidant capacity and phenolic composition
as a function of genetic diversity of wild Tuni-
sian leek (*Allium ampeloprasum* L.). Acad J
Biotech 3:15–26

14. Griffiths G, Trueman L, Crowther T et al
(2002) Onions as global benefit to health.
Phytother Res 16:603–615

15. Hertog MGL, Hollman PCH, Katan MB
(1992) Content of potentially anticarcinogenic
flavonoids of 28 vegetables and 9 fruits commonly consumed in the Netherlands. J Agric
Food Chem 40:2379–2383

16. Ozgur M, Akpinar-Bayaziy A, Ozcan T, Afola-
yan AJ (2011) Effect of dehydration on several
physic-chemical properties and the antioxidant
activity of leeks (*Allium porrum* L.). Not Bot
Hort Agrobot Cluj-Napoca 39:144–151

17. Radovanović B, Mladenović J, Radovanović A
et al (2015) Phenolic composition, antioxidant, antimicrobial and cytotoxic activities of
Allium porrum L. (Serbia) extracts. J Food
Nutr Res 3:564–569

18. Shon MY, Choi SD, Kahng GG et al (2004)
Antimutagenic, antioxidant and free radical
scavenging activity of ethyl acetate extracts
from white, yellow and red onions. Food
Chem Toxicol 42:659–666

19. De Clercq H, Peusens D, Roldán-Ruiz I, Van
Bockstaele E (2003) Causal relationships
between inbreeding, seed characteristics and
plant performance in leek (*Allium porrum*
L.). Euphytica 134:103–115

20. Schweisguth B (1970) Etudes preliminiares a
lamelioration du poireau *A porrum*
L. Proposition dune methode demaliration.
Ann Amel Plant 20:215–231

21. Silvertand B (1996) Induction, maintenance
and utilization of male sterility in leek (*Allium
ampeloprasum* L.) PhD thesis, Department of
Plant Breeding, Agricultural, Wageningen,
Netherlands

22. Smith BM, Crowther TC (1995) Inbreeding
depression and single cross hybrids in leek
(*Allium ampeloprasum* ssp. *porrum*). Euphytica 86:87–94

23. Gray D, Steckel JRA (1986) Self-and open-
pollination as factors influencing seed quality
in leek (*Allium porrum*). Ann Appl Biol
108:167–170

24. Gagnebin F, Bonnet JC (1979) Quelques considerations sur la culture et l'amelioration du
poireau. Revue Hort Suisse 52:112–116

25. Kampe R (1980) Untersuchungen zum aus-
mass von hybrideffekten bei porree. Archiv
Züchtungsforschung 10:123–131

26. Pink DAC (1993) Leek (*Allium ampeloprasum*
L.). In: Kalloo G, Bergh BO (eds) Genetic
improvement of vegetable crops. Pergamon
Press, Oxford, New York, pp 29–34

27. Silvertand B, Jacobsen E, Mazereeuw J et al
(1995) Efficient in vitro regeneration of leek
(*Allium ampeloprasum* L.) via flower stalk segments. Plant Cell Rep 14:423–427

28. De Clercq H, Baert J, Van Bockstaele E (1999)
Breeding potential of Belgian landraces of leek
(*Allium ampeloprasum* L. var *porrum*). Euphy-
tica 106:101–109

29. Alan AR, Kaska A, Celebi-Toprak F (2013)
Edible *Allium* improvement via doubled haploidy technology. Curr Opin Biotech 24:42

30. Celebi-Toprak F, Alan AR (2016) Optimization of gynogenesis induction in leek (*Allium
ampeloprasum* var. *porrum*). J Biotech 230:
S30.   https://doi.org/10.1016/j.jbiotec.
2016.05.126

31. Kaska A, Celebi-Toprak F, Alan AR (2012)
Gynogenesis induction in edible *Allium*s. J
Biotech 161:18

32. Kaska A, Celebi Toprak F, Alan AR (2013)
Gynogenesis induction in leek (*Allium ampe-
loprasum* L.) breeding materials. Curr Opin
Biotech 24:42

33. Schum A, Mattiesch L, Timmann EM, Hof-
mann K (1993) Regeneration of dihaploids
via   gynogenesis   in   *Allium-Porrum*
L. Gartenbauwissenschaft 58(5):227–232

34. Peterson R, Slovin JP, Chen C (2010) A simplified method for differential staining of
aborted and non-aborted pollen grains. Int J
Plant Biology 1:e13

35. Arumuganathan K, Earle ED (1991) Nuclear
DNA content of some important plant species.
Plant Mol Biol Report 9(3):211–215

36. Dolezel J, Bartos J (2005) Plant DNA flow
cytometry and estimation of genome size.
Ann Bot 95:99–110

37. Alan AR, Brants A, Cobb E, Goldschmied PA,
Mutschler MA, Earle ED (2004) Fecund gynogenic lines from onion (*Allium cepa* L.) breeding materials. Plant Sci 167:1055–1066

Part III

Doubled Haploids in Cereals

Chapter 8

Barley Isolated Microspore Culture

Luís Cistué and Begoña Echávarri

Abstract

The production of doubled haploids (DHs) has proved to be a highly valuable tool to obtain new cultivars. Among the cereals, barley (*Hordeum vulgare* L.) is the most successful species in large-scale haploid production. Techniques employed for this purpose are based on either the gynogenetic or the androgenetic pathway. Interspecific cross with *Hordeum bulbosum* L., haploid gene inducer (the *hap* gene), ovary culture, anther culture (AC), and isolated microspore culture (IMC) are the most used methods. Among all of them, IMC is regarded as a particularly effective system owing to the great increase in green plant numbers per spike and also the higher induction of chromosome doubling when compared with other methods. Thus, IMC provides the best way to mass scale production of new varieties.

Key words Androgenesis, Barley, Doubled haploid, Embryo, Microspore

1 Introduction

Doubled haploids have a great potential value in plant breeding programs. Among the several existing methods to produce DHs in barley, the isolated microspore culture (IMC) method is considered the most efficient one [1–3].

Initially, the low response rate for some genotypes, the frequency of albino plants, and the low embryo regeneration percentage limited the use of this method as a practical tool. Over time, IMC has been improved and nowadays, it has become the most widely used technique for haploid production. Several research groups made an effort to enhance the efficiency of this technique.

Some improvements developed by researchers are related to the pretreatment stress. The stress is applied to switch microspores from a gametophyte to a sporophytic pathway. In barley, various stresses have been extensively employed. Cold shock at 4 °C for 28 days has been effectively used for numerous researchers [4]. Osmotic and starvation stress are also frequently used. Anther pretreatment for 4 days on a liquid culture medium containing 0.3 M mannitol prior to transfer to induction medium was effective

Jose M. Seguí-Simarro (ed.), *Doubled Haploid Technology: Volume 1: General Topics, Alliaceae, Cereals*,
Methods in Molecular Biology, vol. 2287, https://doi.org/10.1007/978-1-0716-1315-3_8,
© Springer Science+Business Media, LLC, part of Springer Nature 2021

in promoting microspore division and regeneration of green plant-lets [5]. Cistué et al. [6] found that stressing anthers by pretreat-ment with 0.7 M mannitol solidified with 8 g/l agarose (Sea Plaque, FMC) increased the number of dividing microspores and the ratio of green/albino plants. Recently, it has been shown that the addition of DMSO to the pretreatment medium increases the number of green plants regenerated mainly for recalcitrant genotypes [7].

The composition of the culture medium has undergone many variations over the years. The Hunter's induction medium (FHG) is frequently used to cultivate microspores after pretreatment. The replacement of sucrose by maltose generated notable improve-ments in the results [8].

Another modified component is the source of organic nitro-gen. Microspores are influenced by the amount and type of nitro-gen sources and for proportion inorganic/organic over total nitrogen [9]. Substitution of NO_3NH_4 for glutamine increased the number of DHs regenerated [10, 11].

In addition, Kao [12] demonstrated the efficiency of Ficoll-400 (sucrose polymer) in keeping the dividing microspores floating on the medium, avoiding anaerobic conditions. Kuhlman and Foroughi-Wehr [13] showed that a liquid medium with 20% Ficoll-400 gave the best production of green plantlets. Devaux and Kasha [14] showed that Ficoll-400 favored the production of better quality embryos that led to increased green plantlet regeneration.

Once embryos have developed, they are transferred onto regeneration medium. Manipulation of induction or regeneration medium composition, like the change of macronutrients or micro-nutrients, has proved effective for improving plant regeneration [15, 16].

Several substances are used as growth regulators. Their dose and the type of effect they produce are of crucial importance. The cytokinin benzylaminopurine (BAP) in the induction medium, and the auxin indole acetic acid (IAA) combined with BAP for regener-ation medium are commonly used. Genotypic preferences were found for growth regulator combinations [17].

Barley breeders routinely encounter many low-responding genotypes among hybrids used for DH production. However, depending on the genotype, large differences in green plant regen-eration were observed [3, 18]. Thus, while Davies [19] reports up to 1000 green plants per processed spike for some winter-type genotypes, for spring-type barleys the rate was only 10–20 green plants per spike.

The two main limitations of IMC are the genotype-dependent response in terms of capacity to regenerate plants [20], and the high frequency of albino plants that are regenerated from some genotypes [21]. Six-row barleys typically produce a high proportion of albino plants compared to two-row types [22, 23], and spring-type barleys have also been found to be less responsive than winter types [14]. Considered jointly, six-row spring-type barleys are thus the least productive in IMC.

The methods of Cistué et al. [24], Kasha et al. [25], Davies [19], and Esteves and Belzile [26] with some modifications are used in several breeding programs with success for most of the genotypes used.

2 Materials

2.1 Lab Tools

1. Glassware/beakers made up of borosilicate glass.
2. Wash plastic bottles.
3. Sterile culture tubes and culture tube racks.
4. Parafilm.
5. Forceps and scissors, sterilized by autoclaving.
6. Media filtration system.
7. Millipore membrane filters, 0.22 μm pore size.
8. Sterile glass pipettes and plastic micropipettes.
9. Sterile Petri dishes in different sizes.
10. Microscope slides and cover slips.
11. Magenta boxes.

2.2 Equipment and Facilities

1. Air flow hood.
2. Refrigerators.
3. Centrifuge.
4. Freezer.
5. Autoclave.
6. Oven for dry-sterilizing.
7. Green house.
8. Growth chamber.
9. Stereo microscopes.
10. Inverted microscope.
11. Digital cameras.
12. Imaging software.

2.3 Plant Tissue Media

The artificial culture media supply all the nutrients necessary for growth. The composition of these media has a big impact on the success of the tissue culture.

Table 1
Media composition for isolated microspore culture in barley according to Hunter, 1988) [8] and Murashige and Skoog, 1962 [27]

Medium components (mg·l^{-1})	FGH induction	FHG regeneration	Plant regeneration (MS)
Macronutrients			
KNO$_3$	1900.0	1900.0	1900.0
NH$_4$NO$_3$	165.0	165.0	1650.0
KH$_2$PO$_4$	170.0	170.0	170.0
CaCl$_2$·2H$_2$O	440.0	440.0	440.0
MgSO$_4$·7H$_2$O	370.0	370.0	370.0
Micronutrients			
MnSO$_4$·H$_2$O	16.9	16.9	16.9
ZnSO$_4$·7H$_2$O	8.6	8.6	8.6
H$_3$BO$_3$	6.2	6.2	6.2
KI	0.83	0.83	0.83
NaMoO$_4$·2H$_2$O	0.25	0.25	0.25
CuSO$_4$·5H$_2$O	0.025	0.025	0.025
CoCl$_2$·6H$_2$O	0.025	0.025	0.025
FeNa (EDTA)$_2$	40.0	40.0	37.5
Vitamins and other supplements			
Thiamine–HCl	0.4	0.4	0.1
Inositol	100.0	100	100.0
Glutamine	730.0	–	–
Piridoxine-HCl	–	–	0.5
Nicotinamide	–	–	0.5
Glycine	–	–	2.0
Growth regulators			
BAP	1	1	–
IAA	–	0.5	–
ANA	–	–	2.0
Solidifying agents			
Agarose	–	8000.0	–
Gelrite	–	–	3000.0
Viscosity modifying agent			
Ficoll	20%–40%	–	–

Four different media are mentioned in this barley protocol and their composition, except for pretreatment medium, is listed in Table 1. In all of them, pH is set at 5.8.

1. Pretreatment medium: The basal medium contains 0.7 M mannitol and 40 mM $CaCl_2 \cdot 2H_2O$ with 0.8% (w/v) Sea Plaque agarose [6] supplemented with 1% (*v/v*) of DMSO [7]. Commonly, 6-cm Ø Petri dishes containing 6 ml pretreatment medium are used.

2. Microspore induction medium: (Table 1) FHG medium of Hunter [8] modified (Table 1) is used as microspore induction medium. It is a semi-liquid medium that contains 62 g/l maltose as metabolizable sugar and is supplemented with 730 g/l glutamine and 1 mg/l BAP.
   It contains 200 g/l Ficoll-400.

3. Regeneration medium: (Table 1) FHG medium of Hunter [8] modified (Table 1) is used as embryo regeneration medium. It is a solid medium. It contains 31 g/l maltose as sugar and is supplemented with 1 mg/l BAP and 0.5 mg/l IAA. It contains 8 g/l agarose.

4. MS medium: (Table 1) It is used to grow green plants before transferring to soil. It contains half-strength mineral salts MS medium [27] with 20 g/l sucrose and is supplemented with naphthalene acetic acid (NAA).

3 Methods

3.1 Donor Plants and Growth Conditions

1. Donor plants are grown in controlled environment and under optimal conditions to avoid any kind of stress (*see* **Notes 1** and **2**).

2. Cultivate seedlings of F1 donor plants in paper-pots with a mixture of sand, vermiculite, and peat (1:1:1).

3. Vernalize them in a cold room at 4 °C, 8/16 h light/dark photoperiod for 4 weeks.

4. After vernalization, transfer them to 30 cm pots with the same soil mixture and grow in a growth chamber at 70% humidity at 12 °C constant and 12/12 h light/dark for 4 more weeks.

5. Increase temperature up to 21–18 °C (day/night) and the photoperiod to 16/8 h light/dark.

6. Water and fertilize plants weekly during the whole vegetation period.

7. Fertilized the soil with N:P:K (15:15:15). Besides, apply a foliar fertilizer once a week (*see* **Note 3**).

3.2 Harvesting of Spikes

1. Collect the spikes when most of the microspores are at the mid- to late-uninucleate stage (*see* **Note 4**).

2. Sterilize leaf sheaths with enclosed spikes by spraying with 70% ethanol under aseptic conditions before removing their sheath (*see* **Note 5**).

3.3 Pretreatment of Anthers

1. Extract anthers from the middle zone of the spike in a laminar flow bench and plate them on 6 cm Ø Petri dishes containing 6 ml pretreatment medium. Five to seven flowers (30–42 anthers) from each side of the spike are used (*see* **Notes 6** and **7**).

2. Seal plates with Parafilm and incubate them for 4 days in a growth chamber at 24 °C at the dark.

3.4 Microspore Isolation

1. Following pretreatment, transfer responsive anthers onto glass beakers containing 10 ml cold (4 °C) 0.3 M mannitol using sterilized forceps. Because of the stress pretreatment, responsive anthers absorb water and swell (Fig. 1a).

2. When all anthers have been transferred, pour the mannitol along with the anthers into a sterilized glass rod homogenizer to release the microspores by crushing them with a pestle. Separate the debris from the microspores by filtering through a 100-μm mesh using a small glass funnel.

3. Collect the microspores in graduated centrifuge tubes. Replenish centrifuge tubes with cold (4 °C) fresh mannitol 0.3 M before centrifuging them at 1000 rpm for 5 min.

4. Perform a discontinuous maltose gradient to select viable microspores. Discard the supernatant and suspend the cell pellet in 1 ml mannitol, over a 20% (w/v) cold (4 °C) maltose solution. After centrifuging at 1000 rpm for 5 min, collect the resulting band of living microspores from the mannitol–maltose interface using a sterile Pasteur-pipette and transfer to a new tube with 10 ml cold (4 °C) mannitol 0.3 M.

5. Discard the pellet with the dead microspores. Centrifuge again the tubes at 1000 rpm for 5 min. Discard the supernatant and transfer the viable microspores located in the resulting pellet onto a tube with 100 μl of liquid induction medium.

6. Count the cells using a hemocytometer (Neubauer) under a light microscope. Next, adjust the microspore density to 10^6 cells/ml.

7. Carefully, inoculate microspores on top of the Ficoll induction medium (*see* **Note 8**). The microspores will float on the surface of the medium (Fig. 1b).

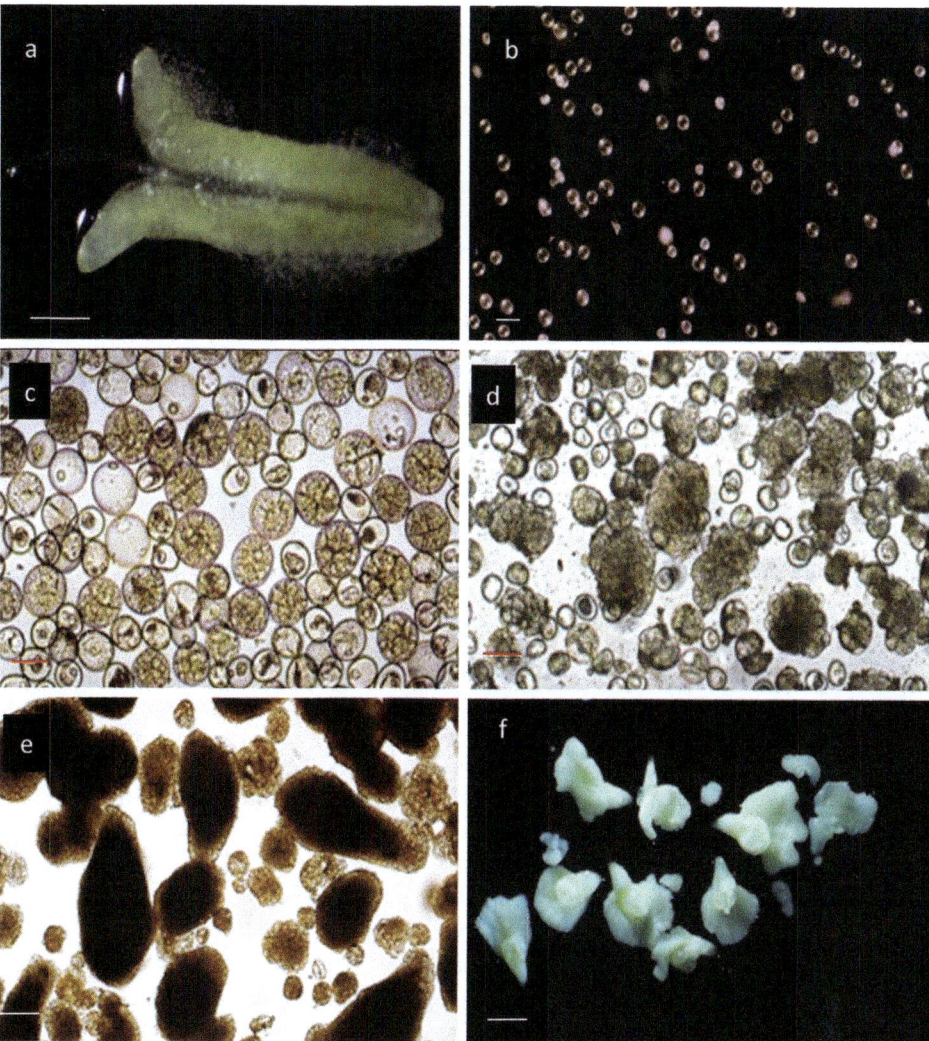

Fig. 1 Process of microspore culture in barley. (**a**) Responsive anther in 0.7 M mannitol medium. (**b–f**) The course of microspore embryogenesis from the microspore until maturing embryo stages. (**b**) Microspores floating on FGH medium just after microspore isolation. (**c**) First divisions observed after 7 days. (**d**) First multicellular aggregates after 12 days. (**e**) Embryo-like structures visible after 3 weeks in FHG medium. (**f**) Well-developed embryos visible after 28 days. Bars: (**a**) 1 cm; (**b–e**) 60 μm; (**f**) 1 cm

3.5 Microspore Culture

1. To induce androgenesis, transfer isolated microspores from previous steps onto 3 cm Ø Petri dishes containing 1.5 ml FHG liquid medium supplemented with 200 g/l Ficoll-400. Keep plates at 25 °C in darkness for a month.

2. After 10–12 days, if the microspores are dividing, replenish dishes with 1.5 ml of the same medium containing 400 g/l Ficoll-400 (*see* **Note 9**).

3. Check microspore development using a stereo microscope. After 1 week, the first divisions can be observed (Fig. 1c), and

by days 10–12, the first multicellular aggregates are seen (Fig. 1d).

3.6 Transfer of Embryos

1. Check plates after 21 days. The first embryo-like structures will be visible on the medium (Fig. 1e).

2. Check plates after 28 days in culture. Well-developed embryos (embryos with scutellum, coleoptile, and coleorhiza) (Fig. 1f) cover the medium surface.

3. Transfer the well-developed embryos to the FHG regeneration medium (6 cm Ø plates) in aseptic conditions. It is advisable to transfer 20 embryos to each plate.

4. Incubate embryos at 25 °C in darkness for 2 days and then transfer them to a growth chamber with 16/8 h light/dark at 25 °C. The embryos will develop green coleoptiles (Figs. 2a, b).

3.7 Plant Transfer to Magenta Boxes

Complete green plants regenerated (Fig. 2c) are transferred to Magenta boxes (Fig. 2d) containing MS medium. Plants are vernalized in the Magenta boxes in a growth chamber at 4 °C and 8/16 h light/dark for 4–6 weeks until a good development of the roots is observed.

3.8 Plant Transfer to Soil

After vernalizing, Magenta boxes are placed in a growth chamber at 16 °C, 12 h photoperiod provided by fluorescent and incandescent lamps and kept for several days in high humidity (90%). Once a good development of the root system is observed, about 3 weeks later, plants are potted (5 cm Ø) in the soil mixture mentioned above. Plants are kept in this chamber for another 3 weeks to be hardened and acclimated to the ex vitro environment.

Finally, plants are transplanted to pots (30 cm Ø), with similar soil mixture, and cultivated in a greenhouse 24 °C and 16/8 h photoperiod for seed production (*see* **Notes 10** and **11**).

4 Notes

1. Growth chambers, tissue culture chambers, or walk-in rooms are used for precise, uniform, and repeatable control of temperature, light, humidity, and other environmental conditions. They are able to create the precise environmental conditions for healthy growth. Greenhouse is used to grow plants from in vitro culture to maturity for seed production. It is also employed to grow donor plants and obtain explants, but it requires facilities to maintain humidity, light, and temperature under controlled conditions.

2. Donor plants of barley grown from October to December produce more responsive microspores. Although a greenhouse

Fig. 2 Process of regeneration of green plants from embryo stage. (**a**) Microspore-derived embryo growing in FHG medium. Embryos developing green coleoptiles after some days at light. (**b**) Embryo-derived plantlets growing in regeneration medium. At the top right of the image, microspores in the process of division, calli, and embryos in induction medium. (**c**) Embryo-derived plantlet germinated in FHG regeneration medium. (**d**) Green plants growing in Magenta boxes containing MS medium. Bars: (**a**, **c**) 1 cm

can be used for it, a growth chamber with controlled light, humidity, and temperature will provided more healthy, uniform, and reproducible results.

3. The vigor of the donor plants, temperature, light intensity, photoperiod, nutrition, and application of pesticides should be carefully controlled before performing an experiment.

4. In barley, the spikes may be selected based upon the distance between the ligules of the flag leaf and the penultimate leaf and on the anther appearance. This distance is about 5–7 cm for

2-row varieties and 3–5 cm for 6-row ones. About 18–20 spikes are required to perform a microspore isolation.

5. Preferably, freshly harvested tillers should be used to carry out this technique. If it is not possible, spikes may be stored in the fridge (2 or 3 days) until they are employed.

6. Stress pretreatment is applied to anthers to achieve efficient androgenic induction. Sugar starvation and osmotic stress are applied to them by using 0.7 M mannitol medium.

7. Mannitol 0.7 M tends to crystallize when using this product at such a high concentration. Precautions shall be taken to prevent that effect. Moreover, the addition of DMSO causes a quicker crystallization which can be avoided by filtering the solution as soon as DMSO is added to it.

8. Ficoll-400 needs to be dissolved in boiling water until the solution runs transparent. After dissolving, it can be added the rest of the components and sterilized by filtering, not autoclaving. The induction medium is sterilized by filtration, and the pH is adjusted to 5.8.

9. After the first 10 days, the concentration of Ficoll is increased to 300 g/l to prevent embryos sinking. Ficoll-400 showed to improve floatation, aeration, and substance exchange of microspores.

10. One advantage of this protocol is that it can be used for spring and winter barley genotypes. Usually winter crosses respond in about 80% and spring crosses around 60%. With responding cultivars (Igri, Cobra), around 150–200 DHs with seeds can be collected from one microspore isolation. For nonresponding cultivars, from 30 to 50 DHs with seeds are collected.

11. A high rate of spontaneous chromosome doubling is obtained using this protocol. Around 80–90% of regenerated plants are doubled haploids. For this reason, it is not necessary to use colchicine treatments to double the chromosome number. Instead, plants are let grow to maturity and the haploid plants are discarded.

References

1. Davies PA, Morton S (1998) A comparison of barley isolated microspore and anther culture and the influence of cell culture density. Plant Cell Rep 17:206–210

2. Kasha KH, Simion E, Oro R, Yao QA, Hu TC, Carlson AR (2001) An improved in vitro technique for isolated microspore culture of barley. Euphytica 120:379–385

3. Li H, Devaux P (2005) Isolated microspore culture over performs anther culture for green plant regeneration in barley (*Hordeum vulgare* L.). Acta Physiol Plant 27:611–619

4. Huang B, Sunderland N (1982) Temperature stress pretreatment in barley anther culture. Ann Bot 49:77–88

5. Roberts-Oehlschlager SL, Dunwell JM (1990) Barley anther culture: pretreatment on mannitol stimulates production of microspore-derived embryos. Plant Cell Tissue Organ Cult 20:235–240

6. Cistué L, Ramos A, Castillo AM, Romagosa I (1994) Production of large number of doubled haploid plants from barley anthers pretreated with high concentrations of mannitol. Plant Cell Rep 13:709–712

7. Echávarri B, Cistué L (2016) Enhancement in androgenesis efficiency in barley (*Hordeum vulgare* L.) and bread wheat (*Triticum aestivum* L.) by the addition of dimethyl sulfoxide to the mannitol pretreatment medium. Plant Cell Tissue Organ Cult 125:11–22

8. Hunter CP (1988) Plant regeneration from microspores of barley, *Hordeum vulgare*. PhD thesis. Wye College, University of London, London

9. Mordhorst AP, Lörz H (1993) Embryogenesis and development of isolated barley (*Hordeum vulgare* L.) microspores are influenced by amount and composition of nitrogen sources in culture media. J Plant Physiol 142:485–492

10. Olsen FL (1987) Induction of microspore organogenesis in cultured anthers of *Hordeum vulgare*. The effects of ammonium nitrate, glutamine and asparagine as nitrogen sources. Carlsb Res Commun 52:393–404

11. Olsen FL (1991) Isolation and cultivation of embryogenic microspores from barley (*Hordeum vulgare* L.). Hereditas 115:255–266

12. Kao KN (1981) Plant formation from barley anther cultures with Ficoll media. Z Pflanzenphysiol 103:437–443

13. Kuhlmann U, Foroughi-Wehr B (1989) Production of doubled haploid lines in frequencies sufficient for barley breeding programs. Plant Cell Rep 8:78–81

14. Devaux P, Kasha KJ (2009) Overview of barley doubled haploid production. In: Touraev A et al (eds) Advances in haploid production in higher plants. Springer, New York

15. Nuutila AM, Hämäläinen J, Mannonen L (2000) Optimization of media nitrogen and copper concentrations for regeneration of green plants from polyembryogenic cultures of barley (*Hordeum vulgare* L.). Plant Sci 151 (1):85–92

16. Wojnarowiez G, Jacquard C, Devaux P, Sangwan RS, Clément C (2002) Influence of copper sulfate on anther culture in barley (*Hordeum vulgare* L.). Plant Sci 162(5):843–847

17. Luckett DJ, Smithard RA (1995) A comparison of several published methods for barley anther culture. Plant Cell Rep 14:763–767

18. Weyen J (2009) Barley and wheat doubled haploids in breeding. In: Touraev A et al (eds) Advances in haploid production in higher plants. Springer, New York

19. Davies PA (2003) Barley isolated microspore culture (IMC) method. In: Maluszynski M, Kasha KJ, Forster BP, Szarejko I (eds) Doubled haploid production in crop plants. Springer, Dordrecht, pp 49–52

20. Lu R, Wang Y, Sun Y, Shan L, Chen P, Huang J (2008) Improvement of isolated microspore culture of barley (*Hordeum vulgare* L.): the effect of floret co-culture. Plant Cell Tissue Organ Cult 93:21–27

21. Torp AM, Andersen SB (2009) Albinism in microspore culture. In: Touraev A et al (eds) Advances in haploid production in higher plants. Springer, New York

22. Cistué L, Ramos A, Castillo AM (1999) Influence of anther pretreatment and culture medium composition on the production of barley doubled haploids from model and low responding cultivars. Plant Cell Tissue Organ Cult 55:159–166

23. Marchand S, Fonquerne G, Clermont I, Laroche L, Huynh TT, Belzile FJ (2008) Androgenic response of barley accessions and F1s with *Fusarium* head blight resistance. Plant Cell Rep 27:443–445

24. Cistué L, Ziauddin A, Simion E, Kasha KJ (1995) Effects of culture conditions on isolated microspore response of barley cultivar Igri. Plant Cell Tissue Organ Cult 42:163–169

25. Kasha KJ, Simion E, Oro R, Shim YS (2003) Barley isolated microspore culture protocol. In: Maluszynski M, Kasha KJ, Forster BP, Szarejko I (eds) Doubled haploid production in crop plants. Springer, Dordrecht, pp 43–47

26. Esteves P, Belzile FJ (2019) Isolated microspore culture in barley. Methods Mol Biol 1900:53–71

27. Murashige T, Skoog F (1962) A revised medium for rapid growth and bio assays with tobacco tissue cultures. Physiol Plant 15 (3):473–497

Chapter 9

Site-Directed Mutagenesis in Barley Using RNA-Guided Cas Endonucleases During Microspore-Derived Generation of Doubled Haploids

Robert Eric Hoffie, Ingrid Otto, Hiroshi Hisano, and Jochen Kumlehn

Abstract

In plant research and breeding, haploid technology is employed upon crossing, induced mutagenesis or genetic engineering to generate populations of meiotic recombinants that are themselves genetically fixed. Thanks to the speed and efficiency in producing true-breeding lines, haploid technology has become a major driver of modern crop improvement. In the present study, we used embryogenic pollen cultures of winter barley (*Hordeum vulgare*) for Cas9 endonuclease-mediated targeted mutagenesis in haploid cells, which facilitates the generation of homozygous primary mutant plants. To this end, microspores were extracted from immature anthers, induced to undergo cell proliferation and embryogenic development in vitro, and were then inoculated with *Agrobacterium* for the delivery of T-DNAs comprising expression units for Cas9 endonuclease and target gene-specific guide RNAs (gRNAs). Amongst the regenerated plantlets, mutants were identified by PCR amplification of the target regions followed by sequencing of the amplicons. This approach also enabled us to discriminate between homozygous and heterozygous or chimeric mutants. The heritability of induced mutations and their homozygous state were experimentally confirmed by progeny analyses. The major advantage of the method lies in the preferential production of genetically fixed primary mutants, which facilitates immediate phenotypic analyses and, relying on that, a particularly efficient preselection of valuable lines for detailed investigations using their progenies.

Key words *cas9*, Cereals, CRISPR, Genome editing, Targeted mutagenesis

1 Introduction

Cereals are the most important source of calories for mankind [1]. Plant breeding is one of the most effective measures for securing and increasing food production for a growing world population and under changing environmental conditions. Methods of haploid production are already well established in cereal breeding and are used for the time-saving production of genetically fixed breeding lines [2].

Jose M. Seguí-Simarro (ed.), *Doubled Haploid Technology: Volume 1: General Topics, Alliaceae, Cereals*,
Methods in Molecular Biology, vol. 2287, https://doi.org/10.1007/978-1-0716-1315-3_9,
© Springer Science+Business Media, LLC, part of Springer Nature 2021

Barley, together with maize, rice and wheat, is one of the four most commonly grown cereals and is used in a variety of ways for the production of food, feed and beverages [1]. Doubled haploid barley lines can be obtained from embryogenic pollen cultures [3]. For this purpose, microspores are isolated from immature spikes and embryogenic development is induced in vitro [4]. By combining this approach with *Agrobacterium*-mediated gene transfer, homozygous transgenic plants can be instantly regenerated [5, 6].

In recent years, site-directed genome engineering has become increasingly important for plant research and breeding. Targeted genome modifications were first achieved using meganucleases, zinc finger nucleases (ZFNs) and transcription activator-like effector nucleases (TALENs), and finally also using RNA-guided endonucleases (RGENs) [7]. The latter originate from microbial immune systems called 'CRISPR/Cas' (clustered regularly interspaced short palindromic repeats/CRISPR-associated), which are particularly directed against viral infections. As biotechnologically used molecular tools, RGENs consist of two components: an endonuclease that is able to induce double-strand breaks in DNA, and an RNA molecule guiding the endonuclease to a genomic target motif to which it was rendered nucleobase-complementary. There, the genomic DNA is cleaved by the endonuclease and repaired by the plant's own mechanisms [8]. If the repair is correct, gRNA and endonuclease are capable of binding repeatedly until a mutation occurs due to an erroneous repair event. In addition to nucleotide insertions and deletions of largely random size, which are used in particular to knock out or knock down target genes, there are already more advanced applications by which not only the location of the mutation but also the resulting DNA sequence can be precisely predefined. These approaches rely either on the use of a synthetic DNA template carrying the nucleotide sequence of choice, which is implemented via homology-mediated DNA repair, or on the use of fusion proteins consisting of modified endonucleases and deaminating enzymes.

In combination with customisable endonucleases, haploid technology was first used for the efficient dissolution of multiple mutations present in chimeric primary mutants, from which populations of genetically fixed doubled haploids were generated [9, 10]. By applying RGENs to haploid microcalli obtained from in vitro cultivated microspores, it is as well possible to directly produce homozygous mutant plants. In the present protocol, we describe the *Agrobacterium*-mediated transfer of Cas9 endonuclease- and gRNA-encoding T-DNAs into embryogenic pollen cultures of winter barley cv. 'Igri' for site-directed mutagenesis in the three individual candidate genes *EIF4E*, *PDIL5-1* and *QSD1* [11–13]. We achieved mutations in all of these target genes and provide evidence for homozygosity of genetic modifications detected in primary mutants, as well as for the correspondingly uniform generative transmission of mutations to progeny.

2   Materials

2.1   Plant Material

This protocol was developed using the winter-type barley cultivar 'Igri' (Saatzucht Ackermann, Irlbach, Germany), which is particularly amenable to pollen embryogenesis and *Agrobacterium*-mediated gene transfer to embryogenic pollen cultures.

2.2   Equipment

1. Refrigerated centrifuge equipped with swing-out baskets.
2. Temperature-controlled incubators equipped with a rotary shaker.
3. Waring blender (heat sterilisable).
4. Hemocytometer (type Rosenthal).
5. Spectrophotometer.
6. Flow cytometer (CyFlow Ploidy Analyser, Partec, Münster, Germany).
7. Gel electrophoresis system (Bio-Rad, Munich, Germany).

2.3   Laboratory Materials

1. Sterile Petri dishes with lid (3.5 cm in diameter).
2. Sterile Petri dishes with lid (10 cm in diameter).
3. Sterile Petri dishes with lid (10 cm in diameter, with two compartments).
4. Safe-lock tubes (1.5 and 2 mL).
5. Sterile screw-cap polypropylene cryotubes (1.5 mL).
6. Sterile screw-cap polycarbonate round-bottomed tubes (12 mL).
7. Sterile screw-cap polypropylene centrifuge tubes (round bottom 12 mL and skirted conical 50 mL tubes).
8. Magenta boxes (autoclaved).
9. Nylon mesh (100μm grid, autoclaved).
10. Nylon mesh filters, for example, CellTrics (30μm mesh diameter, Partec).
11. Wire brush.
12. Sterile filter-stopped tips (1 mL) for standard pipettes.
13. Erlenmeyer flasks (100 mL) with chicane.
14. GeneJET PCR purification kit (Thermo Fisher Scientific, USA).
15. Cloning pGEM-T vector kit (Promega, Madison, WI).

Table 1
Primer sequences for the analysis of regenerated T$_0$ plants. Given are the names and 5′–3′-sequences of DNA oligonucleotides used as primers for PCR amplification of *cas9*, gRNA and *hpt* gene sequences of the T-DNA. Primers for amplification of the target region addressed for site-directed mutagenesis need to be designed specifically; *PDIL5-1* target-specific primers are given here as an example

cas9	
ZmUbi-P-f	5′-TTTAGCCCTGCCTTCATACG-3
Cas9-r	5′-TTAATCATGTGGGCCAGAGC-3
gRNA	
gRNA-forward Oligonucleotide	5′-NNNNNNNNNNNNNNNNNNNNN-3′ (*see* **Note 1**)
OsU3T-r	5′-TCAGCGGGTCACCAGTGTTG-3′
hpt (for hygromycin resistance)	
Hpt-f	5′-GATCGGACGATTGCGTCGCA-3′
Hpt-r	5′-TATCGGCACTTTGCATCGGC-3′
Target-specific primers	
PDIL5-1-f	5′-CATATGACGTCGCGGTCTCC-3′
PDIL5-1-r	5′-GATGCACGACAAGGGGAAAGC-3′

16. Taq DNA polymerase with respective PCR buffer.
17. Primers for amplification of *cas9*/gRNA-constructs, *hpt* gene and target regions (Table 1; *see* **Note 1**).

2.4 Glasshouse and Growth Chamber Materials

1. Germination and vernalisation: 3:1:2 mixture of garden mulch/sand/white and black peat (Substrate 2, Klasmann-Deilmann, Germany).
2. Plant growth after vernalisation: 2:2:1 mixture of compost/Klasmann-Deilmann Substrate 2 and sand.
3. Acclimation of in vitro regenerants: Petuniensubstrat (Klasmann-Deilmann, Germany).
4. Fertiliser: Osmocote (19% N, 6% P and 12% K) (ICL Specialty Fertilizers, Israel).

2.5 Stock Solutions and Chemicals

All stock solutions are to be autoclaved and stored at room temperature, if not stated otherwise.

1. K macro mineral salts (20×): 1.6 g/L NH$_4$NO$_3$, 40.4 g/L KNO$_3$, 4.9 g/L MgSO$_4$·7H$_2$O, 6.8 g/L KH$_2$PO$_4$, 8.8 g/L CaCl$_2$·2H$_2$O.
2. K4N macro mineral salts (20×): 6.4 g/L NH$_4$NO$_3$, 72.8 g/L KNO$_3$, 4.9 g/L MgSO$_4$·7H$_2$O, 6.8 g/L KH$_2$PO$_4$, 8.8 g/L CaCl$_2$·2H$_2$O.

3. K micro mineral salts ($1000\times$): 8.4 g/L $MnSO_4 \cdot H_2O$, 7.2 g/L $ZnSO_4 \cdot 7H_2O$, 3.1 g/L H_3BO_3, 25 mg/L $CuSO_4 \cdot 5H_2O$, 120 mg/L $Na_2MoO_4 \cdot 2H_2O$, 24 mg/L $CoCl_2 \cdot 6H_2O$, 170 mg/L KI.

4. Ferric sodium ethylene diamine tetraacetate (NaFeEDTA; 75 mM, 27.5 g/L): filter-sterilised and stored at 4 °C.

5. $CaCl_2 \cdot 2H_2O$ (1 M, 147 g/L): filter-sterilised and stored at 4 °C.

6. KH_2PO_4 (1 M, 136 g/L): filter-sterilised.

7. K_2HPO_4 (1 M, 174 g/L): filter-sterilised.

8. Phosphate buffer (1 M, pH 5.0): 97.5% 1 M KH_2PO_4, 2.5% 1 M K_2HPO_4, filter-sterilised and stored at 4 °C.

9. Phosphate buffer (1 M, pH 5.9): 90% 1 M KH_2PO_4, 10% 1 M K_2HPO_4, filter-sterilised and stored at 4 °C.

10. $CuSO_4 \cdot 5H_2O$ (10 mM, 2.5 g/L): filter-sterilised and stored at 4 °C.

11. Mannitol (0.4 M, 72.9 g/L), stored at 4 °C.

12. Maltose·H_2O (1 M, standard quality, 360 g/L): filter-sterilised.

13. Maltose·H_2O (1 M, >99% purity, 360 g/L): filter-sterilised.

14. Maltose·H_2O (0.55 M, >99% purity, 198 g/L): filter-sterilised and stored at 4 °C.

15. KM organics ($100\times$, Sigma K-3129): 10 g/L myo-inositol, 200 mg/L L-ascorbic acid, 100 mg/L pyridoxine·HCl, 100 mg/L nicotinamide, 100 mg/L thiamine·HCl, 100 mg/L D-calcium pantothenate, 2 mg/L p-aminobenzoic acid, 40 mg/L folic acid, 20 mg/L riboflavin, 2 mg/L cyanocobalamin, 1 mg/L D-biotin, 1 mg/L retinol, filter-sterilised and stored at −20 °C.

16. Gamborg B5 organics ($1000\times$): 100 mg/L myo-inositol, 10 mg/L thiamine·HCl, 1 mg/L nicotinic acid, 1 mg/L pyridoxine·HCl, filter-sterilised and stored at −20 °C.

17. 2,4-Dichlorophenoxyacetic acid (2,4-D, 1 mM): 221 mg/L dissolved in few drops of 50% ethanol by heating gently, made up to the final volume with hot H_2O, filter-sterilised and stored at 4 °C.

18. 6-Benzylaminopurine (BAP, 1 mM): 224 mg/L dissolved in 30 mL hot H_2O and 3–5 drops 1 M NaOH, made up to the final volume with H_2O, filter-sterilised and stored at 4 °C.

19. Timentin (150 mg/mL), filter-sterilised and stored at −20 °C.

20. Spectinomycin (100 mg/mL), filter-sterilised and stored at −20 °C.

21. Hygromycin (50 mg/mL), filter-sterilised and stored at −20 °C.

22. Tetracycline (10 mg/mL), filter-sterilised and stored at −20 °C.

23. Phytagel (2× stock, 6 g/L): suspended in cold water (at best 1.5 g Phytagel per 250 mL unit).

24. Liquid nitrogen.

25. Ethanol (70%).

26. Glycerol (15%).

27. Isopropyl alcohol (≥99.8%).

28. Agarose gel (1.0–2.0%), freshly prepared (Bio-Rad, Munich, Germany).

29. Phenol (equilibrated, stabilised)/chloroform/isoamyl alcohol 25:24:1 (AppliChem, Darmstadt, Germany).

30. Acetosyringone (1 M, 196 mg/mL): dissolved in dimethyl sulphoxide (DMSO), no sterilisation required, stored at 4 °C.

31. Staining buffer: CyStain UV Ploidy Staining Solution (Partec).

32. Colchicine (1 g/L, 0.1% w/v): dissolved in dimethyl sulphoxide (DMSO), filled up with water, a few drops of Tween 20 added, no sterilisation required, freshly prepared.

33. L-Glutamine (37 mg/mL): dissolved by adding a few drops of 0.1 M KOH and heating in a water bath, filter-sterilised and stored at −20 °C.

34. Morpholinoethanesulphonic acid (MES, 212.2 g/L, pH 5.0, pH 5.5 and pH 5.9, *see* **Note 2**): dissolved in 40 mL water, the pH adjusted with KOH (few pellets), stored at room temperature overnight and the pH readjusted with either 1 M KOH or 1 M HCl, made up to the required final volume, filter-sterilised and stored at 4 °C.

2.6 Culture Media

1. CPY medium for *A. tumefaciens* strain LBA4404: 0.1% (w/v) yeast extract, 0.5% (w/v) pancreatic peptone, 2 mg/L $MgSO_4 \cdot 7H_2O$, 0.5% (w/v) sucrose, pH 7; add 1.2% w/v Bacto Agar to produce solid medium.

2. Barley pollen culture (KBP) medium: 50 mL/L K macro, 1 mL/L K micro, 1 mL/L NaFeEDTA, 12 mL/L L-glutamine, 10 mL/L KM organics, 4 mL/L BAP and 250 mL/L maltose (1 M, >99%) stocks, pH adjusted to 5.9 and stored at 4 °C.

3. Co-culture (CK) medium: 50 mL/L K macro, 1 mL/L K micro, 1 mL/L NaFeEDTA, 10 mL/L KM organics, 250 mL/L maltose (1 M, >99%), 2 mL/L BAP, 0.5 mL/L acetosyringone, 10 mL/L MES (pH 5.9) and 50 mL/L phosphate buffer (pH 5.9) stocks, stored at 4 °C.

4. AgroStop (ASt) medium: 50 mL/L K macro, 1 mL/L K micro, 7 mL/L CaCl$_2$, 1 mL/L NaFeEDTA, 4 mL/L L-glutamine, 10 mL/L KM organics, 250 mL/L maltose (1 M, >99%), 10 mL/L MES (pH 5.0), 25 mL/L phosphate buffer (pH 5.0), 10 mL/L 2,4-D, 2 mL/L BAP, 0.5 mL/L acetosyringone, 250µL/L hygromycin and 1.3 mL/L Timentin stocks, stored at 4 °C.

5. Selection medium (S3): 50 mL/L K macro, 1 mL/L K micro, 1 mL/L NaFeEDTA, 12 mL/L L-glutamine, 10 mL/L KM organics, 1 mL/L BAP, 250 mL/L maltose (1 M, standard quality), 10 mL/L MES (pH 5.5), 1 mL/L hygromycin and 1.3 mL/L Timentin stocks, stored at 4 °C.

6. Solid barley pollen culture (KBP4PT) medium: 50 mL/L K macro, 1 mL/L K micro, 1 mL/L NaFeEDTA, 12 mL/L L-glutamine, 10 mL/L KM organics, 1 mL/L BAP, 250 mL/L maltose (1 M, standard quality), 1 mL/L hygromycin and 1.3 mL/L Timentin stocks; appropriate amounts for a final volume of 0.5 L KBP4PT made up to an intermediate volume of 250 mL (doubled concentrated), adjusted to pH 5.9, heated to about 40 °C and then mixed (1:1) with 250 mL Phytagel stock (melted by heating).

7. Regeneration (K4NBT) medium: 50 mL/L K4N macro, 1 mL/L K Micro, 0.75 mL/L NaFeEDTA, 0.49 mL/L CuSO$_4$, 4 mL/L L-glutamine, 1 mL/L B5 organics, 1 mL/L BAP, 100 mL/L maltose (1 M, standard quality), 500µL/L hygromycin and 1.3 mL/L Timentin stocks; appropriate amounts for a final volume of 0.5 L K4NBT made up to an intermediate volume of 250 mL (double concentrated), adjusted to pH 5.9, heated to about 40 °C and then mixed (1:1) with 250 mL Phytagel stock (melted by heating).

3   Methods

3.1   Vector Construction and Bacterial Strains

1. Design target-specific gRNA sequences, for which online tools such as WU-CRISPR (http://crispr.wustl.edu/) or Benchling (https://www.benchling.com/crispr/) are very helpful, whereby the latter also allows for off-target analyses.

2. Order the corresponding sequences as ssDNA oligonucleotides forward and reverse. For *Bsa*I-based cloning into the vector pSH121 [14], the following overhangs are required: forward: 5′-TGGC-3′, reverse: 5′-AAAC-3′. pSH121 contains ZmUbi-P::Cas9::Nos-T and OsU3-P::gRNA-Scaffold::OsU3-T. *Bsa*I is used to insert pairwise hybridised target-specific gRNA-coding oligonucleotides between OsU3 promoter and gRNA scaffold. *See* also Fig. 1a.

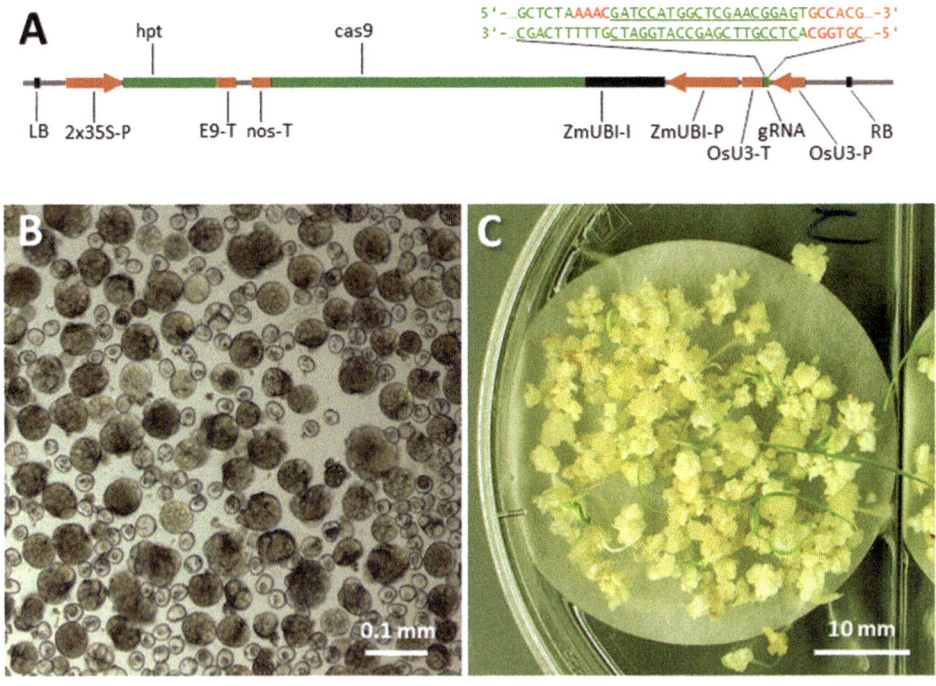

Fig. 1 (**a**) Representative T-DNA carrying expression units for *hpt*, *cas9* and a *HvPDIL5-1*-specific guide RNA (10,400 bp in total), detailed sequence of target-specific guide RNA (green, underlined) with overhangs for *Bsa*I-based cloning (red), (**b**) microspore-derived microcalli ready for agro-inoculation, (**c**) commencing shoot formation from microspore-derived calli under selective conditions, and (**d**) mutated target regions of *HvPDIL5-1* aligned with the wild-type sequence, Cas9/gRNA target motif underlined, protospacer-adjacent motif indicated by green font, Cas9 cleavage site indicated by dashed red line. *P* Promoter, *T* Terminator, *I* Intron, *LB* left border of T-DNA, *RB* right border of T-DNA

3. Clone the fragment containing the gRNA and *cas9* expression units from pSH121 via *Sfi*I into the generic binary vector p6i-2x35S-TE9 (DNA Cloning Service, Hamburg), which is compatible for that step and contains all additional components of the T-DNA (left and right border sequences, as well

as the *hpt* gene under the control of a doubled-enhanced 35S promotor from the Cauliflower Mosaic Virus (CaMV). *See* also Fig. 1a.

4. Transform *Agrobacterium tumefaciens* strain LBA4404/pSB1 via electroporation with the final binary vector. That strain carries the disarmed Ti plasmid pAL4404 and the hypervirulence-mediating plasmid pSB1 [5].

3.2 Growth of Donor Plants

1. Germinate grains of cultivar 'Igri' in a tray filled with germination substrate in a chamber providing a 12-h photoperiod (136μmol m^{-2} s^{-1} photon flux density) with a temperature regime of 14/12 °C for light/dark phases, for 2 weeks (*see* **Note 3**).

2. Vernalise the seedlings for 8 weeks at 4 °C under an 8-h photoperiod.

3. After transfer to 18 cm diameter pots, fertilise the plants by providing 15 g Osmocote, and then return them to the conditions described in **step 1** of Subheading 3.2.

4. At the tiller elongation stage, transfer the pots to a glasshouse maintained at 18/16 °C with a minimum of 16 h photoperiod (170μmol m^{-2} s^{-1} photon flux density).

3.3 Isolation of Microspores

3.3.1 Spike Pretreatment

1. Harvest the spikes when still being encased by the leaf sheath and the awns have just emerged from the flag leaf.

2. Surface-sterilise the boots by spraying with 70% ethanol.

3. Remove the flag leaf sheath under aseptic conditions.

4. Place the dissected spikes in a 10-cm diameter Petri dish with two compartments. Five spikes are placed in one of the compartments and sterile water is added to the other compartment.

5. After sealing, incubate the plates at 4 °C for 4–5 weeks in the dark.

3.3.2 Isolation, Purification and Precultivation of Microspores

All materials that may come into contact with the explants have to be aseptic and cooled down to 4 °C. All solutions are to be kept on ice during the microspore isolation and purification procedure. Liquid transfer is most effectively conducted using a battery-operated pipette with filter-stopped 5-mL or 10-mL pipette tips.

1. Chop 10–15 pre-treated spikes into fragments of approx. 1 cm and macerate them in a blender together with 20 mL of 0.4 M mannitol. Usually, two bursts of 15 s each are conducted using a Waring blender that is set on 'low' speed.

2. Place a 100μm mesh on top of a Magenta box and filter the macerate through it. Rinse the blender with 10 mL of 0.4 M mannitol, which is also passed through the mesh.

3. Gently squeeze with forceps the debris remaining on the mesh and return it to the blender for re-maceration (twice at 10 s) in another 10 mL of 0.4 M mannitol. Pass the macerate through the mesh, followed by rinsing the blender again with 10 mL of 0.4 M mannitol and passing it through the mesh.

4. After being filtered through the mesh, transfer the material collected in the Magenta box to a 50 mL centrifuge tube. Rinse the Magenta box with 5 mL 0.4 M mannitol, which is also added to the same tube. Centrifuge the suspension at $100 \times g$ for 10 min at 4 °C.

5. Resuspend the pellet in a round-bottomed 12 mL tube using 3 mL of 0.55 M maltose. Then, slowly add 2 mL 0.4 M mannitol, which is to be layered carefully over the 0.55 M maltose suspension, thereby forming two liquid layers that remain separated.

6. Perform density gradient centrifugation in swing-out baskets ($100 \times g$, 10 min, 4 °C) with the centrifuge being set to provide slow acceleration and deceleration, which makes sure that the two layers do not become blended. During the centrifugation step, the microspores accumulate in the interphase.

7. Take up the interphase using a pipette, transfer it to a fresh 50-mL tube containing 10 mL of 0.4 M mannitol and resuspend it by gentle shaking.

8. For the estimation of population density, place an appropriate aliquot of the evenly suspended material into a hemocytometer cell.

9. Meanwhile, pellet the remaining microspores by centrifugation ($100 \times g$, 10 min, 4 °C). Before the supernatant is discarded, leave the tube unmoved for approx. 5 min so as to allow still-floating microspores to settle down.

10. Resuspend the pellet in an appropriate volume of KBP medium, by which the density is adjusted to 400,000 microspores per mL.

11. Transfer aliquots of 1 mL microspore suspension into 3.5 cm Petri dishes. The dishes are sealed with Parafilm and incubated in the dark at 25 °C for 7–8 days prior to co-cultivation with *A. tumefaciens*.

3.4 Agrobacterium-Mediated Gene Transfer to Embryogenic Pollen

3.4.1 Preparation of A. tumefaciens Stocks

1. Plate *Agrobacterium* containing the selected binary vector onto solidified CPY medium, containing 50 mg/L tetracycline (500μL/L stock solution) and 250 mg/L spectinomycin (2.5 mL/L stock solution). Incubate the plate at 28 °C for 48 h (*see* **Note 4**).

2. Start liquid cultures from single colonies sampled with a sterile toothpick and grown overnight with shaking at 180 rpm (28 °C) in a tube containing 3 mL of CPY supplemented with the respective antibiotics.

3. Confirm the integrity of the binary vector by restriction analysis of extracted plasmid DNA.

4. Add 20μL of the overnight culture to 5 mL fresh CPY medium containing antibiotics to initiate a fresh culture. Incubate for another 24 h as described above until an OD_{550} of approximately 2 is reached.

5. Cultivate the selected clones in 10 mL of liquid CPY containing antibiotics for another 24 h. Then, divide the culture into two 12 mL round-bottomed tubes and centrifuge ($3,220 \times g$, 12 min, room temperature).

6. Resuspend the pellets by vortexing each in 2.5 mL of fresh CPY without antibiotics and collect the suspensions in one of the tubes. Adjust the cell density by adding an appropriate volume of medium so that the OD_{550} of a 1:10 dilution is 0.45, which corresponds to 10^9 colony-forming units (cfu) per mL medium.

7. Add an equal volume of 15% autoclaved glycerol to the cells, followed by mixing. Transfer 0.5 mL aliquots into 1.5 mL cryotubes, incubate them for 1 h at room temperature and then store the aliquots at −80 °C.

3.4.2 Co-Cultivation of Embryogenic Pollen Cultures and A. tumefaciens

1. Add a glycerol stock of strain LBA4404/pSB1, bearing the binary vector of choice, to 10 mL of CPY medium supplemented with 250 mg/L spectinomycin (2.5μL stock solution) and incubate at 28 °C for 24 h with shaking (180 rpm).

2. Split the liquid culture into two 12 mL round-bottomed tubes. Centrifuge the tubes ($3,220 \times g$, 12 min, room temperature) and resuspend the pellets by vortexing each in 2.5 mL CK medium. Transfer them to a 100 mL Erlenmeyer flask and incubate the flask at 28 °C for 1–3 h with shaking (100 rpm).

3. Remove the KBP medium used to pre-cultivate the microspores (Fig. 1b) by pipetting with a 1-mL pipette tip and add 1 mL of CK medium (*see* **Note 5**). Every tenth dish containing embryogenic pollen remains untouched to later provide a supply of feeder cells to support embryogenic development of the agro-inoculated pollen.

4. Dilute the *A. tumefaciens* cultures tenfold in CK medium. The measured OD_{550} is used to determine the concentration of present colony-forming units (cfu), as well as the volume of inoculum to be added to the pollen cultures.

5. Inoculate each 1 mL of cultivated pollen with 2.5×10^7 cfu of *Agrobacterium*.

3.5 Regeneration of Transgenics

1. After 48 h of co-cultivation, remove the CK medium as described above. Rinse the pollen in 0.4 mL of ASt medium. Make sure to detach the bacterial mucilage from the wall and bottom of the dish using a pipette tip (*see* **Note 6**). Then, replace the medium with a fresh aliquot of 1.1 mL ASt to which 100μL of non-co-cultivated embryogenic pollen is added to serve as feeder cells. Incubate the sealed dishes in the dark at 25 °C for 1 week with gentle shaking (75 rpm).

2. Replace the ASt medium by 1.1 mL of S3 medium taking care to avoid any uptake of embryogenic pollen. Incubate the sealed dishes in the dark at 25 °C for 1 week with shaking (65 rpm), after which the used S3 medium is removed and refreshed by a new aliquot of the same medium. Keep dishes under the same conditions for another week.

3. Transfer the microcalli that are formed 3 weeks after co-cultivation to an ash-free filter paper disk placed over solid KBP4PT medium in a 10-cm Petri dish. After sealing, incubate these dishes in the dark at 25 °C for 2 weeks.

4. Transfer calli that have grown to more than 1 mm in diameter into a 10-cm Petri plate containing K4NBT medium. Calli that have not yet grown to that size can be kept for another week on KBP4PT before being transferred to regeneration medium. After 1 week in the dark at 25 °C, transfer the sealed K4NBT plates into the light for another 2 weeks.

5. Transfer calli along with emerging leaves and shoots (early stage shown in Fig. 1c) to tissue culture boxes with K4NBT. Sub-culture them twice at 3-week intervals.

6. Transfer plantlets with roots to 6 cm diameter pots with substrate (Klasmann-Deilmann Petuniensubstrat), place them in a tray and cover them with a transparent plastic hood to maintain saturated air humidity. Incubate the tray in a chamber with a 12-h photoperiod (136μmol m^{-2} s^{-1} photon flux density) and a temperature regime of 14/12 °C for light/dark phases, respectively.

7. After 1 week, remove the hood and leave the tray uncovered for another week. For vernalisation, hold the plantlets for 8 weeks at 4 °C under an 8-h photoperiod.

8. After vernalisation, cultivate the plants under the same conditions as described for the donor plants.

3.6   Analysis of Putative Transgenic Plants

3.6.1   Ploidy Determination and Colchicine-Induced Whole-Genome Duplication

1. For the determination of ploidy levels, perform flow cytometry of leaf tissue during the vernalisation period.

2. Isolate the nuclei by supplying a leaf sample with ice-cold CyStain UV Ploidy Staining Solution and gently disintegrate the leaf tissue by means of a wire brush.

3. Pour the resulting suspension through a 30-μm CellTrics filter.

4. Analyse the filtered cell suspensions of the regenerants and of controls from glasshouse-grown wild-type plantlets by a flow cytometer such as the CyFlow Ploidy Analyser (Partec) following the manufacturer's instructions.

5. After the vernalisation period, treat the identified haploids with colchicine as follows: Wash off the soil from the roots. Trim the tillers and roots to a length of about 5 cm and 3 cm, respectively. Place the plantlets individually in 50 mL polypropylene conical tubes containing 0.1% w/v colchicine solution. The liquid level is adjusted to reach the bases of the tillers. Cap the tubes and then incubate in the light for 6 h at 21 °C.

6. After being removed from the colchicine, carefully rinse the plantlets in water before transferring them into 9 cm-diameter pots filled with Petuniensubstrat. Saturated air humidity is provided using a tray with plastic hood as described above.

7. More than 80% of plantlets overcome the colchicine treatment, and out of these at least 90% set grains, which indicates that whole-genome duplication of germ-line cells has taken place (*see* **Note 7**).

3.6.2   Molecular Analyses: DNA Isolation, PCR and Sequencing

1. From fresh leaf tissue, 200–400 mg are sampled per plantlet and snap-frozen in liquid nitrogen.

2. Isolate the DNA following the protocol of Palotta et al. [15] (*see* **Note 8**).

3. Perform PCR assays using 50–100 ng genomic DNA as template. Particularly here, four different PCR reactions for the following templates are performed: Cas9 endonuclease, gRNA, selectable marker gene (*hpt*) and target region. Perform all PCRs in reaction volumes of 10μL with the exception of the target region PCR assay, where 25μL volumes are used. For target-specific PCR, *see* **Note 9**.

4. Purify and sequence the target amplicons. Purification is performed using the GeneJET PCR purification kit according to the manufacturer's instructions. Purified PCR samples are Sanger-sequenced.

5. Analyse the sequencing chromatograms and align the resulting sequences to that of the wild-type so as to detect mutations (insertions, deletions or nucleotide exchanges) within the targeted region. Chromatograms with consecutive peaks

commencing around the cleavage site indicate heterozygosity (one mutated and one wild-type allele), biallelic (plants with both alleles carrying different mutations) or chimeric plants (with two or more mutant alleles being present in different sectors) *see* **Note 10**; Fig. 1d.

4   Notes

1. The gRNA-forward oligonucleotide sequences coincide with the specific target regions of individual gRNAs.

2. Separate stocks are required for the various pH values (pH 5.0, pH 5.5 and pH 5.9).

3. The growth conditions of the donor plants strongly influence the quality of the microspore cultures. Therefore, it is of high importance to ensure the optimised conditions as described.

4. The disarmed Ti-plasmid pAL4404 present in the *Agrobacterium* strain LBA4404 harbours the same spectinomycin resistance gene (aadA) as the binary vectors used. Consequently, a higher than standard concentration of spectinomycin is required (250 mg/L) to make sure the binary vectors are maintained. To select for *Agrobacterium* cells carrying the hypervirulence-conferring pSB1 plasmid, tetracycline is used here in addition.

5. This procedure is conducted with particular care so as to avoid taking up any embryogenic pollen. It is important to minimise the time between the removal of the KBP medium and the addition of the CK medium, because cultivated pollen is rapidly damaged when being dried.

6. After being removed from the wall and bottom of the dish, the bacterial mucilage is required to be left inside the culture, because it contains much of the embryogenic pollen.

7. As haploid plants are sterile, only plants that have undergone whole-genome duplication, either spontaneous or colchicine-induced, are capable of setting grains.

8. For molecular analyses (DNA extraction, PCR) of the transgenics, it is recommended to use filter tips, because false-positives can occur owing to cross-contamination of samples, in particular where standard transgenes such as *cas9* and *hpt* are frequently used.

9. For Sanger-sequencing, PCR amplicons of the target regions should be at least of 400 bp in length. Well-established primer combinations previously confirmed not to cause any side products ensure high sequencing quality. The Sanger-sequencing

reaction can be further improved by the use of an additional oligo that binds within the amplified sequence in place of one of the primers that were already used in the PCR reaction.

10. Using this protocol, we targeted three different genes of winter barley. Taken together, 48 T_0 plants were generated; seventeen of these plants carried mutations (indels of 1–22 bp). Out of these, nine plants were chimeric or heterozygous with regard to their mutations, and eight plants carried a mutation at homozygous condition (either being spontaneously doubled haploid or after colchicine treatment). For five of the homozygous primary mutants, progenies were analysed for their mutational status. All of those proved to be homozygous for the very same mutation carried by the primary mutant they derived from.

Acknowledgements

We wish to thank Carola Bollmann and Andrea Müller for her expert technical assistance. We thank the German Federal Ministry for Science and Education for funding our research in frame of the IdeMoDeResBar project (FKZ 031B0199C and 031B0887C). We are also grateful to the Leibniz Institute of Plant Genetics and Crop Plant Research (IPK) Gatersleben for providing our research group with excellent working conditions.

References

1. FAOSTAT. http://www.fao.org/faostat/en/#data. Accessed 22 April 2020

2. Kalinowska K, Chamas S, Unkel K, Demidov D, Lermontova I, Dresselhaus T, Kumlehn J, Dunemann F, Houben A (2019) State-of-the-art and novel developments of in vivo haploid technologies. Theor Appl Genet 132(3):593–605. https://doi.org/10.1007/s00122-018-3261-9

3. Kumlehn J (2014) Haploid technology. In: Kumlehn J, Stein N (Hrsg) biotechnological approaches to barley improvement, Bd 69. Springer, Heidelberg, pp S 379–S 392. https://doi.org/10.1007/978-3-662-44406-1_20

4. Pandey P, Daghma DS, Houben A, Kumlehn J, Melzer M, Rutten T (2017) Dynamics of post-translationally modified histones during barley pollen embryogenesis in the presence or absence of the epi-drug trichostatin a. Plant Reprod 30(2):95–105. https://doi.org/10.1007/s00497-017-0302-5

5. Kumlehn J, Serazetdinova L, Hensel G, Becker D, Loerz H (2006) Genetic transformation of barley (*Hordeum vulgare* L.) via infection of androgenetic pollen cultures with agrobacterium tumefaciens. Plant Biotechnol J 4(2):251–261. https://doi.org/10.1111/j.1467-7652.2005.00178.x

6. Kumlehn J (2009) Embryogenic pollen culture: a promising target for genetic transformation. In: Touraev A, Forster BP, Jain SM (eds) (Hrsg) Advances in haploid production in higher plants. Springer, Netherlands, Dordrecht, pp 295–305. https://doi.org/10.1007/978-1-4020-8854-4_24

7. Koeppel I, Hertig C, Hoffie R, Kumlehn J (2019) Cas endonuclease technology-a quantum leap in the advancement of barley and wheat genetic engineering. Int J Mol Sci 20 (11):2647. https://doi.org/10.3390/ijms20112647

8. Pacher M, Puchta H (2017) From classical mutagenesis to nuclease-based breeding—directing natural DNA repair for a natural

end-product. Plant J 90(4):819–833. https://doi.org/10.1111/tpj.13469

9. Gurushidze M, Trautwein H, Hoffmeister P, Otto I, Müller A, Kumlehn J (2017) Doubled haploidy as a tool for chimaera dissolution of talen-induced mutations in barley. In: Jankowicz-Cieslak J, Till BJ, Kumlehn J, Tai TH (eds) (Hrsg) Biotechnologies for plant mutation breeding: protocols. Springer, New York, pp 129–141. https://doi.org/10.1007/978-3-319-45021-6_8

10. Schedel S, Pencs S, Hensel G, Müller A, Rutten T, Kumlehn J (2017) RNA-guided Cas9-induced mutagenesis in tobacco followed by efficient genetic fixation in doubled haploid plants. Front Plant Sci 7:1995. https://doi.org/10.3389/fpls.2016.01995

11. Stein N, Perovic D, Kumlehn J, Pellio B, Stracke S, Streng S, Ordon F, Graner A (2005) The eukaryotic translation initiation factor 4E confers multiallelic recessive Bymovirus resistance in Hordeum vulgare (L.). Plant J 42(6):912–922. https://doi.org/10.1111/j.1365-313X.2005.02424.x

12. Yang P, Lüpken T, Habekuss A, Hensel G, Steuernagel B, Kilian B, Ariyadasa R, Himmelbach A, Kumlehn J, Scholz U, Ordon F, Stein N (2014) Protein disulfide isomerase like 5-1 is a susceptibility factor to plant viruses. Proc Natl Acad Sci U S A 111 (6):2104–2109. https://doi.org/10.1073/pnas.1320362111

13. Sato K, Yamane M, Yamaji N, Kanamori H, Tagiri A, Schwerdt JG, Fincher GB, Matsumoto T, Takeda K, Komatsuda T (2016) Alanine aminotransferase controls seed dormancy in barley. Nat Commun 7:11625. https://doi.org/10.1038/ncomms11625

14. Gerasimova S, Hertig C, Korotkova A, Kolosovskaya E, Otto I, Hiekel S, Kochetov A, Khlestkina E, Kumlehn J (2020) Conversion of hulled into naked barley by Cas endonuclease-mediated knockout of the NUD gene. BMC Plant Biol 20:255. https://doi.org/10.1186/s12870-020-02454-9

15. Pallotta MA, Graham RD, Langridge P, Sparrow DHB, Barker SJ (2000) RFLP mapping of manganese efficiency in barley. Theor Appl Genet 101(7):1100–1108. https://doi.org/10.1007/s001220051585

Chapter 10

Generation of Doubled Haploid Barley by Interspecific Pollination with *Hordeum bulbosum*

Pooja Satpathy, Sara Audije de la Fuente, Vivian Ott, Andrea Müller, Heike Büchner, Diaa Eldin S. Daghma, and Jochen Kumlehn

Abstract

The generation of doubled haploid barley plants by means of the so-called "Bulbosum" method has been practiced for meanwhile five decades. It rests upon the pollination of barley by its wild relative *Hordeum bulbosum*. This can result in the formation of hybrid embryos whose further development is typically associated with the loss of the pollinator's chromosomes. In recent years, this principle has, however, only rarely been used owing to the availability of efficient methods of anther and microspore culture. On the other hand, immature pollen-derived embryogenesis is to some extent prone to segregation bias in the resultant populations of haploids, which is due to its genotype dependency. Therefore, the principle of uniparental genome elimination has more recently regained increasing interest within the plant research and breeding community. The development of the present protocol relied on the use of the spring-type barley cultivar Golden Promise. The protocol is the result of a series of comparative experiments, which have addressed various methodological facets. The most influential ones included the method of emasculation, the temperature at flowering and early embryo development, the method, point in time and concentration of auxin administration for the stimulation of caryopsis development, the developmental stage at embryo dissection, as well as the nutrient medium used for embryo rescue. The present protocol allows the production of haploid barley plants at an efficiency of ca. 25% of the pollinated florets.

Key words Dicamba, *Hordeum vulgare*, Embryo rescue, Interspecific pollination, *Triticeae*, Uniparental genome elimination

1 Introduction

Over the history of human civilization, barley represented an important source for food, feed, and malt. Plant breeding is one of the most effective measures for securing agricultural production under consideration of actual societal demands. The generation of doubled haploids (DHs) proved to be instrumental to speed up the exceedingly time-consuming process of barley breeding [1]. Barley is an inbreeding species, which is why landraces and approved

Jose M. Seguí-Simarro (ed.), *Doubled Haploid Technology: Volume 1: General Topics, Alliaceae, Cereals*,
Methods in Molecular Biology, vol. 2287, https://doi.org/10.1007/978-1-0716-1315-3_10,
© Springer Science+Business Media, LLC, part of Springer Nature 2021

varieties are in a largely true-breeding condition. Therefore, new cultivars can be directly bred via selection from populations of DH lines.

Various methods are available to produce haploid barley plants. These methods particularly differ regarding the cellular origin of plant regeneration. The stimulation of immature pollen at the microspore stage to undergo cell proliferation and embryogenic development is the most widely used approach in plant research and breeding today, be it through the culture of anthers or of isolated microspores [2, 3]. However, the high efficiency of these methods is at the expense of a comparatively high dependence on the genotype. This dependence has the implication that the recalcitrant portion of the starting material results in too few DH lines. Moreover, not all meiotically recombined individuals of the pollen population used have the capability of conversion into plants. Hence, a percentage of the possible genetic diversity is lost in the process of DH production [4]. The genetic diversity to select from is, however, of utmost importance for the application of haploid technology. The principle of producing haploid plants by uniparental genome elimination after interspecific crossing of barley with *H. bulbosum* is associated with lower genotype dependence than the developmental pathway of pollen embryogenesis [5]. Upon barley × *H. bulbosum* crosses, haploid founder cells for plant development are formed due to the loss of the *H. bulbosum* chromosomes in the hybrid embryo during the first stages of development (Fig. 1). The incompatibility of these foreign chromosomes with the cellular processes occurring in barley is the anticipated reason for this loss. In this context, the failure of *H. bulbosum* chromosomes to condense in time and their incomplete or missing formation of a functional centromere are playing major roles [6, 7]. Consequently, the appropriate processing of the *H. bulbosum* chromosomes in the further course of cell division is prone to fail. Intriguingly, the active degradation of these chromosomes suggests that their aberrant state is recognized by the cells.

The haploid barley genome remaining in the embryo is the basis for the following development of a plant that is likewise haploid. For the utility of haploids in breeding programs, a genome duplication step to reassume the diploid level of somatic cells is still required. However, in contrast to immature pollen-derived embryogenesis, the "Bulbosum" method does not typically involve spontaneous whole-genome duplication. This circumstance further contributes to the relatively low efficiency of this method, because the indispensable, chemically induced genome duplication is not only labor-intensive but also causes losses among the regenerated plants to some extent.

Since the discovery of this method to produce haploid barley plants from crossings of barley with *H. bulbosum* [8], some efficiency-determining factors have been revealed. A particularly

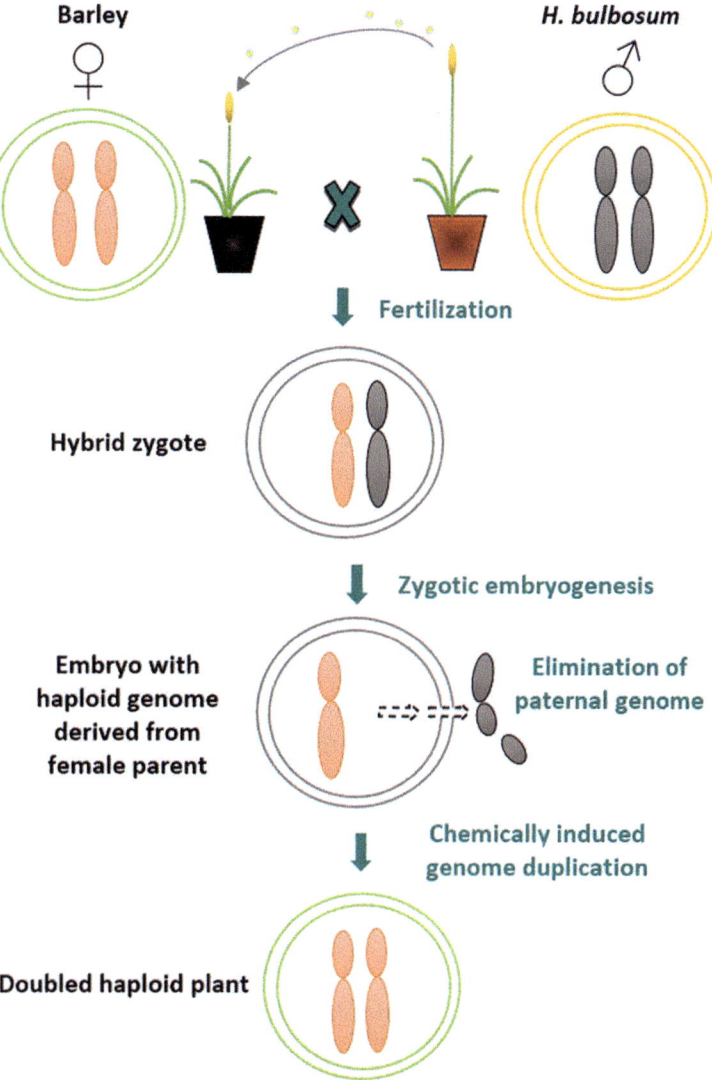

Fig. 1 The "Bulbosum" method involves that barley is pollinated by *Hordeum bulbosum*, which results in the formation of hybrid zygotes. The pollinator's genome gets lost during early embryogenesis and haploid barley plants develop. Upon chemically induced whole-genome duplication, the plants become doubled haploid, entirely homozygous and fertile

important finding was that the limited development of caryopses can be overcome by treatment with strong synthetic auxin analogues such as 2,4-D or Dicamba [9, 10]. After pollination with *H. bulbosum*, the development of the endosperm is at best rudimentary, which is why the resulting grains are hardly capable of germination. Therefore, it is necessary to rescue immature embryos from the caryopses at an appropriate time of their development and to cultivate them in vitro on a suitable nutrient medium. Our experimental work revealed that some methodical variables have a particularly high relevance regarding the efficiency and reproducibility of the protocol. These include the method of emasculation, the temperature at flowering and early grain development, the method, concentration and point in time of auxin administration, the developmental stage at embryo dissection, as well as the nutrient medium used to rescue these embryos. The application of the method described here leads to haploid formation with largely consistent efficiency. Relative to the number of emasculated and *H. bulbosum*-pollinated florets, about 85% of the caryopses respond by increasing in size. Caryopses from about two thirds of the florets produce haploid embryos, about half of which germinate in vitro. Plants established in soil substrate derive from about one quarter of the cross-pollinated florets. Though leaving room for further improvements, the protocol represents a viable means for the generation of haploid plants in the context of biotechnological research and practical breeding in barley.

2   Materials

2.1   Plant Material

The establishment of the present protocol relied on the two-rowed spring-type barley (*Hordeum vulgare* L.) cv. Golden Promise. This British variety is a gamma-ray-induced semidwarf mutant derived from the cv. Maythorpe [11]. Golden Promise is well known for its high tissue culture amenability, and hence it has been widely used in genetic engineering [12]. Golden Promise plants are used as maternal parents in the barley × *Hordeum bulbosum* crosses.

The pollination of barley spikes is conducted using *Hordeum bulbosum* L., which is a wild relative of cultivated barley and the only species representing the secondary gene pool of the latter. Diploid and tetraploid *H. bulbosum* accessions are useful pollinators (*see* **Note 1**).

2.2   Equipment

1. Preparation microscope.
2. pH meter.
3. Rotary shaker.
4. Laminar flow hood.
5. Dark incubator adjusted to 22 °C.
6. Light incubator adjusted to 22 °C.

2.3   Laboratory Materials

1. Forceps, scalpel, needles, brush.
2. Pipettes and disposable autoclaved filter tips (200–1000μL).
3. Sealing tapes Parafilm and Nescofilm.
4. Sterile Petri dishes with lid (10 cm in diameter).

5. Sterile plastic boxes with lid ($107 \times 94 \times 96$ cm).

6. Screw-capped polypropylene tubes (50 mL).

7. Screw-capped polypropylene tubes (15 mL).

8. Cling film (transparent polyethylene).

2.4 Glasshouse and Growth Chamber Materials

1. 3:1:2 mixture of garden soil, sand and Substrate 2 (Klasmann-Deilmann, Germany) for germination of barley grains

2. 2:2:1 mixture of compost, Substrate 2 (Klasmann-Deilmann, Germany) and sand for plant growth.

3. Petuniensubstrat (Klasmann-Deilmann, Germany) for acclimation of regenerants from embryo rescue.

4. Fertilizer: Osmocote (19% N, 6% P, and 12% K) (ICL Specialty Fertilizers, Israel).

2.5 Stock Solutions

Dissolve components in doubled-distilled water if not stated otherwise. All stock solutions are to be autoclaved and then stored at room temperature, if not stated otherwise.

2.5.1 Mineral Salts

1. K-Macro minerals ($20\times$): 6.4 g/L NH_4NO_3, 72.8 g/L KNO_3, 6.8 g/L KH_2PO_4, 8.82 g/L $CaCl_2 \cdot 2H_2O$, 4.92 g/L $MgSO_4 \cdot 7H_2O$, filter-sterilized.

2. K-Micro minerals ($1000\times$): 84 mg/L $MnSO_4 \cdot H_2O$, 31 mg/L H_3BO_3, 72 mg/L $ZnSO_4 \cdot 7H_2O$, 1.2 mg/L $Na_2MoO_4 \cdot 2H_2O$, 0.25 mg/L $CuSO_4 \cdot 5H_2O$, 0.24 mg/L $CoCl_2 \cdot 6H_2O$, 1.7 mg/L KI, filter-sterilized, stored at 4 °C.

3. Chu N6 macro and micro minerals: Ready-to-use product from Sigma-Aldrich (C1416), filter-sterilized [13].

4. Ethylenediaminetetraacetic acid, ferric sodium salt (NaFeEDTA, 75 mM): 2.75 g per 100 mL filter-sterilized, stored at 4 °C (*see* **Note 2**).

5. Copper sulfate (10 mM): 159.6 mg per 100 mL $CuSO_4$, filter-sterilized, stored at 4 °C.

2.5.2 Organic Supplements

1. Maltose (1 M): 360 g/L maltose monohydrate (standard quality), filter-sterilized.

2. Kao and Michayluk vitamins ($100\times$): Ready-to-use product from Sigma-Aldrich (K3129), stored at −20 °C.

3. Gamborg B5 vitamins ($1000\times$): Ready-to-use product from Sigma-Aldrich (G1019), stored at 2–8 °C.

4. L-Glutamine (0.25 M): 3.65 g per 100 mL, dissolved in few drops of 0.1 M KOH, then made up to the final volume using ca. 40 °C double-distilled water, filter-sterilized, stored at −20 °C, defreeze and re-dissolve aliquots in a heated water bath before use.

5. 6-Benzylaminopurine (BAP, 1 mg/mL): stored at 4 °C.

6. Kinetin (10 mM): 215.2 mg per 100 mL dissolved in few drops of 1 M NaOH, made up to 100 mL with hot water, filter-sterilized, stored at 4 °C.

7. Indole-3-butyric acid (IBA, 10 mM): 203.2 mg per 100 mL dissolved in few drops of 50% ethanol, made up to 100 mL with hot water (*see* **Note 3**), filter-sterilized, stored at 4 °C.

8. Dicamba for nutrient medium (100 mg/100 mL): Dissolve 50 mg in 2 mL of absolute ethanol and make volume up to 50 mL by adding heated doubled-distilled water, filter-sterilized.

9. Dicamba for spike treatments (100 mg/L): Diluted with distilled water and mixed with one or two drops of Tween 20 (freshly prepared), filter-sterilized.

2.5.3 Selective Agents

1. Timentin (150 mg/mL): filter-sterilized, stored at −20 °C.

2.5.4 Gelling Agent

1. Phytagel (2×): 8 g/L (for WER1T) and 6 g/L (for K4NBT), autoclaved, cooled down to ca. 50 °C or re-melted before final mixing with other media components.

2.5.5 Other Solutions

1. Sodium hypochlorite (2.4% w/v): 200 mL/L of 12% (w/v) NaOCl, add 0.1% (v/v) Tween 20 (freshly prepared).

2. Ethanol (70% v/v).

3. 1 N NaOH.

4. Autoclaved tap water.

5. Colchicine solution (0.1% w/v, freshly prepared): Dissolved in 0.8% DMSO, made up to final volume by adding tap water, and 1 drop of Tween 20/100 mL added.

2.6 Culture Media

2.6.1 Embryo Rescue

1. Mix all the components described in Table 1 in double-distilled water and fill up to 0.5 L.

2. Adjust the pH to 5.8 by titration with NaOH.

3. Filter-sterilize the solution.

4. Warm up the solution to ca. 40 °C before mixing with the melted Phytagel stock (8 g/L).

2.6.2 Plant Regeneration

1. Mix all the components described in Table 2 in double-distilled water and fill up to 0.5 L.

2. Adjust the pH to 5.8 by titration with NaOH.

3. Filter-sterilize the solution.

4. Warm up the solution to ca. 40 °C before mixing with the melted Phytagel stock (6 g/L).

Table 1
Composition of WER1T medium

Components	Added per L medium	Final concentration
Chu N6 macro and micro minerals	4 g	1×
1 M maltose	0.15µL	0.15µM
75 mM NaFeEDTA	1 mL	75µM
0.25 M glutamine	20 mL	5 mM
KM vitamins (100×)	10 mL	1×
Kinetin (10 mM)	200µL	2µM
IBA (10 mM)	200µL	2µM
Timetin (150 mg/mL)	667µL	100 mg/L

Table 2
Composition of K4NBT medium

Components	Added per L medium	Final Concentration
1 M maltose	100 mL	0.1 M
K-macro minerals (20×)	50 mL	1×
K-micro minerals (1000×)	1 mL	1×
75 mM NaFeEDTA	1 mL	75µM
B5 vitamins (1000×)	1 mL	1×
0.25 M glutamine	4 mL	1 mM
10 mM $CuSO_4$	490µL	5µM
Timetin (150 mg/mL)	1 mL	150 mg/L
BAP (1 mg/mL)	225µL	1µM

3 Methods

3.1 Plant Growth Conditions

3.1.1 Barley

1. Germinate the grains by sowing in trays filled with soil substrate in a growth chamber with a day/night temperature regime of 14/12 °C and a 12-h photoperiod (136µmol m^{-2} s^{-1}).

2. After 3 weeks of germination, pique the seedlings into substrate-filled 2-L pots. At the tillering stage, apply a dressing of 15 g Osmocote to each pot. Transfer the plants during the tiller elongation stage to a glasshouse maintained at 18/16 °C day/night with at least 16 hours of daily photoperiod (170µmol m^{-2} s^{-1}).

3. After pollination, transfer the plants to a glasshouse with 22/20 °C day/night temperature and at least 16-h photoperiod.

3.1.2 Hordeum bulbosum

1. For vernalization, grow the *H. bulbosum* plants first in a cold room at 4 °C with 10-h photoperiod per day provided by cool-white fluorescent tubes for 8 weeks, before being transferred to a glasshouse cabin with 18/16 °C day/night with at least 16 h of photoperiod ($170 \mu mol \, m^{-2} \, s^{-1}$).

3.2 Generation of Haploid Barley

3.2.1 Emasculation

1. Keep the barley plants in a glasshouse maintained at 18/16 °C with a photoperiod of at least 16 h. Emasculate barley ears 2–3 days before anthesis. About 25 florets reside within the synchronously developed central portion of the ear. Slit the outer glumes of these florets longitudinally using surface-sterilized forceps. Then, remove the three anthers per floret (*see* **Note 4**, Fig. 2a). Remove all other florets from the tip and base of the ear.

2. Cover the ears with tagged glassine bags in order to prevent accidental pollination and desiccation. Register the date of emasculation on the tags.

3.2.2 Pollination of Barley with H. bulbosum Pollen

1. Pollinate the ears 2–3 days after emasculation, when the stigmas are spread, which indicates their full receptivity. To this end, collect fresh *H. bulbosum* pollen in a glass Petri dish by gently tapping the ears that are about to undergo anthesis. Gently dust the pollen over the emasculated florets using a paintbrush (Fig. 2b).

2. Then, cover the ears again by the glassine bags and register the pollinator plant and the pollination date on the tag.

3. Transfer the plants to a glasshouse with 22/20 °C day/night temperature right after pollination.

3.2.3 Auxin Treatment

1. Three days after pollination of the barley pistils with *H. bulbosum* pollen, treat the spikes with auxin, which triggers the caryopses to develop, thereby providing room and nutrient supply for the embryos.

2. Before auxin injection, puncture the internode twice in its upper region close to the node residing right below the ear using an injection needle. These venting holes are required to facilitate the release of air from the internode when being filled with Dicamba solution.

3. Tightly seal the lower end of the uppermost internode of the culm using Nescofilm.

4. Then, inject Dicamba solution through the Nescofilm into the base of the uppermost internode of the culm using a syringe. Fill the internode with 0.15 mL or until solution is delivered

Fig. 2 Important steps of the "Bulbosum" method. (**a**) emasculation, (**b**) cross-pollination, (**c**) injection of auxin solution in the uppermost internode of the spike, (**d**) grain development, (**e**) collected caryopses, (**f**) embryo rescue

from the venting holes at the top end of the internode (Fig. 2c). Prevent the solution from running out after removal of the syringe by immediately sealing the injection site with additional layers of Nescofilm.

5. Finally, cover the ear with a glassine pollination bag and register the date of auxin treatment on the tag.

3.2.4 Embryo Rescue

As of point 2, conduct this entire procedure under aseptic conditions.

1. Fourteen days after pollination, harvest the spikes and collect the enlarged caryopses (Fig. 2d, e) in a 15-mL polypropylene tube (*see* **Note 5**).

2. Add 40 mL of 70% (v/v) ethanol to a caryopses-containing tube and keep the closed tube in horizontal orientation on a rotary shaker at 100 rpm for 2 min.

3. Replace the ethanol by 40 mL of autoclaved tap water and wash the caryopses by inverting the tube several times.

4. Replace the water by 2.4% sodium hypochlorite solution supplemented with a drop of Tween 20. Keep the tube on the shaker at 100 rpm for 20 min.

5. Wash the disinfected caryopses five times with autoclaved tap water.

6. Using a scalpel and tweezers, vertically cut the grains open under observation through a preparation microscope. Take the embryos with tweezers out of the caryopses.

7. Transfer the embryos to a Petri dish containing embryo rescue medium WER1T. The embryos are cultivated with their shoot apex facing up for 1 week in the dark at 22 °C (Fig. 2f).

8. Then, transfer the embryo in the same orientation to transparent polypropylene boxes containing K4NB regeneration medium. Incubate these cultures for another 3 weeks in an incubator with 16 h daily light at 22 °C.

9. Once the main shoot and some roots have developed, transfer the regenerated plantlets to pots containing soil substrate.

10. Cover the tray with these pots by a transparent plastic hood for acclimation under the condition of saturated air moisture. Incubate the plantlets in a climate chamber 14/12 °C day/night with 16-h photoperiod.

3.2.5 Colchicine-Induced Whole-Genome Duplication

1. After establishment of plantlets in pots and formation of at least 2–3 tillers, transfer the plantlets to a glasshouse cabin with 18/16 °C day/night for 3 days.

2. Then transfer the plantlets to a cold room (4 °C, 10 h photoperiod) for 48 h in order to accumulate the plants' meristematic cells at the G2 phase of the cell cycle.

3. For the treatment with colchicine, remove the plantlets from the soil substrate and rinse the roots with running water. Cut the roots and shoots to ca. 1.5 cm and 8 cm of remaining length, respectively. Place the plantlets in 50-mL polypropylene tubes containing colchicine solution making sure the tiller bases to immerse, while not submersing the cut tips of the tillers. Close the tubes to prevent any perspiration of the toxic solution. Wrap all tubes together with cling film.

4. Incubate the tubes with plantlets and colchicine solution at 22 °C under strong light for 6 h.

5. Thereafter, discard the colchicine solution and rinse the plantlets carefully using running tap water.

6. Transfer the plantlets back to soil substrate and incubate them in the glasshouse at 18/16 °C day/night till maturity (*see* **Note 6**). For acclimation in saturated air humidity, place the pots on a tray covered by a transparent hood for 1 week.

4   Notes

1. The *H. bulbosum* accessions PB1, FBB, 2929, and 3811b (diploids) as well as 2032, 2920 (tetraploids) were kindly provided by Drs. Frank Blattner and Andreas Houben (IPK Gatersleben, Germany). While all tested accessions were amenable to this protocol, the two diploids FBB and 2929 proved to be the most effective. The accession FBB was previously used to establish microspore-derived haploid formation, and using a DH line derived thereof, the first published sequence-enriched genetic map of *H. bulbosum* was generated [14].

2. If NaFeEDTA stock solution is stored in transparent tubes, wrap these in aluminum foil.

3. To prevent the immediate re-precipitation of the dissolved molecules when being mixed with (doubled-distilled) water, the latter is to be heated prior to the procedure. Precipitation will no longer take place at the final dilution of the stock, even when cooling down the solution.

4. Make sure that the stigma is not injured.

5. Accidentally self-pollinated caryopses can be easily recognized by their much bigger size due to normal grain expansion and endosperm formation.

6. Grain set is a reliable indication of successful whole-genome duplication.

Acknowleddegments

We wish to thank the gardener team headed by Mr. Enk Geyer for the excellent support. We thank the German Federal Ministry for Science and Education for funding our research in frame of the DELITE project (FKZ 031B0550). We are also grateful to the Leibniz Institute of Plant Genetics and Crop Plant Research (IPK) Gatersleben for providing excellent working conditions.

References

1. Kumlehn J (2014) Haploid technology. In: Kumlehn J, Stein N (eds) Biotechnological approaches to barley improvement, series: biotechnology in agriculture and forestry, vol 69. Springer-Verlag Berlin Heidelberg, Berlin, pp 379–392

2. Kumlehn J, Serazetdinova L, Hensel G, Becker D, Loerz H (2006) Genetic transformation of barley (*Hordeum vulgare* L.) via infection of androgenetic pollen cultures with agrobacterium tumefaciens. Plant Biotechnol J 4:251–261

3. Pandey P, Daghma DS, Houben A, Kumlehn J, Melzer M, Rutten T (2017) Dynamics of post-translationally modified histones during barley pollen embryogenesis in the presence or absence of the epi-drug trichostatin A. Plant Reprod 30:95–105

4. Bélanger S, Esteves P, Clermont I, Jean M, Belzile F (2016) Genotyping-by-sequencing on pooled samples and its use in measuring segregation bias during the course of androgenesis in barley. Plant Genome 9:2014.10.0073

5. Kalinowska K, Chamas S, Unkel K, Demidov D, Lermontova I, Dresselhaus T, Kumlehn J, Dunemann F, Houben A (2019) State-of-the-art and novel developments of in vivo haploid technologies. Theor Appl Genet 132:593–605

6. Gernand D, Rutten T, Varshney A, Rubtsova M, Prodanovic S, Brüß C, Kumlehn J, Matzk F, Houben A (2005) Uniparental chromosome elimination at mitosis and interphase in wheat and pearl millet crosses involves micronucleus formation, progressive heterochromatinization, and DNA fragmentation. Plant Cell 17:2431–2438

7. Gernand D, Rutten T, Pickering R, Houben A (2006) Elimination of chromosomes in *Hordeum vulgare* × *H. bulbosum* crosses at mitosis and interphase involves micronucleus formation and progressive heterochromatinization. Cytogenet Genome Res 114:169–174

8. Kasha KJ, Kao KN (1970) High frequency haploid production in barley (*Hordeum vulgare* L.). Nature 225:874–876

9. Matzk F (1991) A novel approach to differentiated embryos in the absence of endosperm. Sex Plant Reprod 4:88–94

10. Pickering RA, Wallace AR (1994) Gibberellic acid + 2,4-D improves seed quality in *Hordeum vulgare* L. × *H. bulbosum* L. crosses. Plant Breed 113:174–117

11. Forster BP (2001) Mutation genetics of salt tolerance in barley: an assessment of Golden promise and other semi-dwarf mutants. Euphytica 120:317–328

12. Gerasimova S, Hertig C, Korotkova A, Kolosovskaya E, Otto I, Hiekel S, Kochetov A, Khlestkina E, Kumlehn J (2020) Conversion of hulled into naked barley by Cas endonuclease-mediated knockout of the NUD gene. BMC Plant Biol 20:255

13. Chu CC, Wang CC, Sun CS, Chen H, Yin KC, Chuc Y, Bi FY (1975) Establishment of an efficient medium for anther culture of rice through comparative experiments on the nitrogen source. Sci Sinica 18:659–668

14. Wendler N, Mascher M, Himmelbach A, Bini F, Kumlehn J, Stein N (2017) A high-density, sequence enriched genetic map of *Hordeum bulbosum* and its collinearity to *H. vulgare*. Plant Genome 10:2017.06.0049

Chapter 11

Bread Wheat Doubled Haploid Production by Anther Culture

Ana María Castillo, Isabel Valero-Rubira, Sandra Allué, María Asunción Costar, and María Pilar Vallés

Abstract

The use of doubled haploid (DH) plants in plant breeding programmes is the fastest route to release new varieties (4–6 years), allowing for a rapid response to end-user needs. Microspore embryogenesis is one of the most efficient methods for DH plant production in bread wheat. In this process, microspores triggered by a stress treatment or by application of bioactive compounds are reprogrammed to follow an embryogenic pathway that leads to the production of haploid or DH plants. In this chapter, we describe a protocol for anther culture of bread wheat. This protocol is based on an osmotic and starvation treatment of the anthers followed by the application of a microtubule disrupting agent. Anthers are cultured in an ovary pre-conditioned medium with mature ovaries from cv. Caramba. This protocol has been applied to a wide range of genotypes and F1s from bread and spelt wheat.

Key words Anther culture, Bread wheat, Colchicine, Doubled haploid, Microspore embryogenesis

1 Introduction

Wheat (*Triticum aestivum* L.) is a major grain cereal, being a key source of calories and protein for humans. Wheat ranks third among the cereals, following maize and rice, in worldwide production, with 734 million tons [1]. It has been estimated that wheat production should increase 38% by 2050 to meet the projected demands of population growth, dietary changes, and increasing biofuel consumption [2]. The development of more nutritious, resilient and productive wheat varieties would help to significantly increase wheat production. Among the different tools available to plant breeders, the use of doubled haploids (DH) is the fastest route for the release of new varieties since DHs are completely homozygous plants that are obtained in one generation, thus saving several generations of selfing. Furthermore, phenotypic selection efficiency performed on DH population is substantially improved compared to conventional methods. Therefore, breeders can

Jose M. Seguí-Simarro (ed.), *Doubled Haploid Technology: Volume 1: General Topics, Alliaceae, Cereals*,
Methods in Molecular Biology, vol. 2287, https://doi.org/10.1007/978-1-0716-1315-3_11,
© Springer Science+Business Media, LLC, part of Springer Nature 2021

release a new variety in 4–6 years, responding more rapidly to end-user needs.

Of the different methods for DH plant production, intergeneric crosses and microspore embryogenesis (ME) are the most efficient in wheat. In ME, microspores are reprogrammed changing their developmental fate to follow an embryogenic pathway. The initial trigger for this reprogramming is generally a stress treatment, which is followed by an in vitro culture phase that leads to the development of embryos and finally haploid or DH plants [3]. The haploid plants should be treated with a chromosome doubling agent to obtain DH plants. One of the main advantages of ME vs. interspecific crosses with maize is the recovery of a percentage of spontaneous DH plants [4]. However, the main drawback of ME is a genotype dependence greater than in interspecific crosses, and the regeneration of albino plants [5–7]. Both ME and interspecific crosses have been used for DH production by public institutions and seed private companies, and around 263 bread wheat DH varieties had been released until 2016 (for review *see*, [7]).

Microspore embryogenesis can be achieved by anther or isolated microspore culture. Since the first anther culture-derived wheat DH plantlets were reported in 1973 [8], different methodological improvements have been succeeded. [9] published an excellent book of protocols for DH production of different species, including anther and isolated microspore culture in bread wheat. After the publication of these protocols, different strategies have been used to increase ME efficiency by modification of treatments before culture and/or modification of culture media composition. Thus, the application of bioactive compounds, among which we can highlight microtubule disrupting agents (colchicine, *n*-butanol); a membrane permeabilizer agent (DMSO), and epigenetic modifiers (histone deacetylase inhibitors, Trichostatin A), alone or in combination with conventional stress treatments (temperature, osmotic stress, starvation) increased the efficiency of microspore reprogramming [10–14]). Also, the incorporation of antioxidant agents (glutathione, dimethyl tyrosine conjugate peptide), a promoter of cell proliferation (phytosulfokine alpha), as well as the use of medium pre-conditioned with mature ovaries, enhanced embryo formation and green plant regeneration [4, 15–17]. Most of the assays with new compounds have been applied to a reduced number of genotypes. However, robust and efficient wheat anther culture protocols for a wide range of genotypes that are managed in a plant-breeding programme are needed.

In this chapter, we describe a detailed protocol for anther culture of bread wheat, from donor plant growth conditions to plant regeneration, acclimatization and chromosome doubling. This protocol has been successfully used for a wide range of genotypes and F1s of bread wheat, spelt wheat and advanced breeding lines of spelt x bread wheat. This protocol is based on an osmotic and starvation anther stress treatment (0.7 M mannitol), followed by a microtubule disrupting agent (*n*-butanol) application. Anthers are cultured in an ovary pre-conditioned medium with mature ovaries from cv. Caramba, supplemented with Ficoll Type 400, and ovaries are maintained along culture (ovary co-culture). Colchicine is applied to haploid plantlets at 3–4 tillers stage.

2 Materials

2.1 Plant Material

This protocol was developed primarily with two different *Triticum aestivum* L. cultivars: Pavon and Caramba, a high and a mid-low responding cultivar for microspore embryogenesis, respectively. However, the protocol has been successfully used with several cultivars, F1 crosses and wheat transgenic lines, as well as with landraces of spelt wheat, and advanced breeding lines derived from spelt x bread wheat crosses.

2.2 Equipment

1. Laminar flow hood.

2. Incubators at 25 °C without light and with 120 $\mu E/m^2/s^1$ supplied by Philips Master TLD 58 W/865 and GE LED BrightStick 15 W 3000 K.

3. Growth chambers at 6 °C with 100 $\mu E/m^{-2}/s^{-1}$ provided by Philipps Master TLD 36 W/865, 8 h light.

4. Growth chambers at 12/14 °C (day/night) with 12 h light and 18/21 °C (day/night) and 16 h light. Both growth chambers with 450 $\mu E/m^2/s^1$ supplied by high-pressure metal halide lamps (Phillips Powertone HPI-T plus 400 W/645) and 60% RH.

5. Growth chamber at 12/14 °C (day/night) with 200 $\mu E/m^2/s^1$ provided by OSRAM Lumilux Cool White L58W/840 and OSRAM Halolux Ceram Eco 100 W 2900, 12 h light and 70% RH.

6. Inverted Microscope (Nikon TE300 or similar) with epifluorescence and filters EX 365/DM 400/BA and digital sight DS-Ri1 and NIS-Elements F software for DAPI analysis and image processing.

7. Stereo microscope (Nikon SMZ750 or similar) with a digital camera and NIS-Elements F software.

8. Flow cytometer (Partec-Sysmex PAS or a similar device).

9. General laboratory equipment: precision balance, pH meter, autoclave, agitator, refrigerators, vacuum pump, etc.

1. Peat, vermiculite (size 3), and fine sand.

<table>
<tbody>
<tr><td>

2.3 Mother Plant Growth/Disinfection/ Sterilization, Culture, Acclimation and Ploidy Analysis

</td><td>

2. Soil Fertilizers: Osmocot Pro (17:11:10) (Projar) and Nutri-chem (20:20:20) (Agrichem) or equivalent.

3. Plastic seedbeds (trays with 96 sockets) and pots (17 cm Ø × 15.5 cm height).

4. Transparent plastic.

5. Plastic spray bottles with 96% (v/v) ethanol.

6. Sterile Petri dishes of 3, 6 and 9 cm Ø.

7. Sterile 5 and 10 ml disposable pipettes.

8. Sterile glass Pasteur pipettes.

9. Automatic or manual pro-pipettes.

10. Sterile 0.22 μm pore size membrane filters of 150 ml or 500 ml.

11. Bunsen-type burner, or similar, for flame sterilization.

12. Magenta box with ethanol 96% for sterilization of scissors, tweezers, etc.

13. Parafilm.

14. Microscope slides and coverslips.

15. Sterile glass bottles for culture media.

16. Scissors and gauze.

17. Beakers and magnetic stirrer.

18. Fine point and blunt tweezers.

19. Sterile Magenta boxes.

20. Small plastic boxes with wet Whatman filter paper (to keep samples for flow cytometry analysis).

21. Non-sterile Petri dishes cover 3 cm Ø (for flow cytometry).

22. Plastic tubes 3.5 ml (for flow cytometry).

23. 30 μm pore size filters (for flow cytometry).

24. Razor blades.

</td></tr>
</tbody>
</table>

2.4 Culture Media

1. Stress treatment medium (SM medium) (Table 1): macroelements from FHG [18] supplemented with 127.5 g/l mannitol, 40 mM $CaCl_2$ and 8 g/l Sea Plaque agarose. To prepare the medium:

   (a) Add macroelements stock solution (×10) and agarose to MilliQ water and make up to half of the final volume, and sterilize it at 121 °C for 23 min in autoclave (Solution A).

   (b) Dissolve mannitol by heating at 50 °C (*see* **Note 1**) in half of the final volume and sterilize it by filtration through a sterile 0.22 μm pore size membrane filter inside a sterile glass bottle, in the laminar hood (Solution B).

Table 1
Media composition for wheat anther culture (concentration expressed in mg/l)

	Stress treatment	Embryo induction	Plant regeneration	Plant rooting
Compounds	SM	MSMI/MSMIF2/MSMIF4	J25-8	J25-8 R
Macroelements				
KNO_3	1900	1400	2200	2200
NH_4NO_3	165	300	600	600
$KH_2 PO_4$	170	170	150	150
$CaCl_2 \cdot 2H_2O$	1028	440	295	295
$MgSO_4 \cdot 7H_2O$	370	370	310	310
$(NH_4)_2SO_4$			67	67
$NaH_2PO_4 \cdot H_2O$			75	75
Microelementss				
$MnSO_4 \cdot H_2O$		14.2	5	5
$ZnSO_4 \cdot 7H_2O$		8.6	5	5
H_3BO_3		6.2	3	3
KI		0.83		
$CuSO_4 \cdot 5H_2O$		0.025	0.025	0.025
$CoCl_2 \cdot 6H_2O$		0.025	0.025	0.025
$Na_2MoO_4 \cdot 2H_2O$		0.25	0.25	0.25
Iron source				
$FeNa_2$ EDTA		40	40	40
Vitamins				
Tiamine–HCl		0.4	10	10
Piridoxine–HCl		0.5	1	1
Nicotinic acid		0.5		
Glicine		1		
Biotine		0.25		
Ascorbic acid		0.5		
Pantothenic acid		0.25		
Nicotinamide			1	1
Amino acids				
L-Glutamine			120	120
L-Asparagine			50	50

(continued)

Table 1
(continued)

	Stress treatment	Embryo induction	Plant regeneration	Plant rooting
L-Threonine			25	25
L-Arginine			25	25
L-Proline			50	50
Organic compounds				
Myo-inositol		100	100	100
Glutamine		1000		
Casein			125	125
Malic acid			100	100
Coconut water			25[a]	25[a]
Carbohydrates				
Mannitol	127,000			
Maltose		90,000		
Sucrose			20,000	20,000
Glucose			7000	7000
Growth regulators				
2,4-D		1		
BAP		1		
ANA				2
Agarose	8000			
Ficoll type 400		0/200,000/400,000[b]		
Gelrite			3000	3000
pH = 5.8				

[a]25 ml of coconut water
[b]0/200,000/400,000 mg/l of Ficoll correspond to MSMI, MSMIF2 and MSMIF4 media, respectively

> (c) Mix both solutions inside the hood and pour 10 ml in each 6 cm Ø Petri dishes.

2. Embryo induction and development culture media (MSMI, MSMIF2 or MSMIF4; Table 1). These are MMS3-modified media [19] supplemented with 90 g maltose, 100 mg myo-inositol, 1 g glutamine, 1 mg of 2,4-dichlorophenoxiacetic acid (2,4-D), 1 mg of benzylaminopurine (BAP) (MSMI, used for n-butanol treatment) and with 200 or 400 g/l of Ficoll-Type 400 (MSMIF2 medium for embryo induction and MSMIF4, media for embryo

development, respectively). For medium preparation (*see* **Note 2**), pH is adjusted to 5.8 with 1 N NaOH, and sterilized by filtration through 0.22 μM pore size membrane filter inside a sterile glass bottle, in the laminar hood.

3. Plantlet regeneration medium (J25-8) [20] (Table 1). To prepare the medium:
   (a) Dissolve all the components with the exception of activated charcoal and gelrite in half of the final volume (*see* **Notes 3** and **4**), adjust pH to 5.8 with KOH and sterilize by filtration through a 0.22 μM pore size membrane filter inside a sterilized glass bottle (Solution A).
   (b) Add activated charcoal and gelrite to half of the final volume and autoclave at 121 °C for 23 min (Solution B).
   (c) Mix solutions A and B in the laminar hood and pour 10 ml in 6 cm Ø Petri dishes.

4. Rooting medium (J25-8R): This is a J25-8 medium (Table 1) supplemented with 2 mg of naphthaleneacetic acid (NAA). The medium is prepared as J25-8 with the incorporation of 2 mg/l of NAA in solution A (*see* Subheading 2.4, **item 3**).

2.5 Chemicals and Solutions for Anther Culture, Ploidy Analysis and Watering of the Plants

1. 96% Ethanol (v/v).

2. 4,6-Diamidino-2-phenylindole (DAPI) stock solution: 1 mg/ml in MilliQ quality water, stored at −20 °C in the dark.

3. DAPI working solution (4 μg/ml): Dilute the DAPI stock solution 250 times in MilliQ water and add 6 drops of Tween 80 to facilitate the diffusion of DAPI inside the microspores. Cover the Eppendorf tube with aluminum foil. This working solution can be stored at 4 °C for several weeks.

4. 2,4-dichlorophenoxiacetic acid (2,4-D) stock solution 1 mg/ml: Weight 50 mg of 2,4-D in an analytical balance and put in a beaker with a magnetic bar. Add 25 ml ethanol 96%, stir until it gets dissolved and then add MilliQ water, very slowly to avoid precipitation, to make up the volume to 50 ml.

5. Benzylaminopurine (BAP) stock solution 1 mg/ml: Weight 50 mg of BAP in an analytical balance, and put it in small beaker with a magnetic bar, stir and add HCl 1 N, drop by drop, until it is completely dissolved, and then add MilliQ water very slowly to make up the final volume to 50 ml.

6. Naphthaleneacetic acid (NAA) stock solution 1 mg/ml: Weight 50 mg of NAA in an analytical balance, put it in small beaker with a magnetic bar, stir and add NaOH 1 N, drop by drop, until it is completely dissolved and then add MilliQ water very slowly to make up the volume to 50 ml.

7. Commercial Cystain UV ploidy (Partec-Sysmex) containing lysis buffer and DAPI.

8. Hoagland nutrients solution [21] for watering during acclimation.

3   Methods

3.1   Growth Conditions of the Mother Plants

1. Seeds of donor plants are sown in a seedbed with a mixture of peat:vermiculite:fine sand (40 kg:8 kg:50 kg). Soil is fertilized with 285 g of Osmocote Pro (17:11:10).

2. Spring and winter materials are vernalized for 4 and 6 weeks, respectively, in a growth chamber at 6 °C.

3. After vernalization, two plantlets are transplanted to an individual 17 cm Ø × 15.5 cm height pot with the same soil mixture described above, and cultivated in a growth chamber at 14/12 °C (day/night) with 12 photoperiod. After 3 weeks, the temperature is increased to 21/18 °C (day/night) and the photoperiod lengthened to 16 h light. Each pot is fertilized three times during plant development with 200, 200 and 250 ml of 30 g/l of Nutrichem, 4 days before changing the photoperiod and once per week after that, respectively. Plants are watered every 2–3 days as required (*see* **Note 5**).

3.2   Harvest of Spikes and Sterilization

1. Spikes are harvested early in the morning (1–2 h after lights are on) when 60–70% of the microspores are at mid-to-late uninucleate stage (Fig. 1). Spikes at different developmental stages are harvested to establish a correlation between morphological characters of the tiller and spike with microspore developmental stage (Fig. 1a, *see* **Notes 6** and **7**). The microspore developmental stage is determined by DAPI staining (Fig. 1b). Six anthers are taken from the central flowers of a spike and squashed in 15 μl of DAPI working solution. After 10 min of incubation, nuclei are visible in an epifluorescence microscope under a UV lamp.

2. The flag leaf and the rest of the leaves are cut with scissors, the spikes are placed in a plastic container with water and transferred to the laboratory for sterilization.

3. Spikes inside the sheath are disinfected by spraying with ethanol 96% in a flow hood. After drying on a sterile gauze, spikes are removed from sheath and placed in a 9 cm Ø Petri dishes (4–5 spikes/Petri dish), sealed with Parafilm and use directly for anther culture, or store at 4 °C in the dark for no more than 2–3 days. A selection of the spikes is performed at this step taking into account their stiffness and rachilla length (*see* **Note 7**).

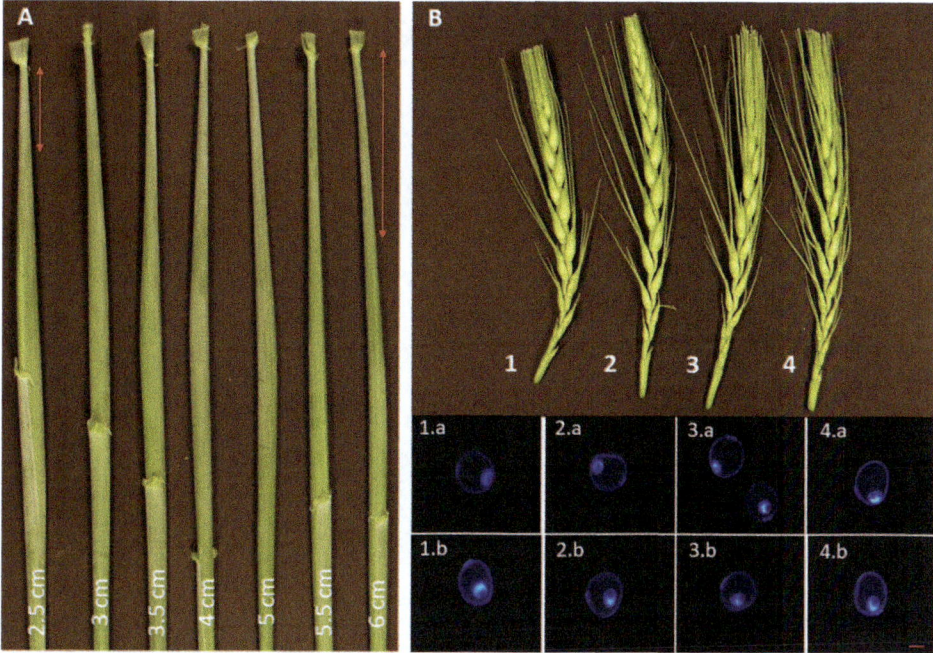

Fig. 1 Correlation between the stage of microspore development and the morphological characters of tillers and spikes in cv. Pavon. (**a**) Tillers with 2.5–6.0 cm length from the distal part of the spike to basal part of the flag leaf are selected from each genotype. (**b**) Upper panel: Spikes from tiller with 3 cm (1), 4 cm (2), 5 cm (3) and 5.5 cm length (4). Note the differences in spike stiffness and rachilla length. Lower panel: DAPI analysis from spikes with 3 cm (1a, 1b), 4 cm (2a, 2b), 5 cm (3a, 3b) and 5.5 cm length (4a, 4b). Tillers with 3–5 cm length contained late uninucleate microspores (1a, 1b, 2a, 3a) and mid-late uninucleate microspores (2b, 3b). Tiller with 5.5 cm length contained early to mid uninucleate microspores (4a, 4b). Bars, B: 20 μm

*3.3 Ovary
Pre-conditioned
Medium*

1. Spikes inside the sheath leaf from cv. Caramba are covered with a sterilized cellophane bag (Fig. 2a) until at least 2/3 of the spike is outside the sheath (around 4–5 days later) (Fig. 2b, *see* **Note 8**). At this stage, ovaries are mature (stigma opened) (Fig. 2d) and microspores are at late binucleate stage (Fig. 2e) [4].

2. Spikes are collected and transferred to the laboratory for sterilization. The cellophane bag is sprayed with ethanol 96% in the laminar hood. The bag is removed and spikes are sprayed thoroughly with ethanol 96% and let dry for 10 min inside the hood (Fig. 2c).

3. Six mature ovaries are excised with forceps under the stereoscopic microscope and cultured in a 3 cm Ø Petri dish containing 2 ml of MSMIF2 (Table 1) for 5–6 days (Fig. 2f).

*3.4 Anther
Pre-Treatment
and Culture*

1. Anthers are extracted from the spike under a stereoscopic microscope and placed on 6 cm Ø Petri dish containing SM medium (Table 1). Cultures are kept at 25 °C in darkness for 5 days (Fig. 3a).

Fig. 2 Preparation of Caramba ovary preconditioned medium. (**a**) Spike inside the sheath is bagged with a sterilized cellophane bag. (**b**) Spike before harvesting for ovary pre-conditioned medium. (**c**) Sterilized spike for ovary excision. (**d**) Mature ovary before culture in MSMIF2 medium. (**e**) Flowers with microspore at late binucleate stage are used for medium preconditioned (DAPI staining). (**f**) Ovaries after 5 days in culture, right before anther culture. Bars, (**d**) 1 mm; (**e**) 10 μm; (**f**) 3 mm

2. After mannitol treatment, white or yellowish swollen anthers are selected for *n*-butanol treatment (Fig. 3b, *see* **Note 9**). Thirty swollen anthers are plated in a 3 cm Ø Petri dish containing 2 ml of MSMI medium (Table 1) with 4 μl of *n*-butanol (0.2% v/v) for 2–3 h at room temperature [11]. Petri dishes are shaken gently every 30 min [11, 22].

3. MSMI medium is removed with a sterile glass Pasteur pipette and replaced with 2 ml of MSMIF2 pre-conditioned medium and 6 ovaries (Table 1, *see* Subheading 3.3, *see* **Note 10**) [4]. Cultures are incubated in darkness at 25 °C.

4. Culture development is followed under an inverted microscope (*see* **Note 11**). The first microspores with star-like morphology can be observed already at 0 days of culture and the number

Fig. 3 Anther culture and DH plant production. (**a**) Anthers at 0 days in stress medium (SM). (**b**) Anthers after 5 days in SM medium, white or yellowish swollen anthers (arrow) are selected for n-butanol treatment. (**c**) Star-like microspores after 5 days in culture. (**d**) Some microspore-derived structures break the exine at 10 days of culture. (**e**) Mature embryos in MSMIF2 medium at 35 days of culture before transferring to regeneration medium. (**f**) Green and albino plant regeneration after 15–20 days in J25-8 medium. (**g**) Acclimation process in a growth chamber with 70% RH. (**h**) Ploidy analysis by flow cytometry. (**i**) Growth of DH plant in the greenhouse. (**j**) Spikes from a colchicine-treated plant with a low number of seeds (arrow). (**k**) Spikes from spontaneous DH plant with 100% seed set. Bars, (**c, d**) 50 μm; (**e**) 2.5 mm

increased after 3–4 days (Fig. 3c). Exine rupture takes place after 10–12 days of culture depending on the genotype (Fig. 3d).

5. After 10–12 days of culture, 2 ml of MSMIF4 medium containing 400 g/l Ficoll is added to each Petri dish (*see* **Note 12**).

6. After 30–35 days of culture, depending on the genotype, mature embryos are developed and transferred to regeneration medium J25-8 [20] (Fig. 3e, *see* **Notes 13** and **14**). Around 20 embryos are transferred to one Petri dish and placed at

25 °C in darkness for 2–3 days and then transferred to light (120 $\mu E/m^2/s^1$ and 16 h light).

3.5 Vernalization and Transfer to Soil

1. Green plantlets, developed after 15–20 days (Fig. 3f), are transferred to Magenta boxes containing J25-8R to strengthen the root system, and placed at 25 °C with 16 h light for 5–7 days before vernalization (*see* **Notes 15** and **16**).

2. Depending on the vernalization requirements of the genotypes, Magenta boxes are transferred to a growth chamber at 6 °C for 4–6 weeks, with the same conditions described in the Subheading 3.1, **step 2**.

3. Magenta boxes are transferred to a growth chamber at 12/14 °C with 200 $\mu E/m^2/s^1$ and 70% humidity for 1 week.

4. Rooted plants are transferred to a seedbed with a mixture of peat:vermiculite (1 kg:0.4 kg), watered with a diluted Hoagland nutrient solution (1:4) [21], and covered with a transparent plastic for 1 week. Plastic is removed and plants are watered with the same solution as needed (Fig. 3g).

3.6 Ploidy Analysis, Chromosome Doubling and Seed Production

1. Small pieces of young leaves are cut out of plantlets 7 days after transferring to soil and placed in a plastic box containing a wet filter paper and kept at a 4 °C. Small pieces of young leaves from seed-derived plants are also cut out to use as controls (*see* **Note 17**).

2. Ploidy analysis is performed by flow cytometry. Turn on the cytometer and the UV light and clean it up before use.

3. Cut a leaf piece of 0.5 cm^2 and place it on a non-sterile Petri dish cover 3 cm Ø. Add 100 μl of Cystain UV ploidy solution and chop it thoroughly using a razor blade until obtaining small chunks in order to get as many nuclei as possible. Add 400 μl of the same solution to homogenize the sample.

4. Filter the homogenate through a 30 μm pore size filter and collect the nuclei in a plastic tube compatible with the loading port of the cytometer. Add 2 ml of MilliQ water and mix thoroughly.

5. Insert the vial in the loading port and analyse DNA content. Fix the fluorescence peak of control samples at X axis at 150 and infer the ploidy level of the samples by comparison with control. A minimum of 8000 counts per sample should be performed (Fig. 3h).

6. Haploid plants with 3–4 tillers are removed from the soil, roots are washed and cut back to about 2 cm. Plants are then immersed, with aeration, to a depth of 4 cm in a 0.12% colchicine aqueous solution containing 2% DMSO for 4 h at 24 °C (modified from [22], *see* **Note 18**).

7. After colchicine treatment, the area of the plantlet exposed to colchicine solution is watered thoroughly with tap water for 15 min before transferring to soil.

8. Colchicine-treated plants and DH plants are transferred to pots with the same soil mixture as indicated in the Subheading 3.1 and placed in a greenhouse at 20 °C for seed production (Fig. 3i–k). Colchicine-treated plants are watered with a diluted Hoagland solution (1:4) [21]) for the first month, and spontaneous DH plants with water as needed. Plants are fertilized twice with Nutrichem before heading (*see* **Notes 19** and **20**).

4 Notes

1. Dissolve the mannitol in less than half of the final volume. For example, to prepare 1 l of SM, add 127.54 g mannitol to 400 ml of MilliQ water, heat it at 50 °C (avoid overheating, otherwise mannitol structure will be disrupted), and then add MilliQ water to make the volume to 500 ml.

2. Stock solution containing all macroelements, microelements, iron source and vitamins (×2) is prepared and kept at −20 °C. A small volume of medium is prepared since anthers are cultured in 3 cm Ø Petri dishes. For example, to prepare 400 ml of MSMIF2 or MSMIF4:

   (a) Defrost 200 ml of stock solution and add 36 g maltose, 40 mg myo-inositol, 400 mg glutamine, 400 μl 2-4-D (1 mg/ml) and 400 μl BAP (1 mg/ml) (Solution A).

   (b) Add 40 or 80 g of Ficoll in 140 ml MilliQ water (for MSMF2 and MSMF4, respectively). Due to the high amount of Ficoll, the volume of the solution increases significantly, so it is important to dissolve Ficoll in a reduced volume of water. To dissolve Ficoll, it is necessary to apply heat until it starts boiling and the solution turns transparent. Over boiling should be avoided, otherwise Ficoll will be burnt and turn brownish. Let it cool down at room temperature (Solution B).

   (c) Mix solutions A and B and make up the volume to 400 ml.
       To prepare 400 ml of MSMI, solution A should be made up to 400 ml with MilliQ water. After sterilization, the three media are kept at room temperature for 4 days and then placed at 4 °C until use.

3. Stock solutions containing macroelements (×20), microelements (×100), iron source (×100), vitamins (×100), and aminoacids (×100) are prepared and kept at −20 °C. Macroelements are prepared in two stock solutions: the first contains $(NH_4)_2SO_4$ and $NaH_2PO_4 \cdot H_2O$ and the second

contains the rest of the macroelements (Table 1) to avoid precipitation. To prepare 1 l of J25-8 medium: defrost 50 ml of macroelement stock solution and 10 ml of iron source, vitamins and aminoacids stock solutions, and 25 ml of coconut water (*see* **Note 4**). Add the stock solutions to 250 ml MilliQ water and mix, then add 100 mg myo-inositol, 125 mg casein, 20 g sucrose and 7 g glucose. Dissolve the malic acid in 50 ml of MilliQ water by adding 3 drops of NH_4OH and add it to the solution. Make up the volume to 500 ml.

4. This medium contains coconut water (the liquid endosperm of coconuts), which is prepared in the laboratory from mature coconuts, acquired in the supermarket, as follows:

    (a) Select heavy coconuts with a high content of water (it is advisable to shake them and hear the water sloshing around).

    (b) Poke a hole through the two softest of the three eyes of the coconut with a clean screwdriver or a big nail and a hammer.

    (c) Invert the coconut and place over a glass container to drain the juice (pour the liquid from each coconut separately, it should smell well, otherwise the juice should be thrown away).

    (d) Mix the water from different coconuts and boil it for 10 min for protein precipitation, let it cold down at room temperature.

    (e) Filter through a Whatmam filter before freezing at $-20\,°C$.

5. Plants should be kept free from pests and fungal infections. The application of pesticides or fungicides before or after heading decreases significantly the number of viable microspores and thus the efficiency of embryo production.

6. The selection of the spikes at the right stage of microspore development is one of the most critical points to achieve a high efficiency of microspore embryogenesis. Since it is not possible to determine this parameter from each spike, correlations between the stage of microspore development and morphological characters such as the length from the distal part of the spike to basal part of the flag leaf (Fig. 1a) are established. From each batch of plants and genotype, 6–8 spikes with the different lengths described above are selected for DAPI staining of the microspores (Fig. 1b). The stage of microspore development is recognized by the position of the nucleus in the microspore: in late uninucleate microspores, the nucleus is located at the opposite side of the pore (Fig. 1b: 1a, 1b, 2a, 3a); in mid uninucleate microspores, the nucleus is half way between the pore and the opposite side (Fig. 1b: 2b, 3b).

Spike 4 will be eliminated since microspores are at early to mid uninucleate stage (nucleus located close to the pore; Fig. 1b: 4a, 4b). Therefore, spikes with a 3–5 cm distance between the basal part of the flag leaf and the distal part of the spike will be harvested from this genotype.

7. Spikes which have a very flabby or very rigid palea, lemma and low glume and spikes with a long or a very short rachilla are eliminated (Fig. 1b). Spike 4 (on the right side of Fig. 1b) with a long rachilla and very soft palea and lemma will not be used for anther culture since microspores are at the early to mid-uninucleate stage.

8. Ovary genotype, as well as ovary developmental stage are important factors to be considered since they affect the percentage of embryo formation. Ovaries from cv. Caramba are used in our laboratory since they induce microspore embryogenesis more efficiently than ovaries from other genotypes. Furthermore, mature ovaries from flowers containing microspores at late binucleate stage produced higher percentages of embryogenesis than young ovaries (Fig. 2d, e). After 5 days in culture medium, ovaries grow, stigmas open and lodicules swell (Fig. 2f).

9. During stress treatment, anthers change their colour and morphology. They turn white to yellow and swell at different degrees. White or yellowish swollen anthers, containing microspores that have changed their developmental pathway, are selected for n-butanol treatment (Fig. 3b).

10. Ovaries are maintained along anther culture until the embryos are developed (30–35 days of culture).

11. During mannitol treatment, a degradation of the tapetum and opening of anther wall is produced in some anthers. Thus, microspores are released from the anther into the culture medium and its development can be followed under an inverted microscope.

12. Ficoll is a high-molecular-weight polysaccharide that increases the density of the medium avoiding sinking of pro-embryos and dying due to anoxia. As divisions take place, the pro-embryos have a tendency to sink into the medium, which it usually happens at the time of exine rupture. Therefore, the same medium with a higher concentration of Ficoll is added to each Petri dish after 10–12 days of culture.

13. Mature embryos are characterized by a white colour due to starch accumulation in their cells. A large variation in the quality of the embryos can be observed, and this is a genotype-dependent character. Preferably, well-developed embryos are selected since the percentage of plant regeneration from these embryos is quite high. However, if one genotype

produces only poor-quality embryos, all embryo-like structures are transferred for regeneration, but not calli.

14. Embryos develop asynchronously. Therefore, mature embryos are transferred to regeneration medium every 12–14 days (two or three times per plate).

15. Green and albino plants are produced, but only green plants are transferred to J25-8R medium since albino plants are not able to survive after transferring to soil.

16. Plants are transferred to the same medium with 2 mg/l NAA to strengthen the root system, which is critical for plant acclimation.

17. Leaves of young plantlets should be used to obtain a low variation of fluorescence coefficient in samples.

18. Several factors are critical to achieve a higher percentage of plant survival to colchicine treatment, including the stage of development of the plantlet and the light intensity in the greenhouse. Plantlets with 1–2 tillers have a high rate of mortality and plants with 5–6 tillers have a low percentage of seed set. The application of the colchicine treatment should be performed in a sunny day to achieve a higher percentage of doubling.

19. Colchicine-treated plants develop slower than DHs. Death of some of the colchicine-treated tillers usually takes place, but new tillers develop healthy. These plants are more sensitive to fungal infection than spontaneous DH plants. The percentage of seed set in these plants varies from 0% to 50% (Fig. 3j), while in spontaneous doubling it reaches 100% (Fig. 3k).

20. If possible, plants should be transferred to the greenhouse at the end of winter to avoid high temperature during the filling stage of seeds, which would promote the occurrence of shrivelled grains.

Acknowledgements

This research was supported by Projects AGL2013-46698-R and AGL2016-77211-R from State R&D Program Oriented to the Challenges of the Spanish Society, and funds for Reference Research Groups recognized by Diputación General de Aragón, Spain (Groups A06 and A08-20R: Genetics, Genomics, Biotechnology and Plant Breeding). Isabel Valero-Rubira was funded by the Spanish Ministry of Science and Innovation grant no. BES-2017-080970 (linked to Project AGL2016-77211-R).

References

1. FAOSTAT (2020). http://faostat.fao.org. Accessed March 2020

2. Ray DK, Mueller ND, West PC, Foley JA (2013) Yield trends are insufficient to double global crop production by 2050. PLoS One 8 (6):e66428. https://doi.org/10.1371/jour nal.pone.0066428

3. Soriano M, Li H, Boutilier K (2013) Microspore embryogenesis: establishment of embryo identity and pattern in culture. Plant Reprod 26(3):181–196. https://doi.org/10.1007/s00497-013-0226-7

4. Castillo AM, Sánchez-Díaz R, Vallés MP (2015) Effect of ovary induction on bread wheat anther culture: ovary genotype and developmental stage, and candidate gene association. Front Plant Sci 6:402. https://doi.org/10.3389/fpls.2015.00402

5. Lantos C, Weyen J, Orsini JM, Gnad H, Schlieter B, Lein V, Kontowsky S, Jacobi A, MihÁly R, Broughton S, Pauk J (2013) Efficient application of in vitro anther culture for different European winter wheat (Triticum aestivum L.) breeding programmes. Plant Breed 132(2):149–154. https://doi.org/10.1111/pbr.12032

6. Castillo AM, Allué S, Costar MA, Alvaro F, Vallés MP (2019) Doubled haploid production from Spanish and central European spelt by anther culture. J Agric Sci Technol 21 (5):1313–1324

7. Devaux P, Cistué L (2016) Wheat doubled haploids: production to sequencing. What makes them so appealing? In: Bonjean AP, Angus WJ, van Ginkel M (eds) The world wheat book. A History of Wheat Breeding. Lavoisier Tec & Doc Publishers, Paris, pp 885–938. http://hdl.handle.net/10261/132172

8. Picard E, De Buyser J (1973) Obtention de plantes haploïdes de Triticum aestivum L. à partir de cultures d'anthères in vitro. C R Acad Sci 277:1463–1466

9. Maluszynski M, Kasha K, Forster BP, Szarejko I (eds) (2003) Doubled haploid production in crop plants: a manual. Springer Science & Business Media, Berlin, pp 1–415

10. Soriano M, Cistué L, Vallés MP, Castillo AM (2007) Effects of colchicine on anther and microspore culture of bread wheat (Triticum aestivum L.). Plant Cell Tissue Organ Cult 91:225–234. https://doi.org/10.1007/s11240-007-9288-2

11. Soriano M, Cistué L, Castillo AM (2008) Enhanced induction of microspore embryogenesis after n-butanol treatment in wheat (Triticum aestivum L.) anther culture. Plant Cell Rep 27:805–811. https://doi.org/10.1007/s00299-007-0500-y

12. Echavarri MB, Cistué L (2016) Enhancement in androgenesis efficiency in barley (Hordeum vulgare L.) and bread wheat (Triticum aestivum L.) by the addition of dimethyl sulfoxide to the mannitol pretreatment medium. Plant Cell Tissue Organ Cult 125(1):11–22. https://doi.org/10.1007/s11240-015-0923-z

13. Jiang F, Ryabova D, Diedhiou J, Hucl P, Randhawa H, Marillia EF, Foroud NA, Eudes F, Kathiria P (2017) Trichostatin a increases embryo and green plant regeneration in wheat. Plant Cell Rep 36(11):1701–1706. https://doi.org/10.1007/s00299-017-2183-3

14. Wang HM, Enns JL, Nelson KL, Brost JM, Orr TD, Ferrie AMR (2019) Improving the efficiency of wheat microspore culture methodology: evaluation of pretreatments, gradients, and epigenetic chemicals. Plant Cell Tissue Organ Cult 139(1):1–11. https://doi.org/10.1007/s11240-019-01704-5

15. Asif M, Eudes F, Goyal Randhawa H, Spaner D (2013) Organelle antioxidants improve microspore embryogenesis in wheat and triticale. In Vitro Cell Dev Biol Plant 49:489–497. https://doi.org/10.1007/s11627-013-9514-z

16. Asif M, Eudes F, Randhawa H, Admusen E, Spaner D (2014) Phytosulfokine alpha enhances microspore embryogenesis in both triticale and wheat. Plant Cell Tissue Organ Cult 116:125–130. https://doi.org/10.1007/s11240-013-0379-y

17. Sinha RK, Eudes F (2015) Dimethyl tyrosine conjugated peptide prevents oxidative damage and death of triticale and wheat microspores. Plant Cell Tissue Organ Cult 122:227–237. https://doi.org/10.1007/s11240-015-0763-x

18. Hunter CP (1988) Plant regeneration from microspores of barley, Hordeum vulgare L. PhD thesis, Wye College University of London, London

19. Hu TC, Kasha KJ (1997) Improvement of isolated microspore culture of wheat (Triticum aestivum L.) through ovary co-culture. Plant Cell Rep 16:520–525. https://doi.org/10.1007/BF01142316

20. Jensen CJ (1977) Monoploid production by chromosome elimination. In: Reinert J, Bajaj

YPS (eds) Applied and fundamental aspects of plant cell, tissue and organ culture. Springer-Verlag, Berlin, pp 299–330

21. Hoagland DR, Arnold DI (1950) The water culture method for growing plants without soil. Cir 347. California Agr Expt Stn. University of California, Berkeley

22. Devaux P (2003) The Hordeum bulbosum (L.) method. In: Maluszynski M, Kasha KJ, Forster BP, Szarejko I (eds) Doubled haploid production in crop plants. Kluwer Academic Publisher, Dordrecht, pp 15–19. https://doi.org/10.1007/978-94-017-1293-4_3

Chapter 12

Unpollinated Ovaries Used to Produce Doubled Haploid Lines in Durum Wheat

Olfa Ayed Slama and Hajer Slim Amara

Abstract

The use of doubled haploid lines improves the efficiency of cultivar development and homozygous genotypes can be obtained in one generation, as opposed to conventional line production, which requires several cycles of self-pollination. However, in durum wheat (*Triticum turgidum* subsp. *durum* Desf.), the low efficiency of green plant regeneration and the very high frequency of albino plants hinder the application of this technique.

We observed the success of using gynogenesis for durum wheat and the significant influence of growing conditions on ovary and callus development, and on plant regeneration. Our results suggested that the cold pretreatment for 2 weeks is efficient for durum wheat. Furthermore, the addition of 2,4-D, vitamins and glutamine, and the use of maltose as sugar source in media improved the ovary regeneration. We describe in this work an efficient method to regenerate green plants from in vitro durum wheat gynogenesis.

Key words Doubled haploid lines, Durum wheat, Gynogenesis, In vitro culture, Unpollinated ovary culture

1 Introduction

One goal of recent biotechnology research is to obtain doubled haploid lines, which offers complete homozygosity and phenotypic uniformity in one generation [1]. This method has several advantages such as a substantial reduction in the cost and the time required to produce breeding lines by conventional methods [2, 3]. Production of durum wheat doubled haploid lines in large number would be interesting considering the importance of this culture in the world and in Tunisia. The successful use of doubled haploid in most crops to develop commercial cultivars highly relies on an efficient protocol for inducing haploids. Haploids can be induced from male (androgenesis) or female (gynogenesis) germ line cells. In bread wheat (*Triticum aestivum* L.), androgenesis including anther and isolated microspore culture is one of the most efficient methods for doubled haploid production [4–

Jose M. Seguí-Simarro (ed.), *Doubled Haploid Technology: Volume 1: General Topics, Alliaceae, Cereals*,
Methods in Molecular Biology, vol. 2287, https://doi.org/10.1007/978-1-0716-1315-3_12,
© Springer Science+Business Media, LLC, part of Springer Nature 2021

6]. The efficiency of this technique has reached an adequate level and has led to the creation of commercial bread wheat varieties such as "Florin" [7]. However, in durum wheat, the low efficiency of green plants regeneration and the very high frequency of albino plants hinder the application of this technique [8–10].

In vitro gynogenesis induced by unpollinated ovary culture is also used to develop haploid plants. This technique has been used in a few species, such as onion [11] and sugar beet [12] where it has proven to be an efficient method. Unpollinated ovary culture is practiced more rarely in wheat although unlike androgenesis, gynogenesis of cereal species does not produce albino plants [9, 13]. The major problem of this technique is the lack of established and efficient protocols for most species, especially durum wheat [14].

In order to use the durum wheat doubled haploid production system into a routine technique, many problems have been solved and then optimized (pretreatment, induction medium, genotype dependence, and growth conditions). We define in this work an optimized protocol to produce doubled haploid durum wheat lines using unpollinated ovary in vitro culture.

2 Materials

2.1 Equipment and Laboratory Materials

1. Laminar air-flow cabinet.
2. Autoclave.
3. Adjustable growth room (for tissue culture).
4. Glasshouse (for donor plants and doubled haploid lines).
5. 40 cm diameter plastic pots with peat soil mix.
6. Small pots (5 cm diameter) containing a mixture of sand and peat (2:1).
7. Big pots (30 cm diameter) containing a mixture of sand and peat (1:2).
8. Balance.
9. pH meter.
10. Stereoscope.
11. Light microscope.
12. Magnetic agitator.
13. Agitator.
14. Refrigerator (4 °C).
15. Double-distiller of water.
16. Sterile fine-tipped forceps.
17. Sterile filter paper.
18. Parafilm.

19. 45 μm pore size filter.

20. Bain marie.

2.2 Glassware

1. Sterile glass beakers (100–1000 ml).

2. Sterile Erlenmeyer (100–1000 ml).

3. Sterile flasks (100–1000 ml).

4. Burette (100–1000 ml).

5. 5.5 or 9 cm diameter sterile plastic Petri dishes.

6. Sterile jars (500 ml).

2.3 Solutions

Prepare all solutions using bidistilled water.

1. Carmine stain solution: 45 ml of cold acetic acid, 5 g of carmine, 55 ml of distilled water. Mix and boil for 30 min, then filter the solution.

2. Disinfection solution: 12% sodium hypochlorite.

3. Saturated α-bromonaphthalene solution: 1 drop in 1000 ml ordinary water for 5 h at room temperature (21 ± 1 °C) or 16 h at 4 °C.

4. Colchicine solution: 1 g colchicine, 20 ml dimethylsulfoxide (DMSO; optional), 10 drops Tween 20 (optional), add 1000 ml distilled water.

5. Pectinase 2%: 2 g pectinase in 100 ml distilled water.

6. Feulgen staining solution: 0.9 g Fuchsin, 4.8 g $Na_2S_2O_5$, 3.2 ml HCl. Adjust to 250 ml with distilled water. Store in the dark at 4 °C.

2.4 Culture Media

Sterilize all the glassware used. Mix the media components according to the final concentration of each chemical. Start with less water than required. Once all media components are added, make up to final volume. These procedures must be done in a laminar air-flow hood.

1. Stock solutions: Prepare each stock at the concentration described in Table 1, using the components described in Table 2. Dispense each stock solution into 500 ml bottles and store them in the refrigerator (*see* **Note 1**).

2. For 1000 ml of final volume of induction, differentiation, and development media, mix stock solutions in a 1000 ml beaker (*see* **Note 2**):

    (a) 100 ml of 10× concentrated macroelement stock solution.

    (b) 10 ml of 100× concentrated micro element stock solution.

    (c) 10 ml of 100× concentrated iron stock solution.

Table 1
Preparation of stock solutions

Component	Concentration of stock solution with respect to the concentration of each component shown in Table 2
Macroelements	10×
Microelements	100×
Iron source	100×
Vitamins	100×
Amino acids	100×
Growth regulators	100×

(d) 10 ml of 100× concentrated vitamin stock solution (*see* **Note 3**).

(e) 10 ml of 100× concentrated amino-acid stock solution (*see* **Note 3**).

(f) 10 ml of 100× concentrated growth regulator stock solution (*see* **Notes 3** and **4**).

3. Add 60 g maltose/30 g sucrose.

4. Bring up to the volume of 1000 ml with distilled water.

5. Adjust the pH of all components of the media to 5.8 using 0.1 N HCl or 0.1 N NaOH, and solidify them with 7 g purified agar.

6. Sterilize the medium by autoclaving at 120 °C for 20 min, cool it and distribute the medium into sterile Petri dishes respectively for induction and differentiation medium (*see* **Notes 5** and **6**).

3 Methods

3.1 Donor Plants and Growth Conditions

1. Sow and germinate durum wheat seeds into a peat soil mix (10 seeds in each 40 cm plastic pot; *see* **Note 7**).

2. Place the pots in a glasshouse at 20–25 °C day/10–15 °C night temperatures and 12–14 h day length and natural light. Water plantlets with tap water every 2–3 days.

3.2 Spike Selection and Pretreatment

1. Check the developmental stage of the microspore based on the location of nucleus relative to the microspore pore [15]. Use the staining of microspores from the oldest anthers on a single spike in acetocarmine 5% and observe them under microscope.

Table 2
Composition of induction medium (Ind M), differentiation medium (Diff M), and development medium (Dev M) for in vitro unpollinated ovary culture based on [10]. 2,4-D: 2,4 Dichlorophenoxyacetic acid, 2iPA: 6-γ, γ- Dimethylamino purine. NAA: Naphthaleneacetic acid

Component	Ind M	Diff M	Dev M
Macroelements (g/l)			
NH_4NO_3	0.160	0.160	0.160
$CaCl_2 \cdot 4H_2O$	0.440	0.440	0.440
$MgSO_4 \cdot 7H_2O$	0.370	0.370	0.370
KH_2PO_4	0.170	0.170	0.170
KNO_3	1.900	1.900	1.900
Microelements (mg/l)			
KI	0.83	0.83	0.83
H_3BO_3	6.20	6.20	6.20
$MnSO_4 \cdot 2H_2O$	22.30	22.30	22.30
$ZnSO_4 \cdot 2H_2O$	8.60	8.60	8.60
$Na_2MO_4 \cdot 4H_2O$	0.25	0.25	0.25
$CuSO_4 \cdot 5H_2O$	0.025	0.025	0.025
$CoCl_2 \cdot 6H_2O$	0.025	0.025	0.025
Iron source (mg/l)			
$FeNa_2EDTA$	40	40	40
Vitamins (mg/l)			
Myo-inositol	100	100	100
Nicotinic acid	1	0.5	0.5
Pyridoxine HCl	1	0.5	0.5
Thiamine HCl	1	0.1	0.1
Pyruvate Na	–	–	5.0
Amino acids (mg/l)			
Glutamine	750	146	146
Glycine	–	2.25	2.25
Growth regulators (mg/l)			
2,4-D	2	1	–
NAA	–	1	–
Kinetin	0.5	–	–
2iPA	–	0.1	–

(continued)

Table 2
(continued)

Component	Ind M	Diff M	Dev M
Maltose g/l	60	–	–
Sucrose g/l	–	30	30
Purified Agar g/l	7	7	7
pH	5.8	5.8	5.8

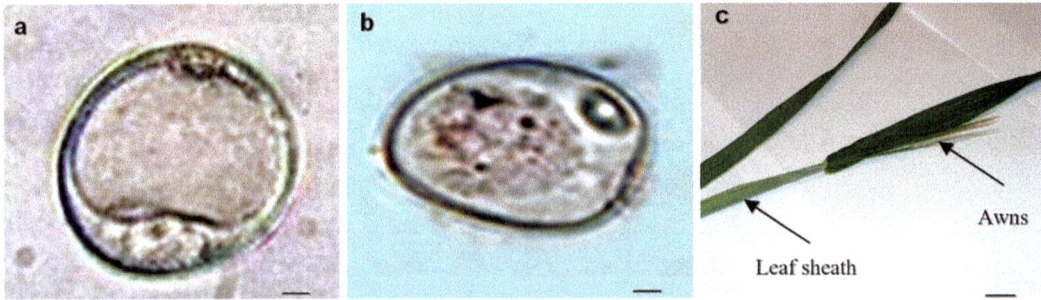

Fig. 1 Microspore at the late uninucleate stage (**a**); Microspore at the binucleate stage (**b**). Optimal morphological stage for maturity of durum wheat ovaries (**c**). Bars: (**a, b**) 5μm. (**c**) 2 cm

2. Confirm the maturity of ovaries: the best stage to collect young spikes from donor plants is when microspores are at the late uninucleate or binucleate stage (Fig. 1a, b; *see* **Note 8**).

3. Select the young spikes at the proper development stage and keep at 4 °C in ordinary water in the dark for 14 days for pretreatment (*see* **Note 9**).

3.3 Ovary Culture and Plant Regeneration

1. In the laminar flow hood, remove the remaining foliage encasing spikes and then take out the boot for disinfection.

2. Disinfect the spikes by immersion in disinfection solution for 15 min.

3. Wash spikes in sterile distilled water three times.

4. Remove the glumes and the lemma with sterilized forceps. Forceps should be dipped in 80% alcohol, flamed and cooled before re-use.

5. Carefully extract the ovaries of 1 to 1.5 mm length by the fine-tipped forceps and placed them into 5.5 cm diameter Petri dishes containing induction medium, no more than 20 ovaries per Petri dish.

6. Seal the Petri dishes with Parafilm and incubate the cultures in the incubator in the dark at 27 °C for 5 to 6 weeks.

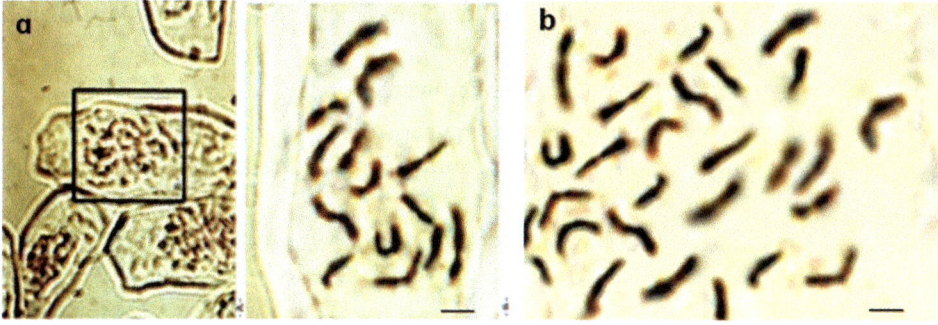

Fig. 2 Chromosome count from mitotic cells of haploid plants (**a**) and doubled haploid plants after colchicine treatment (**b**). Bars (**a**) 7μm, (**b**) 0.7μm

7. When their diameter is 5–6 mm, transfer ovary-derived calli to 9 cm diameter Petri dishes with differentiation medium (*see* **Note 10**).

8. Place the cultures in the growth room for 6 weeks at 25 °C with a 16 h photoperiod at light intensity of 80–100μE m^{-2} s^{-1}.

9. Transfer the calli with emerging shoots to development medium and keep in the same conditions for regeneration (*see* **Note 11**).

10. For better root formation, transfer green regenerated plantlets of about 2 to 3 cm in height into jars containing 125 ml of development medium.

11. Once plantlets reach the 3 to 4 leaf stage, count their chromosome number (Fig. 2a), prior to chromosome doubling (if needed) and transfer to soil.

3.4 Chromosome Studies

1. Determine the ploidy level before (Fig. 2a) and after colchicine treatment (Fig. 2b) for all regenerated plantlets, looking for cells undergoing mitosis [16] (*see* **Notes 12** and **13**).

2. Collect the mitotic cells from root tips in active growth (*see* **Note 14**).

3. Pretreat root tips in saturated solution of α-bromonaphthalene (*see* **Note 15**).

4. Fix cells at the metaphasic stage with acetic acid 90%, 30 min at room temperature (21 ± 1 °C) or 10–12 h at 4 °C.

5. Rinse in ethanol 95°, 5 min each then store in ethanol 70° (*see* **Note 16**).

6. Rehydrate in ordinary water, 3 successive rinses, 5 min each.

7. Hydrolyze in HCl 1 N for 10–12 min at 60 °C (*see* **Note 17**).

8. Use pectinase 2% (in distilled water) for 43 min at room temperature for maceration of root tips (*see* **Note 18**). Then rinse with distilled water.

9. Stain with Feulgen staining solution for 60 min at room temperature.

10. Rinse with distilled water for 10 min.

11. Transfer gently the root tip onto a microscope slide and add a drop of acetocarmine. Squash the meristematic tissue with a cover slip.

12. Observe and count the chromosome number under a light microscope (objective ×100 with immersion oil; *see* **Note 19**).

3.5 Chromosome Doubling

1. At the three-leaf stage, remove the agar from plant roots by gently washing with tap water and carefully wipe them with filter paper.

2. Cut away excess leaves and trim tips of leaves.

3. Immerse the plants in 0.1% colchicine solution during 4–5 h in the greenhouse under a lamp (*see* **Note 15**).

4. Rinse carefully with running water during 0.5–1 h.

5. Check chromosome doubling (Fig. 2b).

3.6 Transfer of Plants to Soil

1. Place doubled haploid plantlets in small pots (5 cm diameter) containing a mixture of sand and peat (2:1).

2. Place pots into a growth room at $25° \pm 1$ °C day/night temperature, a 16-h photoperiod, at light intensity of $350–450\mu E\ m^{-2}\ s^{-1}$.

3. Cover with glass caps for about 1 week to prevent plants from water stress.

4. When the plants establish a vigorous growth and roots over-grow the soil, transplant plants into bigger pots (30 cm diameter) with (1:2) mixture of sand and peat.

5. Place the pots in the glass room under a light regime of 16/8 h light/dark and at 25 ± 1 °C for seed production.

6. Collect and store the produced doubled haploid seeds at 4 °C.

4 Notes

1. To facilitate media preparation, most media components (macro and micro elements, iron, growth regulators, vitamins...), can be prepared in stock solutions 10–100× concentrated and stored in a refrigerator at 4 °C for 1 or 2 months.

2. Dissolve components one at a time in distilled water and make sure each component is completely dissolved before adding the next component.

3. Thermolabile components such as some hormones, vitamins and amino acids can be filter-sterilized using 45μm pore size filter and added to the autoclaved media.

4. To facilitate the preparation of growth regulator stock solution, dissolve auxins first in a few drops of absolute ethanol.

5. The differences between the differentiation and development media and the induction medium are the nature and concentration of the carbon source, the lower concentration of vitamins and amino acids, the addition of sodium pyruvate and the lack of growth regulators for the development medium.

6. Culture media are essential for the successful of unpollinated ovary culture and require improvements for each species. For durum wheat, our best induction response was obtained with culture media characterized by a low concentration of macro nutrients, a high level of organic nitrogen (glutamine), the presence of vitamins, and the use of 2,4-D as a growth hormone (Table 2).

7. Durum wheat seeds can be sown in the experimental field. Donor plants grown in the field produced generally higher production and more vigorous spikes than in glasshouse.

8. The developmental stage of microspores is determinant for the success of gynogenesis. It is essential to collect the durum wheat spikes from donor plants at the optimal stage (Fig. 1a, b). However, tillers containing spikes can be preselected on the basis of their morphology. This morphological stage is obtained when the awns have a length of 5 or 6 cm outside the flag leaf and the leaf sheath begin to split (Fig. 1c).

9. The pretreatment is also a critical step for the achievement of durum wheat unpollinated ovary culture. Flag leaves are not removed during the pretreatment period. The collected material is placed in a beaker containing ordinary water. The spikes are wrapped with aluminum foil, labeled, and refrigerated at 4 °C for 2 weeks. This cold treatment improves the ovary culture response.

10. Cultures are transferred every 4 weeks for the different media.

11. The development of the ovaries can be monitored using a stereoscope.

12. The protocol of chromosome count from mitotic cells of root tips used in this study is valid for all species of Triticeae.

13. Flow cytometry can be also used for chromosome count.

14. The removal of tissue with mitotic cells of root tips is better in the late afternoon, due to the very turgid tissue at that time. It is also possible to use young leaves.

15. Wear a mask and gloves for the preparation of colchicine and α-bromonaphthalene solutions and use an extractor fume hood.

16. Root tips can be preserved in ethanol 70° at 4 °C during 1 year.

17. It is very important to respect duration and temperature for hydrolysis. These conditions depend on species.

18. The maceration step of chromosome count is optional.

19. Chromosome count must be carried out on at least 3 complete and well spread plates.

Acknowledgments

The authors would like to thank the Laboratory of Genetics and Cereal Breeding of National Agronomic Institute of Tunisia (INAT) for his financial support.

References

1. Zheng MY, Liu W, Polle E et al (2001) Culture of freshly isolated wheat (*Triticum aestivum* L.) microspores treated with inducer chemicals. Plant Cell Rep 20:685–690

2. Meenakshi S, Nii A, Dipak KS et al (2012) An improved wheat microspore culture technique for the production of doubled haploid plants. Crop Sci. https://doi.org/10.2135/cropsci2012.03.0141

3. Lu R, Chen Z, Gao R et al (2016) Genotypes-independent optimization of nitrogen supply for isolated microspore cultures in barley. Biomed Res Int. https://doi.org/10.1155/2016/1801646

4. Soriano MC, Cistué L, Valles MP et al (2007) Effects of colchicine on anther and microspore culture of bread wheat (*Triticum aestivum* L.). Plant Cell Tissue Org Cult 91:225–234. https://doi.org/10.1007/s11240-007-9288-2

5. Castillo AM, Sanchez-Diaz RA, Vallés MP (2015) Effect of ovary induction on bread wheat anther culture: ovary genotype and developmental stage, and candidate gene association. Front Plant Sci 6:1–12

6. Lantos C, Pauk J (2016) Anther culture as an effective tool in winter wheat (*Triticum aestivum* L.) breeding. Russ J Genet 52:794–801. https://doi.org/10.1134/S102279541608007X

7. Picard E, Crambers E, Mihamou-Ziyyat A (1994) L'haplodiploïdisation: un outil multiusage pour la génétique et l'amélioration des céréales. In: Aupelf-uref (ed) Quel avenir pour l'amélioration des plantes? John Libbey Eurotext, Paris

8. Cistué L, Romagosa I, Batlle F et al (2009) Improvements in the production of doubled haploids in durum wheat (*Triticum turgidum* L.) through isolated microspore culture. Plant Cell Rep 28:727–735

9. Slama-Ayed O, Bouhaouel I, Ayed S et al (2019) Efficiency of three haplomethods in durum wheat (*Triticum turgidum* subsp. *durum* Desf.): isolated microspore culture, gynogenesis and wheat × maize crosses. Czech J Genet Plant Breed 55(3):101–109

10. Sibi ML, Kobaissi A, Shekafandeh A (2001) Green haploid plants from unpollinated ovary culture in tetraploid wheat (*Triticum durum* Desf.). Euphytica 122:351–359

11. Campion B, Alloni C (1990) Induction of haploid plants in onion (*Allium cepa* L.) by *in vitro* culture of unpollinated ovaries. Plant Cell Tissue Org Cult 20:1–6

12. Gürel S, Gürel E, Kaya Z (2000) Doubled haploid plant production from unpollinated ovules of sugar beet (*Beta vulgaris* L.). Plant Cell Rep 19:1155–1159

13. Slama-Ayed O, Slim-Amara H (2007) Production of doubled haploids in durum wheat (*Triticum durum* Desf.) through culture of unpollinated ovaries. Plant Cell Tissue Org Cult 91:125–133

14. Hadziabdic D, Wadt PA, Reed SM (2018) Haploid cultures. In: Trigiano RN, Gray DJ

(eds) Plant tissue culture, development and biotechnology, 1st edn. CRC Press, Boca Raton, Florida

15. Kasha KJ, Simion E, Oro R et al (2001) An improved *in vitro* technique for isolated microspore culture of barley. Euphytica 120:379–385

16. Jahier J, Chevre AM, Eber F et al (1992) Techniques de cytogénétique végétale. In: Jahier (ed) . INRA, France

Chapter 13

In Vitro Anther Culture for Doubled Haploid Plant Production in Spelt Wheat

Csaba Lantos and János Pauk

Abstract

Doubled haploid (DH) plant production belongs to modern biotechnology methods of plant breeding. The main advantage of DH plant production methods is the development of genetically homozygous lines in one generation, whilst in conventional breeding programmes, the development of homozygous lines requires more generations. The present chapter describes an efficient protocol for DH plant production in spelt wheat genotypes using in vitro anther culture.

Key words In vitro androgenesis, Anther culture, Doubled haploid, Spelt wheat

1 Introduction

In the last decades, spelt wheat (*Triticum spelta* L.) has become an attractive species in research and breeding programmes of cereals because of its many beneficial traits; high nutritional value, wide adaptability, abiotic stress tolerance, high tillering ability and biomass production [1–5]. In organic farming system, spelt is one of the most preferred species due to these attributes. The interest in spelt wheat is increasing in human consumption because of high protein content, minerals (Zn, Cu and Fe) and other bioactive compounds [3, 6–11].

In research and breeding of crop plants, many effective methods can be applied to reduce the long process of breeding and increase the efficiency of breeding programmes. The doubled haploid (DH) plant production techniques belong to these biotechnological methods of modern plant breeding [12–17]. One of the main advantages of DH plant production methods is the production of homozygous lines during the time of one winter cereal season. Homogeneity is one of the essential requirements in the breeding of new varieties and hybrids. Furthermore, the DH plant production methods can be combined with other breeding

Jose M. Seguí-Simarro (ed.), *Doubled Haploid Technology: Volume 1: General Topics, Alliaceae, Cereals*,
Methods in Molecular Biology, vol. 2287, https://doi.org/10.1007/978-1-0716-1315-3_13,

strategies, such as marker assisted selection, mutation or transgenic technologies [18, 19].

In crop plants, there are three frequently applied DH plant production methods, namely (1) chromosome elimination, (2) anther culture and (3) isolated microspore culture. Recently, the in vitro anther culture has proved to be an efficient method for DH plant production in spelt wheat via more tested spelt genotypes and F_1 combinations [20–23]. In the last years, our purposes were to screen the responsivity of spelt genotypes in anther culture and to establish an efficient DH plant production method for breeding of this species.

2 Materials

2.1 Equipment

1. Conditioned glasshouse for growing of donor plants and regenerated, transplanted plantlets (*see* **Note 1**).

2. Volldünger chemical fertilizer (N:P:K:Mg = 14:7:21:1, plus 1% microelements: B, Cu, Fe, Mn and Zn).

3. Cold chamber for vernalization of winter type spelt genotypes.

4. Inverted microscope (*see* **Note 2**).

5. Clean bench (*see* **Note 3**).

6. Incubators (32 °C and 28 °C).

7. Autoclave for media preparation.

8. TissueLyser II.

9. Flow cytometer.

10. Standard laboratory equipments: balance, pH meter, fridge, freezer, pipettes, tweezers, scissors, Eppendorf tubes.

2.2 Culture Containers

1. Petri dishes (90 mm) for anther culture.

2. Petri dishes (90 mm) for plant regeneration of embryo-like structure (ELS).

3. Sterile plastic containers (100 × 140 × 103 mm) for rooting of in vitro green plantlets.

4. Plastic racks (6 × 11 places) for acclimatization of anther culture-derived plantlets in glasshouse.

2.3 Solutions and Culture Media

1. Sterilizing solution: 300 mL 2% NaOCl solution with one drop Tween-80.

2. All the media used for androgenesis induction, plantlet regeneration and rooting are described in Table 1 (*see* **Note 3**).

3. Galbraith buffer for flow cytometry 100 mL [24]: 912.5 mg $MgCl_2 \cdot 6H_2O$, 887.5 mg Na-citrate, 412.5 mg MOPS, 100 μL Triton X-100, adjust pH to 7.

Table 1
Ingredients of media for induction of in vitro androgenesis, plantlet regeneration and rooting in spelt wheat

Ingredients	'W14mf' induction medium mg/L	'190-2Cu' regeneration medium mg/L	'190-3Cu' rooting medium mg/L
KNO_3	2000	1000	1000
K_2SO_4	700	–	–
$NH_4H_2PO_4$	380	–	–
$CaCl_2 \cdot 2H_2O$	140	–	–
$MgSO_4 \cdot 7H_2O$	200	200	200
KH_2PO_4	–	300	300
$(NH_4)_2SO_4$	–	200	200
$Ca(NO_3)_2 \cdot 4H_2O$	–	100	100
KCl	–	40	40
Na_2EDTA	37.3	37.3	37.3
$FeSO_4 \cdot 7H_2O$	27.8	27.8	27.8
$MnSO_4 \cdot 4H_2O$	8	8	8
$ZnSO_4 \cdot 7H_2O$	3	3	3
H_3BO_3	3	3	3
KI	0.5	0.5	0.5
$CuSO_4 \cdot 5H_2O$	0.025	0.5	0.5
$CoCl_2 \cdot 6H_2O$	0.025	–	–
$Na_2MoO_4 \cdot 4H_2O$	0.005	–	–
Thiamine HCl	2	1	1
Pyridoxine HCl	0.05	0.5	0.5
Nicotonic acid	0.05	0.5	0.5
Myo-inositol	–	100	100
Glycine	–	2	2
Maltose	80,000	–	–
Sucrose	–	30,000	30,000
2,4-D	2	–	–
Kinetin	0.5	0.5	–
NAA	–	0.5	2
pH	5.8	5.8	5.8
Ficoll	100,000	–	–
Gelrite	–	2800	2800

4. 1 mg/mL RNase solution for degradation of RNA content.

5. 1 mg/mL propidium iodide solution for DNA staining.

3   Methods

3.1   Growing of Plant Materials

1. Sow the seeds of the selected donor genotypes in greenhouse (*see* **Notes 1** and **4**).

2. Transfer the germinated winter type genotypes at 4 °C for 8 weeks with continuous artificial dim light during the vernalization period.

3. Transplant the vernalized donor plants to 2 L plastic pots containing a 1:1 ratio of peat and sandy soil mixture.

4. Fertilize the donor plants with Volldünger chemical fertilizer once in a fortnight.

5. Adjust 20/15 °C day/night temperature for plants, respectively. Natural light can be supplemented with 3 h artificial light until the collection of donor materials.

6. Apply the required fungicides for protecting donor plants and remove the weeds manually.

3.2   Pre-treatment of Donor Tillers

1. Check the developmental stages of microspores with the inverted microscope (*see* **Note 2**). Crush the anthers of donor genotypes in a drop of tap water to monitor the microspore population under the microscope.

2. Collect the tillers of donor genotypes containing anthers with early- and mid-uninucleate, vacuolated microspores (Fig. 1a).

3. Place the collected tillers into Erlenmeyer flasks containing tap water and cover by PVC bags to keep high humidity.

4. Pre-treat these tillers at 3–4 °C for 14 days (*see* **Note 5**).

3.3   Preparation of Anther Cultures

1. Check again the developmental stages of microspores under the inverted microscope.

2. Sterilize the spikes containing anthers with early- and mid-uninucleate, vacuolated microspores in sterilizing solution for 20 min on a horizontal shaker.

3. Rinse the spikes three times with autoclaved sterile distilled water in laminar air-flow cabinet (*see* **Note 3**).

4. Isolate 300 anthers manually with two tweezers in each 90 mm diameter glass Petri dish containing 12 mL 'W14mf' induction medium [25, 26].

5. Place the anther cultures at 32 °C to apply heat shock during the first 3 days of culture.

6. Keep the Petri dishes at 28 °C for 8 weeks in an incubator in darkness (*see* **Note 5**).

Fig. 1 (**a**) Uni-nucleate, vacuolated microspore for the induction of in vitro androgenesis in spelt wheat. (**b**) Microspore-derived ELS develop in anther culture after 4–5 weeks of cultivation. (**c**) The ELS produce green- and albino plantlets onto regeneration medium. (**d**) The green plantlets produce roots in plastic boxes (15 green plantlets/box), (**e**) which acclimatize to the greenhouse conditions. Bars: (**a**) 10μm; (**b**) 5 mm; (**c**) 10 mm

3.4 Plant Regeneration of Anther Culture-Derived ELS

1. Transfer the well-developed ELSs (Fig. 1b) with a size of 1–2 mm to 90 mm diameter plastic Petri dishes (*see* **Note 5**) containing '190-2Cu' plantlet regeneration medium [27]. In this regeneration medium (Table 1), the microspore-derived ELSs regenerate green and albino plantlets within 2 weeks (Fig. 1c).

2. Transfer the well-developed green plantlets with 2–3 leaves into sterile plastic containers (Fig. 1d) containing '190-3Cu' medium for rooting (Table 1) [21]. Up to fifteen green plantlets can be placed in each plastic container.

3. Adjust the temperature to 24 °C and photoperiod to 16/8 h day/night during the plantlet regeneration period.

3.5 Acclimatization of Anther Culture-Derived Green Plantlets

1. In glasshouse, transplant the well-rooted green plantlets from plastic containers into plastic racks (66 plantlets/rack) containing the above-mentioned 1:1 peat and sandy soil mixture (Fig. 1e).

2. Cover the plantlets with a transparent plastic cover during the 3–4 day-long acclimatization period (*see* **Note 6**). The acclimatized plants are grown in the glasshouse following the above-mentioned growing protocol for donor plants.

3.6 Flow Cytometry

The ploidy level of the plantlets can be determined by flow cytometric analysis.

1. Collect leaf samples from plantlets grown in the glasshouse.

2. Isolate the nuclei from the young leaf of plantlets.

3. Disrupt the samples in Eppendorf tubes containing 1 mL Galbraith buffer using a TissueLyser II at 20 Hz for 1.5 min [24].

4. Purify the suspension using 20μm sieves.

5. Add 10μL RNase solution to each sample for 60 min at room temperature to degrade the RNA content.

6. Stain DNA with 40μL PI solution for 30 min at room temperature.

7. After the preparation of samples (*see* **Note** 7), determine the DNA content of samples based on histograms of flow cytometric analyses (Fig. 2).

3.7 Growing of Anther Culture-Derived Plantlets in DH Nursery

1. Transplant the acclimatized anther culture-derived, winter type plantlets manually in the nursery in October (*see* **Note 8**).

2. Water the plantlets as needed to support the development of new roots and acclimatization of the plantlets to field conditions.

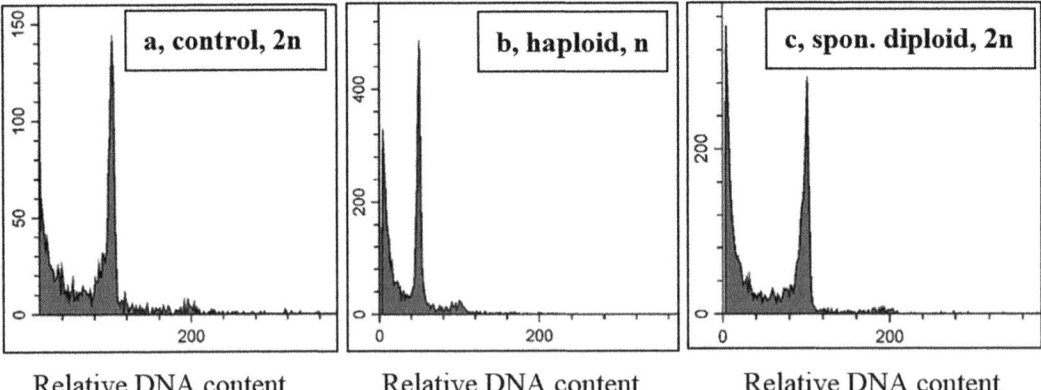

Fig. 2 Ploidy level determination of anther culture-derived plantlets based on flow cytometric analysis. Histograms of (**a**) seed-derived control, anther culture-derived (**b**) haploid and (**c**) spontaneous diploid plantlets

3. Harvest the spikes of the fertile DH plants next July.

4. Determine the percentage of spontaneous genome doubling based on seed production.

3.8 Integration of DH Lines into Breeding Program

In the following generation, sow the seeds of each DH line in one or more row system or microplot depending on the breeding system used (*see* **Note 9**).

4 Notes

1. The quality of donor plants is one of the most important critical factors which influence the efficiency of in vitro androgenesis in anther culture. The healthy (grown under ideal growing condition) donor plants, tillers and spikes are important for large-scale DH plant production. According to the needs of spelt genotypes, the donor plantlets can also be grown in well-controlled glasshouse or in breeding nursery.

2. The developmental stage of the microspores is one of the most critical factors in the induction of in vitro androgenesis of spelt wheat. Early- and mid-uninucleate, vacuolated microspores are ideal for efficient androgenesis induction in spelt wheat anther cultures, which can be checked quickly and easily in a drop of tap water under inverted microspore.

3. The anther culture method requires in vitro manual work (isolation of anthers, transfer of ELS and green plantlets). So, the sterile work practice is a key factor in the implementation of tissue culture. However, the application of antibiotics is not necessary in anther culture of spelt wheat, accurate sterile work is sufficient.

4. The genotype influences the efficiency of in vitro anther culture of spelt wheat genotypes [21–23]. Based on the data of 26 spelt wheat genotypes, significant numbers of in vitro green plantlets (6.3–85.0 green plantlets/100 anthers) can be produced using our protocols, whilst the number of albino plantlets is moderate. It is worthwhile using more genotypes in the adaptation of the method.

5. Following the time table (time of donor collection and pre-treatment, ELS and green plantlets transfer etc.) is also critical for the efficient application of method. The significant delays of different steps can decrease the efficiency of green plantlets production.

6. In vitro plantlets are sensitive to the quick changing of humidity. The plantlets should be transplanted into plastic racks as quickly as possible and covered by a transparent plastic cover, which can be removed after the acclimatization period (3–4 days).

7. The preparation of samples is a critical step in flow cytometric analysis. Optimization of sample preparation may be required depending on the growing conditions of the plantlets (in vitro plantlets, glasshouse-grown plantlets or field nursery-grown acclimatized plantlets).

8. The ploidy level of anther culture-derived plants can be determined by flow cytometric analysis. The haploid and doubled haploid plantlets can be separated based on the histograms of analyses to transplant the doubled haploids and save field in the nursery. Generally, acclimatized plantlets are grown in the nursery.

9. The homogeneity of DH lines can be checked visually in DH_{1-2} generations in the nursery. The segregating lines (somatic tissue-derived plants, cross pollination, seed mixing, volunteer plants) can be identified based on their phenotype. Molecular marker analyses are well-founded in scientific research programmes such as QTL analyses.

Acknowledgements

This project was supported by the János Bolyai Research Scholarship of the Hungarian Academy of Sciences. The experiments were interlocked with scientific programmes (project code: OTKA-K_16-K119835; name of the project: Improvement of spelt wheat lines with low fermentable carbohydrate content (FODMAP) using modern and classical research methods), Thematic Excellence Programme 2019 (project code: TUDFO/51757/2019-ITM, supporter: National Research, Development and Innovation Office) and GINOP project (project number: GINOP-2.2.1-15-2016-00026). The authors thank the conscientious work of Ferenc Markó, Krisztina Kéri and Sándor Vajasdi-Nagy. Furthermore, the authors also thank László Láng (Centre for Agricultural Research, Hungarian Academy of Sciences, Martonvásár, Hungary) and Center for Plant Diversity (Tápiószele, gene bank of Hungary) for supplying the tested spelt wheat varieties ('Franckenkorn', 'Mv Martongold' and 'Oberkulmer Rotkorn') and gene bank germplasms (RCAT056296, RCAT058694, RCAT060960) for experiments, respectively.

Funding: This research was funded by the Hungarian Academy of Sciences, grant number 'János Bolyai Research Scholarship'; National Research, Development and Innovation Office, grant number 'OTKA-K_16-K119835', 'GINOP-2.2.1-15-2016-00026' and 'TUDFO/51757/2019-ITM'.

References

1. Raman H, Rahman R, Luckett D, Raman R, Bekes F, Lang L, Bedo Z (2009) Characterisation of genetic variation for aluminium resistance and polyphenol oxidase activity in genebank accessions of spelt wheat. Breed Sci 59:373–381

2. Koutroubas SD, Fotiadis S, Damalas CA (2012) Biomass and nitrogen accumulation and translocation in spelt (*Triticum spelta*) grown in Mediterranean area. Field Crop Res 127:1–8

3. Arzani A, Ashraf M (2017) Cultivated ancient wheats (*Triticum* spp.): a potential source of health-beneficial food products. Compr Rev Food Sci Food Safe 16:477–488

4. Andruszczak S (2018) Spelt wheat grain yield and nutritional value response to sowing rate and nitrogen fertilization. J Anim Plant Sci 28:1476–1484

5. Pauk J, Lantos C, Ács K, Gell G, Tömösközi S, Hajdú Búza K, Békés F (2019) Spelt (*Triticum spelta* L.) *in vitro* androgenesis breeding for special food quality parameters. In: Al-Khayri JM, Mohan Jain S, Johnson DV (eds) Advances in plant breeding strategies: Cereals. Springer Nature Switzerland AG, Cham, Switzerland, pp 525–557

6. Fan MS, Zhao FJ, Fairweather-Taitc SJ, Poultona PR, Dunhama SJ, McGrath SP (2008) Evidence of decreasing mineral density in wheat grain over the last 160 years. J Trace Elem Med Biol 22:315–324

7. Zielinski H, Ceglinska A, Michalska A (2008) Bioactive compounds in spelt bread. Eur Food Res Technol 226:537–544

8. Gomez-Becerra HF, Erdem H, Yazici A, Tutus Y, Torun B, Ozturk L, Cakmak I (2010) Grain concentration of protein and mineral nutrients in a large collection of spelt wheat grown under different environments. J Cereal Sci 52:342–349

9. Escarnot E, Aguedo M, Agneessens R, Wathelet B, Paquot M (2011) Extraction and characterization of water-extractable and water-unextractable arabinoxylans from spelt bran: study of the hydrolysis conditions for monosaccharides analyses. J Cereal Sci 53:45–52

10. Guzman C, Medina-Larque AS, Velu G, Gonzalez-Santoyo H, Singh RP, Huerta-Espino J, Ortiz-Monasterio I, Pena RJ (2014) Use of wheat genetic resources to develop biofortified wheat with enhanced grain zinc and iron concentration and desirable processing quality. J Cereal Sci 60:617–622

11. Hlisnikovsky L, Hejcman M, Kunzova E, Mensik L (2019) The effect of soil-climate conditions on yielding parameters, chemical compositions and baking quality of ancient wheat species *Triticum monococcum* L., *Triticum dicoccum* Schrank and *Triticum spelt* L. in comparison with modern *Triticum aestivum* L. Arch Agron Soil Sci 65:152–163

12. Dunwell JM (2010) Haploids in flowering plants: origins and exploitation. Plant Biotechnol J 8:377–424

13. Germana MA (2011) Gametic embryogenesis and haploid technology as valuable support to plant breeding. Plant Cell Rep 30:839–857

14. Hensel G, Oleszczuk S, Daghma DES, Zimny J, Melzer M, Kumlehn J (2012) Analysis of T-DNA integration and generative segregation in transgenic winter triticale (×*Triticosecale* Wittmack). BMC Plant Biol 12:171

15. Niu Z, Jiang A, Abu Hammad W, Oladzadabbasabadi A, Xu SS, Mergoum M, Elias EM (2014) Review of doubled haploid production in durum and common wheat through wheat × maize hybridization. Plant Breed 133:313–320

16. Shchukina LV, Pshenichnikova TA, Khlestkina EK, Misheva S, Kartseva T, Abugalieva A, Borner A (2018) Chromosomal location and mapping of quantitative trait locus determining technological parameters of grain and flour in strong-flour bread wheat cultivar Saratovskaya 29. Cereal Res Commun 46:628–638

17. Sharma P, Chaudhary HK, Manoj NV, Kumar P (2019) New protocol for colchicine induced efficient doubled haploidy in haploid regenerants of tetraploid and hexaploid wheats at *in vitro* level. Cereal Res Commun 47:356–368

18. Testillano PS (2019) Microspore embryogenesis: targeting the determinant factors of stress-induced cell reprogramming for crop improvement. J Exp Bot 70:2965–2978

19. Kalinowska K, Chamas S, Unkel K, Demidov D, Lermontova I, Dresselhaus T, Kumlehn J, Dunemann F, Houben A (2019) State-of-the-art and novel developments of in vivo haploid technologies. Theor Appl Genet 132:593–605

20. Lantos C, Jenes B, Bona L, Cserháti M, Pauk J (2016) High frequency of doubled haploid plant production in spelt wheat. Acta Biol Cracov Ser Bot 58:107–112

21. Lantos C, Bóna L, Nagy É, Békés F, Pauk J (2018) Induction of *in vitro* androgenesis in

anther and isolated microspore culture of different spelt wheat (*Triticum spelta* L.) genotypes. Plant Cell Tiss Org 133:385–393

22. Lantos C, Purgel S, Ács K, Langó B, Bóna L, Boda K, Békés F, Pauk J (2019) Utilization of *in vitro* anther culture in spelt wheat breeding. Plan Theory 8:436

23. Castillo AM, Allue S, Costar A, Alvaro F, Valles MP (2019) Doubled haploid production from Spanish and central European spelt by anther culture. J Agric Sci Tech Iran 21:1313–1324

24. Galbraith DW, Harkins KR, Maddox JM, Ayres NM, Sharma DP, Firoozabady E (1983) Rapid flow cytometric analysis of the cellcycle in intact plant-tissues. Science 220:1049–1051

25. Ouyang JW, Jia SE, Zhang C, Chen X, Fen G (1989) Annual report, a new synthetic medium (W_{14}) for wheat anther culture; Institute of Genetics. Academia Sinica, Beijing, China, pp 91–92

26. Lantos C, Pauk J (2016) Anther culture as an effective tool in winter wheat (*Triticum aestivum* L.) breeding. Russ J Genet 52:794–801

27. Pauk J, Mihály R, Puolimatka M (2003) Protocol of wheat (*Triticum aestivum* L.) anther culture. In: Maluszynski M, Kasha KJ, Forster BP, Szarejko I (eds) Doubled haploid production in crop plants. A manual. Kluwer Academic Publishers, Dordrecht, The Netherlands, pp 59–64

Chapter 14

Production of Wheat Doubled Haploids Through Intergeneric Hybridization with Maize

Pierre Devaux

Abstract

The intergeneric hybridization of wheat (*Triticum aestivum* L.) with maize (*Zea mays* L.) enables the production of doubled haploids (DHs) of wheat from all wheat hybrids with high efficiencies. Wheat and maize donor plants are raised in environmentally controlled greenhouses until crossing. Before anthesis, wheat spikes are emasculated and then pollinated with maize. Auxin is applied to each individual wheat floret 1 day after pollination. About 2 weeks after crossing, in vitro embryo culture is performed, enabling the regeneration of haploid wheat plantlets after maize chromosome elimination. Haploid plantlets are transferred to the greenhouse and after recovery, their genome is doubled with colchicine. Haploid plantlets can be sampled for DNA extractions and molecular analyses to aid the rapid discard of undesirable plantlets. Doubled haploid plants are raised in a greenhouse until maturity. Seeds of each fertile DH are harvested and often sown the same year. Several cycles of multiplication and evaluation in replicated plot trials and different geographical locations are then done to select the best candidate(s) for varietal registration.

Key words Breeding, Cultivar, Doubled haploid, Haploid, Homozygous lines, *Triticum aestivum* L, Wheat

1 Introduction

Practical breeding programs in self-pollinating crops such as wheat (*Triticum aestivum* L., $2n = 6x = 42$) aim to combine favorable alleles of different parents after crossing and meiotic recombination. During this time, many traits are evaluated using phenotypic, marker-assisted and more recently genomic selection while several generation cycles of selfing are done to achieve homozygosity for DUS (Distinctiveness, Uniformity, and Stability)—a requirement for granting Plant Breeding Rights. Current breeding methods, such as pedigree, bulk, single-seed descent inbreeding, are laborious and time-consuming. The instant production of true breeding lines, for example, doubled haploid (DH) lines in crop species saves several generations in the breeding program [1, 2]. Consequently, the time needed before a new cultivar can be released onto the

Jose M. Seguí-Simarro (ed.), *Doubled Haploid Technology: Volume 1: General Topics, Alliaceae, Cereals,*
Methods in Molecular Biology, vol. 2287, https://doi.org/10.1007/978-1-0716-1315-3_14,
© Springer Science+Business Media, LLC, part of Springer Nature 2021

market can be reduced by 3–4 years. Selection is more effective on homozygous lines than on early generation segregating lines, and the potential value of quantitative traits such as yield and quality can be predicted earlier in the breeding program. In addition, DHs can be used to construct genetic maps, locate genes of importance, and identify molecular markers linked to useful traits [3] and sequencing. Several methods for DH production in crops have been reported ([4], Chapter 1 of this volume), but in wheat, the most successful method so far is the intergeneric hybridization between wheat and maize, first reported by Laurie and Bennett [5]. The wheat × maize hybridization also referred to as the "maize method" recently accounted for more than 75% of the DHs released as cultivars worldwide [6]. The protocol described here is for the DH production of wheat using the intergeneric hybridization between wheat and maize. It has been used for more than 25 years in the breeding programs of our Company, and has contributed to the release of many cultivars internationally. Over the years, refinements of the protocol were carried out in order to ensure DH production from every hybrid of the breeding programs with higher efficiencies and in the shortest time possible. Critical factors for success include proper growing conditions for the wheat and maize plants, the use of suitable maize hybrids, emasculation and pollination techniques, growth substance treatments, embryo and plantlet culture conditions, chromosome doubling treatment, and culture conditions of DHs until maturity.

2   Materials

2.1   Greenhouse

1. Lighting: Green Power Philips Master SON-T Agro 400.

2. PG-Mix: 50% Irish peat 0–10, 20% Medium Baltic peat, 20% clay, 30 kg/m^3 of Swedish clay, 10% medium perlite, 7 kg/m^3 lime, 1.5 kg/m^3 PG-mix, 0.2 kg/m^3 trace elements, pH > 6.5 (*see* **Note 1**).

3. Nitrogen fertilizers: 50% ammonium nitrate +50% Blaukorn Classic 12-8-16 + 3 MgO.

4. 4 × 4 × 6 individual plastic containers (length 4 cm × width 4 cm × height 6 cm).

5. 7 × 7 × 6.5 individual plastic containers (length 7 cm × width 7 cm × height 6.5 cm).

6. Plastic pots (Ø 16 cm × H 16 cm) for wheat.

7. Plastic pots (Ø 20 cm × H 20 cm) for maize.

8. Bug-scan blue (thrips) and yellow (flying insects) from Biobest (http://www.biobestgroup.com, *see* **Note 2**).

9. Phytosanitary products: Acetamiprid, Pyrimicarbe, Deltamethrin, sulfur rolls (*see* **Note 3**).

10. Ø 90 mm Petri dishes.

11. Ø 55 mm Petri dishes.

12. 11 cm stainless steel forceps.

13. Stainless steel spatula.

14. 20 ml syringe.

2.2 Laboratory

Prepare all solutions using ultrapure water and high-quality grade reagents. Stock solutions are stored at 4 °C for up to 2 weeks.

1. Seed disinfecting solution: a 1:1 mixture of 70% ethanol and a 50% commercial chlorhexidine gluconate + benzalkonium chloride solution.

2. Mercuric chloride solution: 100 ml of 0.3% (w/v) mercuric chloride + one drop of Tween 20.

3. 95% ethanol.

4. Modified Gamborg's B-5 medium: *see* Table 1.

5. Colchicine solution: 0.1% aqueous colchicine + 2% dimethyl sulfoxide.

6. 30 ml glass tubes.

7. 11 cm stainless steel forceps.

8. Needle handle + needle.

9. 250 ml beakers.

10. 100 ml Erlenmeyers.

11. Glass bead sterilizer, glass beads.

12. Portable and stand-alone tag printers.

13. Industrial thermal printers for plastic labels of 120 mm × 17 mm.

14. Cold culture room (4 °C) with Osram cool white L58W/840 fluorescent tubes.

15. Warm culture room (20 ± 2 °C) with a 1:1 mixture of Osram cool white L58W/840 and Fluora L58W/77 fluorescent tubes.

16. Tillering room (14 ± 1/10 ± 1 °C day/night, 12 h day length) with a 1:1 mixture of Green Power Philips Master SON-T Agro 400 and HPI-T Plus 400 lamps.

Table 1
Modified Gamborg's B-5 medium [7]

Medium components	Concentration (mg/L)
Macro nutrients	
$NaH_2PO_4 \cdot H_2O$	150
KNO_3	2500
$(NH_4)_2SO_4$	134
$MgSO_4 \cdot 7H_2O$	250
$CaCl_2 \cdot 2H_2O$	150
Micro nutrients	
$MnSO_4 \cdot H_2O$	10
H_3BO_3	3
$ZnSO_4 \cdot 7H_2O$	2
$Na_2MoO_4 \cdot 2H_2O$	0.25
$CuSO_4 \cdot 5H_2O$	0.039
$CoCl_2 \cdot 6H_2O$	0.025
KI	0.75
FeNa EDTA	40
Vitamins	
Nicotinic acid	1
Thiamine HCl	10
Pyridoxine HCl	1
Myo-inositol	100
Miscellaneous	
Agar	8000
Sucrose	20,000
pH	5.5

3   Methods

3.1   Donor Plants and Growth Conditions

Healthy plants are a prerequisite for success and much care should be taken to control diseases and insects inside the vernalization room and particularly in the greenhouses. As the number of plant protection products authorized for use on crops is decreasing, it is important to observe the following recommendations:

1. First, wheat and maize plants are grown in separate greenhouses, especially as the growing conditions are different for these two species.

2. Second, all plants are grown in pots in which the growing medium is destroyed after each growing cycle.

3. Third, between each cycle, the greenhouses are completely cleaned, disinfected and left at least 2 weeks empty.

With these precautions, the development of pests and diseases is reduced and thus limits the use of phytosanitary products.

1. Hybrid wheat seeds are sown in $4 \times 4 \times 6$ individual plastic containers filled with commercial PG-mix, watered, and transferred to the greenhouse.

2. When seedlings have reached 5–7 cm height, they are left in a cold room at 4 °C for 4–10 weeks depending on their vernalization requirements with 8 h day length ensuring a photoactive radiation (PAR) of 50μmol m^{-2} s^{-1} at canopy level, and occasional watering. Spring wheat is placed in the same cold room but only for 2 weeks (see **Notes 4** and **5**).

3. Then, the plantlets are transplanted to wheat plastic pots containing the same PG-mix and raised in the greenhouse at $20 \pm 4/16 \pm 2$ °C (day/night, 16 h day length) and regularly watered (see **Note 6**). Natural light is supplemented when necessary with the additional lighting system to maintain a minimum PAR of 140μmol m^{-2} s^{-1} at canopy level. A high nitrogen fertilizer is applied about 3 weeks before heading.

4. Maize (*Zea mays* L.) seeds are planted directly in maize plastic pots, 3 seeds per pot containing the PG-mix, and raised in a greenhouse with the same photoperiod and thermoperiod conditions as for wheat but with at least a PAR of 200μmol m^{-2} s^{-1}.

5. Maize pollen donors are chosen from F_1 hybrids commercially available according to the following requirements: they must be short-strawed, easy to raise in greenhouse conditions and produce high amounts of pollen. As there are slight genotypic differences for haploid production [8], each individual maize F_1 hybrid is tested on a range of 5–10 different wheat hybrids and three to four of the best inducers are introduced for routine production.

6. A mixture of pollen from these selected maize hybrids is used for decreasing the genotypic variation in wheat haploid production. To ensure continuous crossing year around, wheat and maize seeds are sown every week.

Fig. 1 Steps for production of haploid plants and molecular analyses. (**a**) Wheat spike emasculation. (**b**) Wheat spike pollination with maize pollen. (**c**) Overview of donor wheat plants for DH production. (**d**) Removal of seeds from wheat spike. (**e**) Excision of haploid embryo. (**f**) Transfer of haploid embryo to culture medium. (**g**) In vitro haploid plantlets prior to transfer to soil. (**h**) Leaf sampling for DNA extraction. (**i**) Single nucleotide polymorphism analyses of haploid plants. (All images are from Florimond Desprez)

3.2 Hybridization and Post-pollination Treatment

1. Once wheat spikes arise, they are emasculated immediately after the central florets of each spikelet have been removed with a surgical pair of forceps (Fig. 1a). A cellophane bag is then placed on each emasculated spike with 25–30 remaining florets for 2–3 days before pollination with maize.

2. For this, pollen of maize is collected in the morning by shaking the male inflorescence at anthesis above a folded cardboard. After anthers removal with the pair of forceps, the freshly collected pollen is poured into a Ø 90 cm Petri dish. Maize pollen is then applied immediately to each emasculated wheat floret with a small spatula (Fig. 1b; *see* **Note 7**). After pollination, the same cellophane bag is replaced on the pollinated spike.

3. One day after pollination, an application of auxin, for example, 2,4-D [9] is essential for the successful recovery of wheat haploid embryos. For this, a drop of an aqueous solution of 100 mg/l 2,4-D is manually added to each pollinated floret using a 20 ml syringe (*see* **Note 8**). Alternatively, Dicamba can be used for durum wheat [10] and various combinations of an auxin with 6-benzylaminopurine or gibberellic acid [11].

4. After the treatment with the aqueous auxin solution has dried, spikes are covered with a brown paper bag (15 × 4 cm) and identified by a bar-coded label printed directly from a portable printer.

5. Spikes are left on the plant (Fig. 1c) until they are collected for seed dissection and embryo culture. Most florets treated that way produce plump seeds, but these seeds do not always contain an embryo. Attempts to identify embryo-containing seeds by illumination and visual observation [12] and more sophisticated methods that we tried, for example, X-ray radiography and 3D-computed tomography prior to dissecting them have not been satisfactory.

3.3 Embryo Culture

1. About 2 weeks after pollination, spikes +15–20 cm tillers with their paper bags and identification labels are cutoff from the plants and immersed to a depth of 5–6 cm in a beaker containing tap water (*see* **Note 9**).

2. Embryo culture is essential to obtain haploid wheat plantlets, as seeds from such crosses do not have a solid endosperm. Instead, they are filled with an aqueous solution that does not allow the seed to develop to maturity and germinate (*see* **Note 10**). Wheat seeds from several spikes of each wheat hybrid are removed from spikes (Fig. 1d) and gathered in a Ø 55 mm Petri dish.

3. Seeds are transferred to a 250 ml beaker containing seed disinfecting solution and shaken continuously for 20 min. Then, seeds are quickly rinsed in 95% ethanol and transferred into a 100 ml Erlenmeyer containing mercuric chloride solution and left for 20 min with occasional manual shaking (*see* **Notes 11** and **12**). They are subsequently rinsed three times in sterile distilled water.

4. 20–30 aseptic seeds (*see* **Note 13**) are placed on a sterile Ø 90 cm Petri dish (*see* **Note 14**) and embryos are excised under a ×20 binocular microscope (Fig. 1e) in a pre-disinfected laminar flow hood that has been operating for at least 15 min.

5. Aseptically excised embryos are transferred to 30 ml glass tubes (Fig. 1f), three per tube, containing 8 ml of modified Gamborg's B-5 medium (*see*, **Notes 15** and **16**). Newly printed labels are pasted on the corresponding tubes.

6. Numbers of cultured embryos per wheat hybrid are recorded each working day in a spreadsheet.

7. Tubes are incubated in darkness at 4 ± 0.5 °C.

8. When coleoptiles have reached 0.5 cm in length, place them in the warm culture room (20 ± 2 °C) with a 16 h day length and a PAR of 75 µmol m^{-2} s^{-1} at shelf level.

9. Between 2 and 3 weeks later, young plantlets (Fig. 1g) are directly transferred to $7 \times 7 \times 6.5$ individual plastic containers filled with PG-mix, identified by a plastic bar-coded label and placed on shelves in a greenhouse with the same environmental conditions as for raising the donor plants of wheat.

3.4 Sampling and Chromosome Doubling

As all regenerated plantlets are haploid, there is no need for a ploidy level check: all the plantlets must be subjected to colchicine (*see* **Note 17**) treatment for genome doubling.

1. Prior to colchicine treatment, a 30 mg piece of leaf of each plantlet can be sampled in a 96-well deep well plate (Fig. 1h) for DNA extraction and molecular marker analyses (Fig. 1i; *see* **Note 18**). There are numerous advantages of using haploid material to perform molecular and genomic analyses [6].

2. For colchicine treatment (*see* **Note 19**), plantlets at the 2–3 tiller stage are removed from the substrate, roots are washed in running tap water for 2 min and pruned to about 1 cm.

3. A single incision of ~1 cm length is made at each tiller base with a scalpel [13].

4. All the plants are then tied in a bundle in the same order as the identification tags with metal wire and then submerged colchicine solution (Fig. 2a) for 5 h at 25 ± 1 °C.

Fig. 2 Production and field management of DHs. (**a**) Colchicine treatment of haploid plants. (**b**) Recovered wheat haploid plants after colchicine treatment. (**c**) Transfer of colchicine-treated plants after vernalization. (**d**) DH plant covered with paper bag to prevent cross-pollination. (**e**) Harvest of each individual DH. (**f**) Replicated wheat DH trials. (All images are from Florimond Desprez)

5. After the treatment, the lower part of the plants and the roots are rinsed in running tap water for a few minutes, potted in PG-mix and returned to the same greenhouse.

6. After a recovery period of 2 weeks, plantlets have usually developed new tillers. Dead leaves and tillers are removed, and the top part of the plantlets is cutoff to leave 6–7 cm of aerial tillers and leaves (Fig. 2b).

3.5 Seed Collection, Multiplication and Breeding

1. Winter and facultative plantlets are vernalized according to their requirements between 4 and 10 weeks in the same cold room and environmental conditions as for the donor plants. Spring material is placed in the cold room for 2 weeks and then transferred to the tillering room for an additional 2–3 weeks.

2. In the meantime, molecular analyses have been done so that it is possible to eliminate those with undesirable alleles.

3. After vernalization and tillering, plantlets are repotted in wheat plastic pots containing the PG-mix (Fig. 2c) and transferred to an unheated greenhouse without any artificial lighting. In most cases, this is carried out in January to promote high tillering and the natural temperature and day length encourage optimum plant growth until maturity.

4. Before anthesis, each plant is covered with a paper bag (Fig. 2d) commonly used for wrapping bread to prevent cross-pollination (*see* **Note 20**).

5. Immediately after seed set, the bags are removed to allow insect and pest control if necessary. In addition, plants that have a suboptimal phenotype such as tall straw are discarded.

6. At maturity, usually in June and July, fertile spikes are collected manually (Fig. 2e) (*see* **Notes 21** and **22**) from the remaining DH plants and placed in a paper bag closed by a staple with the plastic label of the pot.

7. Harvested seed is then dispatched to the designated trial site for evaluation. In the following autumn, seeds of each DH of winter and facultative types are sown in the field for agronomic assessment while spring DH would be sown in the next spring.

8. The most suitable lines are harvested and tested further in replicated plots, and when possible, in several locations.

9. The best selections are entered into replicated trials in the subsequent year (Fig. 2f) and considered for registration. They can also be used as parents in the breeding program.

4   Notes

1. When a new batch of soil substrate is delivered by the provider, two samples are randomly taken; one is sent to an independent analytical laboratory for chemical and physical analyses and the second retained as a backup. Deviations in the quality of the growing medium may occur and lead to significant decreases in plant production efficiency.

2. As a preventative action, sticky plates for detecting and trapping thrips (Bug-scan blue) and flying insect pests (Bug-scan yellow) are regularly dispersed in all greenhouses.

3. Sulfur is occasionally evaporated at night in greenhouses using a sulfur evaporator to limit powdery mildew development.

4. A 2-week 4 °C treatment with 8 h day length provided by Osram cool white L58W/840 fluorescent tubes giving a photoactive radiation (PAR) of 50μmol m^{-2} s^{-1} at canopy level and occasional watering is recommended for spring wheat. This treatment removes a possible, albeit low need for vernalization in some spring wheat types.

5. When the number of hybrid seeds of wheat is low (<6) and for spring types, the young plantlets are placed in a tillering room for 2–3 weeks at $14 \pm 1/10 \pm 1$ °C (day/night, 12 h day length) provided by a 1:1 mixture of Green Power Philips Master SON-T Agro 400 and HPI-T Plus 400 yielding a minimum PAR of 800μmol m^{-2} s^{-1} at canopy level to enhance tillering and obtain a higher number of spikes for crossing.

6. For best results, it is important not to overwater the plants and therefore to adapt the volume and frequency of watering to the environmental conditions. In practice, the potting soil has to dry somewhat before being rehydrated without the plant suffering from a lack of water.

7. Depending on the temperature inside the greenhouse, pollination of the wheat spikes should be done within 10–15 min from maize pollen collection. As soon as the pollen grains clump in the Petri dish, they should be replaced by freshly collected pollen. Thus, it is recommended to split the pollen collection to ensure efficient fertilization.

8. To avoid contact of fingers with growth substances, for example, auxins, safety gloves must be worn when the spike treatments are carried out.

9. Prior to embryo rescue, collected spikes can be stored up to 2 weeks at 4 °C in the dark. However, there is a progressive degradation of embryo quality with storage time, resulting in poorer embryo development rates and increased in vitro contamination by microorganisms. Therefore, it is recommended to process them within 2–4 days.

10. Wheat seeds derived from intergeneric crosses with maize are smaller than seeds resulting from selfing. The two categories can be easily differentiated visually so that selfed seeds can be left on the spike and thrown away when the seeds from intergeneric cross have been collected for embryo rescue.

11. It is important to respect the safety rules in force in laboratories and greenhouses, especially when handling toxic products such as mercuric chloride and colchicine. Wearing a cotton gown, gloves, safety glasses, and safety shoes is mandatory.

12. All disinfection and toxic solutions are collected in a dedicated container and sent to a waste collection company to be disposed of.

13. No more than 20–30 seeds should be placed in a Ø 90 cm Petri dish for dissection in order to limit the risk of contamination by microorganisms. As soon as the embryos have been removed from the seeds, the used Petri dish should be replaced by a new, sterile one.

14. Plastic sterile Ø 90 cm Petri dishes can be used for seed dissection, but glass Petri dishes are preferred because they are heavier and consequently more stable under the binocular microscope. After use, these are cleaned in a washing machine and sterilized in an oven at 180 °C for at least 1 h.

15. Most compounds for in vitro culture and colchicine treatments are from Sigma Aldrich. This is not mandatory, but it is better to purchase "suitable for plant cell culture" grade products when available.

16. After sterilization and before the Gamborg's B-5 medium solidifies, the tubes are placed in a rack inclined at ~30° until the agar medium solidifies. This allows the condensation of water at the bottom of the tube to drain off and the embryos are placed on a surface of water-free medium.

17. Although caffeine and trifluralin may be used in wheat anther culture to improve plant regeneration and genome doubling [14], colchicine has remained the most efficient antimitotic compound in many plant species.

18. When leaf fragments are sampled, the 96-well plates are held on crushed ice to avoid possible DNA degradation by enzymes inside the tissues and to allow quality molecular analyses.

19. Aqueous colchicine solution can be used as many as five different times without any loss of activity. The solution is stored at 4 °C in the dark and brought to ~25 °C before each use.

20. To bag DH and ensure self-fertilization, it is preferable that the bags have a transparent window. This window makes it possible to check whether the plant is being infested with fungal and insect pests such as aphids and to carry out a phytosanitary treatment accordingly.

21. All haploid plants produced bear the same identification as that of the hybrid from which they are derived. The individual identification of each fertile and harvested DH is denoted by an additional number after that of the hybrid.

22. When harvesting, each DH is physically isolated from the others to avoid a possible mixture of ears from different plants.

Acknowledgments

The author gratefully acknowledges contributions of colleagues and technical support to the development over the years of the described protocol. Many thanks are due to Dr. R.A. Pickering for critically reviewing this manuscript. This article is dedicated to the memory of Mr. Michel Desprez, breeder and scientist, former Research Director at Florimond Desprez.

References

1. Kasha KJ, Reinbergs E (1975) Utilization of haploidy in barley. In: Gaul H (ed) Barley genetics III, proc 3rd Int barley genet Symp, Garching. Springer, Berlin, pp 307–315

2. Pickering RA, Devaux P (1992) Haploid production: approaches and use in plant breeding. In: Shewry PR (ed) Barley: genetics, Molecular Biology and Biotechnology. CAB Int Publ, Wallingford, pp 511–539

3. Forster BP, Thomas WTB (2005) Doubled haploids in genetics and plant breeding. In: Janick J (ed) Plant breeding reviews, vol 25. John Wiley & Sons, New York, pp 57–88

4. Palmer CE, Keller WA (2005) Overview of haploid. In: Palmer CE, Keller WA, Kasha KJ (eds) Biotechnology in agriculture and forestry, Vol 56. Haploids in crop improvement II. Springer-Verlag, Berlin, Heidelberg, pp 3–9

5. Laurie DA, Bennett MD (1988) The production of haploid wheat plants from wheat x maize crosses. Theor Appl Genet 76:393–397

6. Devaux P, Cistué L (2016) Wheat doubled haploids: production to sequencing – what makes them so appealing? In: Bonjean A, Angus WJ, Ginkel van M (eds) The world wheat book, a history of wheat breeding, vol 3. Lavoisier Tec&doc, Cachan Cedex, pp 885–938

7. Gamborg OL, Miller RA, Olima K (1968) Nutrient requirements of suspension cultures of soybean root cell. Exp Cell Res 50:151–158

8. Lefebvre D, Devaux P (1996) Doubled haploids of wheat from wheat x maize crosses: genotypic influence, fertility and inheritance of the 1BL-1RS chromosome. Theor Appl Genet 93:1267–1273

9. Laurie DA, Reymondie S (1991) High frequencies of fertilization and haploid seedling production in crosses between commercial hexaploidy wheat varieties and maize. Plant Breed 106:182–189

10. Knox RE, Clarke JM, DePauw RM (2000) Dicamba and growth condition effects on doubled haploid production in durum wheat crossed with maize. Plant Breed 119:289–298

11. Niu Z, Jiang A, Hammad WA, Oladzadabbasabadi A, Xu SS, Mergoum M, Elias EM (2014) Review of doubled haploid production in durum and common wheat through wheat x maize hybridization. Plant Breed 133:313–320

12. Bains NS, Mangat GS, Singh K, Nanda GS (1998) A simple technique for the identification of embryo-carrying seeds from wheat x maize crosses prior to dissection. Plant Breed 117:191–192

13. Pickering RA (1980) Use of the doubled haploid technique in barley breeding at the welsh plant breeding station. Rep Welsh Pl Breed Stn 1979:208–226

14. Broughton S, Castello M, Liu L, Killen J, Hepworth A, O'Leary R (2020) The effect of caffeine and trifluralin on chromosome doubling in wheat anther culture. Plants 9:105. https://doi.org/10.3390/plants9010105

Chapter 15

Isolation of Staged and Viable Maize Microspores for DH Production

Philippe Vergne and Antoine Gaillard

Abstract

Isolated microspore culture systems have been designed in maize by several groups, mainly from the late 1980s to early 2000s. However, even with optimized protocols, microspore embryogenesis induction has remained very dependent on the genotype in maize, with elite germplasm generally displaying no response or very low response. Yet, these last few years, significant progress has been accomplished in understanding and controlling microspore embryogenesis induction in model dicot and monocot species. This knowledge may be transferred to maize, and isolated microspore culture may gain new interest in this crop, at least for embryogenesis research. The methods we hereby present in detail permit the purification of 3–12×10^5 viable microspores per maize tassel, at the favorable stage for microspore embryogenesis. When cultured in appropriate liquid media, microspores from responsive genotypes give rise to androgenic embryos, which can then be regenerated into fertile doubled haploid plants.

Key words Androgenesis, Isolated microspores, Maltose, Microspore culture, Microspore embryogenesis, Sucrose, Percoll, *Zea mays* L

1 Introduction

In maize, the use of isolated microspore culture for inducing in vitro microspore embryogenesis (or androgenesis) has been developed by several research groups mainly over a period of ca. 15 years, from late 1980s to early 2000s (reviewed in [1–4]). This progress had been allowed by the previous construction by these groups of maize stocks displaying a high responsiveness in anther culture experiments [1, 5, 6]. Indeed, androgenesis responsiveness in maize displays a strong dependence on genotype, with elite germplasm generally displaying no response or very low response [1, 5]. Some progress has been made in introducing microspore embryogenesis ability into elite material [6] and in approaching the precise genetic determinants of responsiveness [7]. But still, microspore culture has not become a major tool in practical maize breeding. Also, it is to mention that besides

Jose M. Seguí-Simarro (ed.), *Doubled Haploid Technology: Volume 1: General Topics, Alliaceae, Cereals*,
Methods in Molecular Biology, vol. 2287, https://doi.org/10.1007/978-1-0716-1315-3_15,
© Springer Science+Business Media, LLC, part of Springer Nature 2021

embryogenic induction per se, there are other critical steps in maize DH plant development through androgenesis, like regeneration of plants from androgenic embryos and the level of spontaneous chromosome doubling [5]. Actually, over the last two decades, interest has been mainly focused on obtaining maize DH plants through the use of *in planta* gynogenesis induction [8] (*see* Chapter 2 of Volume 2).

In parallel, a fair amount of knowledge has been gained on cellular and molecular events during the early stages of androgenesis, mainly in model species for this developmental pathway, namely tobacco, wheat, and particularly rapeseed and barley with highly responsive varieties and refined experimental systems [9–11]. It appears that findings of the past few years may pave the way for efficient application of microspore embryogenesis in recalcitrant species and genotypes, through isolated microspore culture [11]. This may hold true for both maize model genotypes and agronomically relevant germplasm, at least for embryogenesis and regeneration research. This perspective is also to consider with respect to unprecedented progress made these last few years in understanding plant cell fate reprogramming and regeneration [12, 13], using *Arabidopsis* as model, even if it is likely that some differences will be found in this domain between eudicots and monocots.

In monocot cereals, including maize, the issue of isolating microspores from whole inflorescences or whole spikelets for further culture and DH plant production comprises three consistent requirements: (a) to sufficiently dilacerate anthers so as to free high numbers of microspores, (b) to purify a homogeneous population of microspores at the convenient cytological stage, given the fact that developmental gradients occur along spike axis or tassel branches, and between the first and the second flower of each spikelet, and (c) that this purified population displays high rates of cell integrity and viability. In maize, several satisfactory protocols have been designed in different laboratories, combining mechanical blending or pulverizing with differential filtration steps and purification by density gradient centrifugation [1–4].

The protocol presented in this chapter is based on our experience with maize microspores [3, 14–16]. It has enabled the production of fertile DH plants [15], studies of the microspore heat shock response [16, 17] and investigation of cellular and molecular events during the early stages of microspore embryogenesis [18–22]. It would also be amenable to implementing chemical screens [23] or pharmacological approaches [11].

2   Materials

2.1   Plant Material

1. Healthy maize plants growing either in the field during summer season, or under optimal artificial conditions in a greenhouse or a growth chamber.

2.2   Equipment and Materials

1. Laminar flow hood for sterile work.

2. 1-mL, 2-mL, 5-mL, 10-mL, 25-mL, and 50-mL pipettes.

3. Pipette controller with adjustable speed.

4. Low-speed centrifuge mounted with a swing-out rotor with buckets fit for 15 and 50-mL conical centrifugation tubes.

5. 15- and 50-mL conical centrifugation tubes (so-called "Falcon tubes").

6. Dispersing tool (IKA Ultra-Turrax T25 with N18-G dispersing element; *see* **Note 1**).

7. Stackable 200-µm and 50-µm sieves (200-µm sieve above) arranged on a 100-mL beaker (*see* **Note 2**).

8. 100-mL GL 45 bottle for sterilizing tassel branch segments.

9. Several forceps, of sizes and design convenient to handle tassel branch segments, spikelets, spikelet debris, and 0.5–2 mm embryos.

10. Malassez counting chamber, or any cell-counting device adapted to maize microspore size.

11. Microscope with ×10/×20/×40 lenses (an epifluorescence microscope, mounted with UV and blue illumination, is preferable).

12. 60 × 15 mm Petri dishes, cell culture treated (*see* **Note 3**).

13. 140 × 20 mm Petri dishes.

14. 100 × 20 mm Petri dishes.

15. Commercial transparent film to seal dishes; cut the roll up into ≈1 cm wide spools with a sharp knife. Used to seal Petri dishes.

16. Culture tubes or ventilated vessels, ca. 15 cm high, suitable for maize plantlet development.

2.3   Reagents and Stocks

It is recommended to use cell culture grade or plant tissue culture grade reagents for preparing culture media.

1. Purified water of suitable quality for plant tissue/cell culture.

2. Bleach solution: 10% commercial bleach solution with approx. 3 drops of commercial dishwashing liquid (detergent) in a GL 45 100-mL bottle.

3. Percoll®, sterile.

4. Yu-Pei major salt [24] $10\times$ stock solution (g/L): KNO_3 25, NH_4NO_3 1.65, KH_2PO_4 5.1, $MgSO_4 \cdot 7H_2O$ 3.7, $CaCl_2 \cdot 2H_2O$ 1.76, stored at 4 °C.

5. N6 major salts [25] $10\times$ stock solution (g/L): $(NH_4)_2SO_4$ 4.63, KNO_3 28.3, KH_2PO_4 4.0, $MgSO_4 \cdot 7H_2O$ 1.85, $CaCl_2 \cdot 2H_2O$ 1.66 stored at 4 °C.

6. MS major salts [26], $10\times$ stock solution, stored at 4 °C.

7. MS minor salt [26] $1000\times$ stock solution, stored at 4 °C.

8. MS Fe-EDTA [26] $200\times$ stock solution, stored at 4 °C, protected from light.

9. B5 minor salt [27] $1000\times$ stock solution, stored at 4 °C.

10. Straus vitamins [28], $100\times$ stock solution (g/L): glycine 0.77, nicotinic acid 0.13, thiamine HCl 0.025, pyridoxin HCl 0.025, Ca pantothenate 0.025, sterilized by filtration and stored as aliquots at -20 °C.

2.4 Sucrose-Based Media for Microspore Isolation and Culture (See Note 4)

1. Isolation medium, ISO_S: Yu-Pei major salts/2, MS minor salts/2, MS Fe-EDTA, 50 g/L sucrose, 1 g/L myo-inositol, 1 g/L glutamine, 0.1 g/L serine, 10 mM MOPS, pH 7.0, sterilized by autoclave (115 °C, 20 min), Straus vitamins and 1% (v/v) Percoll aseptically added after autoclaving once the medium has cooled down.

2. ISO_S/Percoll mixture (7:3, v/v).

3. High density sucrose medium, S_{150}: Yu-Pei major salts, MS minor salts, MS Fe-EDTA, 150 g/L sucrose, 0.1 g/L myo-inositol, 1 g/L glutamine, 0.1 g/L serine, pH 5.8, sterilized by autoclave (115 °C, 20 min), Straus vitamins aseptically added after autoclaving once the medium has cooled down.

4. Dilution medium, S_{30}: Yu-Pei major salts, MS minor salts, MS Fe-EDTA, 30 g/L sucrose, 0.1 g/L myo-inositol, 1 g/L glutamine, 0.1 g/L serine, pH 5.8, sterilized by autoclave (115 °C, 20 min), Straus vitamins aseptically added after autoclaving once the medium has cooled down.

5. Microspore culture medium, S_{120}: Yu-Pei major salts, MS minor salts, MS Fe-EDTA, 120 g/L sucrose, 0.1 g/L myo-inositol, 1 g/L glutamine, 0.1 g/L serine, Straus vitamins, pH 5.8, sterilized by filtration.

2.5 Maltose-Based Media for Microspore Isolation and Culture (See Note 4)

It has been reported that maltose in place of sucrose in maize microspore culture medium promotes a higher induction rate of androgenesis [2–4]. Therefore, depending on the aim of the experiment, if it is desirable to evaluate maltose-based media, then the following media should be used instead of sucrose-based media. They have equimolar concentrations of maltose as compared to sucrose in the sucrose-based media [3].

1. Isolation medium, ISO_M: Yu-Pei major salts/2, MS minor salts/2, MS Fe-EDTA, 52 g/L maltose monohydrate, 1 g/L myo-inositol, 1 g/L glutamine, 0.1 g/L serine, 10 mM MOPS, pH 7.0, sterilized by autoclave (115 °C, 20 min), Straus vitamins and 1% (v/v) Percoll aseptically added after autoclaving once the medium has cooled down.

2. ISO_M/Percoll mixture (7:3, v/v).

3. High density sucrose medium, M_{157}: Yu-Pei major salts, MS minor salts, MS Fe-EDTA, 157 g/L maltose monohydrate, 0.1 g/L myo-inositol, 1 g/L glutamine, 0.1 g/L serine, pH 5.8, sterilized by autoclave (115 °C, 20 min), Straus vitamins aseptically added after autoclaving once the medium has cooled down.

4. Dilution medium, M_{31}: Yu-Pei major salts, MS minor salts, MS Fe-EDTA, 31 g/L maltose monohydrate, 0.1 g/L myo-inositol, 1 g/L glutamine, 0.1 g/L serine, pH 5.8, sterilized by autoclave (115 °C, 20 min), Straus vitamins aseptically added after autoclaving once the medium has cooled down.

5. Microspore culture medium, M_{126}: Yu-Pei major salts, MS minor salts, MS Fe-EDTA, 126 g/L maltose monohydrate, 0.1 g/L myo-inositol, 1 g/L glutamine, 0.1 g/L serine, Straus vitamins, pH 5.8, sterilized by filtration.

2.6 Media for Plantlet Regeneration from Embryos [29]

1. NBM Maltose solid medium for embryo culture [30]: N6 major salts, B5 minor salts, MS Fe-EDTA, 90 g/L maltose, 0.1 g/L myo-inositol, pH 5.8, sterilized by filtration in a pre-autoclaved bottle containing Gelrite to reach 2.5 g/L. Gelrite is melted by heating in a microwave oven for a minimum time, then Straus vitamins are aseptically added once the medium has cooled down. Medium is poured in 100 × 20 mm Petri dishes (30 mL/dish).

2. Medium for embryo root development: MS major salts, minor salts and Fe-EDTA, MS vitamins, 30 g/L sucrose, 0.1 g/L myo-inositol, pH 5.8, solidified by 2 g/L. Gelrite, sterilized by autoclave (115 °C, 20 min). Medium is poured in 100× 20 mm Petri dishes (30 mL/dish).

3. Medium for plantlet development: MS major salts, minor salts, and Fe-EDTA, MS vitamins, 10 g/L sucrose, 0.1 g/L myo-inositol, pH 5.8, solidified by 2 g/L. Gelrite, sterilized by autoclave (115 °C, 20 min). Medium is poured in culture tubes or ca. 15 cm high design culture vessels.

2.7 Reagents for Cytological Assays

1. 0.1 M–0.2 M citrate–phosphate buffer, pH 4.0, 1% Triton X-100: mix 30 mL 0.1 M citric acid with 20 mL 0.2 M Na_2H-$PO_4 \cdot 12H_2O$, add 500μL Triton X-100, mix well, and store as 1-mL aliquots at −20 °C.

2. DAPI stock solution: 1 mg/mL DAPI dissolved in ultrapure water, store at 4 °C protected from light.

3. DAPI working solution: dilute 1µL of stock solution in a 1-mL aliquot of citrate–phosphate buffer, 1% Triton X-100.

4. Fluorescein diacetate (FDA) stock solution: 2 mg/mL FDA dissolved in acetone, store at −20 °C. To perform FCR viability test [31], dilute stock solution 1/1000 in microspore suspension.

5. Alexander staining solution: 10 mL 96% ethanol, Malachite green (1 mL of 1% solution in 96% ethanol), 50 mL purified water, 25 mL glycerol, 5 g phenol, 5 g chloral hydrate, acid fuchsin (5 mL of 1% solution in water), Orange G (0.5 mL of 1% solution in water), 2 mL glacial acetic acid (*see* **Note 5**).

3   Methods

3.1   Tassel Sampling and Pretreatment

1. Harvest tassels just before emergence from the leaf whorl (*see* **Note 6**). Practically, insert fingers at the base of the 'spindle' made by the tassel and the few leaves surrounding it, and grab vertically and sharply, so that the stem breaks at the base of the tassel.

2. Cold pretreatment: remove leaves to get a 'naked' tassel and wrap it with aluminum foil. Store it at 7 °C for 18 days (*see* **Note 7**).

3. After cold pretreatment, cytologically determine the portions of tassel branches where microspores of the first florets of the spikelets are at the mid-late uninucleate/early bicellular stage (*see* **Note 8**), using the staining method of Alexander [32] (bright field illumination) or DAPI staining [33] (UV illumination with epifluorescent microscope) (*see* **Note 9**).

3.2   Microspore Isolation and Purification

Use cold (4 °C) media at all steps, unless otherwise stated.

1. Cut the determined segments of the tassel and sterilize them: immerse them for 15 min in bleach solution at room temperature (Fig. 1a). Manually invert the bottle several times during the sterilization step. Then rinse tassel branch segments three times for 5 min each with sterile deionized water (at room temperature). Manually shake the bottle at each rinsing step, so that the tassel portions are thoroughly rinsed (*see* **Note 10**).

2. Using forceps, strip off approx. 150 spikelets from tassel portions and transfer them into a 50-mL tube containing 25 mL ISO$_S$ (or ISO$_M$). Spikelets will represent a ca. 20 mL 'volume' in the tube (Fig. 1b).

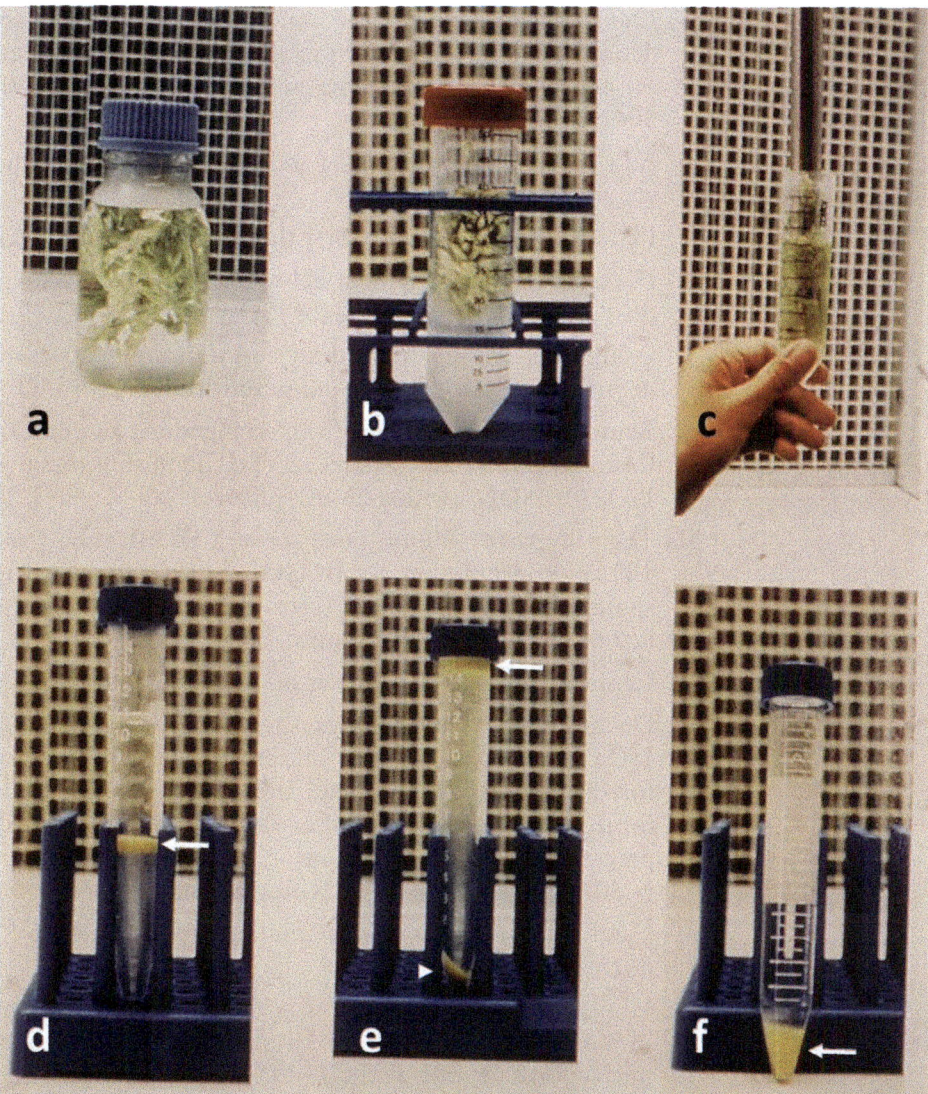

Fig. 1 Overview of microspore isolation procedure: (**a**) sterilization of tassel branch segments; (**b**) spikelet sample, just prior to homogenization; (**c**) homogenization of spikelets with the Ultra-Turrax dispersing tool; (**d**) result of the Percoll gradient centrifugation. The arrow indicates the band of staged microspores; (**e**) result of the centrifugation in S_{150}. The arrow indicates the ring of viable microspores floating at the top of the tube; the arrowhead indicates the pellet of dead microspores; (**f**) fraction of purified microspores gathered from several gradient tubes (arrow), after the final centrifugation in S_{30}

3. Pulverize spikelets for approx. 3 s at 8000 rpm with the Ultra-Turrax dispersing tool while moving the tube upwards and downwards several times (Fig. 1c).

4. Pour the resulting slurry through the sieve arrangement, and quickly rinse spikelet debris retained on the 200-µm sieve with 25 mL ISO$_S$ (or ISO$_M$), using a 25-mL pipette.

5. Using forceps, collect the remaining macroscopic spikelet debris retained on the 200-μm sieve as well as debris retained on the dispersing element, and put them back into the 50-mL tube.

6. Add 25 mL ISO_S (or ISO_M) and immediately repeat the pulverizing step once.

7. Pass the resulting second slurry through the sieve arrangement.

8. Place the sieve arrangement below the dispersing element and rinse it with a few mL ISO_S (or ISO_M).

9. Using a 25-mL pipette, quickly and thoroughly rinse spikelet debris retained on the 200-μm sieve with ISO_S (or ISO_M).

10. Remove the 200-μm sieve from arrangement and quickly and thoroughly rinse microspores retained on the 50-μm sieve with ISO_S (or ISO_M) (with a 25-mL pipette).

11. Then invert the 50-μm sieve above a 50-mL tube (*see* **Note 11**), and rinse the sieve with ISO_S (or ISO_M), so that microspores are transferred in ISO_S (or ISO_M) into the 50-mL tube, in a final volume of 20–40 mL.

12. Centrifuge the 50-mL tube at $65 \times g$ for 3 min.

13. Discard supernatant and resuspend the pellet in 2.5–5 mL ISO_S (or ISO_M). Using a 5-mL pipette and low-speed setting for dispensing, carefully load the resulting suspension onto 5 mL of a mixture of ISO_S (or ISO_M) and Percoll (7:3, v/v) prepared in advance in a 15-mL tube, to form a discontinuous 1–30% Percoll gradient (*see* **Notes 12** and **13**).

14. Carefully handle this gradient and centrifuge it at $225 \times g$ for 3 min.

15. Using a 5-mL pipette and low-speed setting in aspirate mode, carefully collect into a fresh 15-mL tube, the band of staged microspores floating at the 1–30% Percoll interface (Fig. 1d); this results in a 2–3 mL volume, mainly made of ISO_S (or I-SO_M) (*see* **Note 14**).

16. Directly pour from a GL45 bottle 10 mL of S_{150} (or M_{157}) in the tube containing collected microspores (*see* **Note 15**) and centrifuge at $100 \times g$ for 3 min.

17. Using a 1-mL pipette and low-speed setting in aspirate mode, collect in a minimal volume, the viable microspores floating as a yellow ring at the top of the tube (Fig. 1e), and transfer them into a fresh 15-mL tube. The pellet is made of dead microspores, which are sucrose (or maltose) permeable (Fig. 1e).

18. Homogeneously dilute the purified fraction of live microspores with 10 mL S_{30} (or M_{31}) using a 10-mL pipette. Centrifuge the resulting suspension at $80 \times g$ for 3 min (Fig. 1f).

19. Discard supernatant and resuspend well microspore pellet in 4 mL of S_{120} (or M_{126}) culture medium (at room temperature).

20. Quickly, take a small aliquot ($\approx 50\mu L$) of purified fraction, using a 1-mL pipette, and assess yield (number of microspores) and viability rate (FCR test [31]) in a Malassez counting chamber on the epifluorescent microscope. Depending on the genotype, plant growing conditions, and the stage of tassel sampling, a yield of 3×10^5 to 1.2×10^6 microspores is generally achieved per tassel, with a viability rate (FCR-positive) routinely in the range 85–95%.

3.3 Microspore Culture

1. Adjust microspore density to 60.000–80.000 microspores/mL with S_{120} or M_{126} (see **Note 16**).

2. Culture microspores in sealed 60 × 15 mm Petri dishes, with 2–4 mL of microspore suspension per dish.

3. Incubate dishes in darkness at 28 °C in the following arrangement: three dishes of microspore culture are placed in a large 140 × 20 mm dish, with one open 60 × 15 mm dish containing purified water to prevent dehydration.

4. In these conditions, and in responsive genotypes, multinuclear structures can be observed 3–7 days after culture initiation [18] (see **Note 17**).

5. Then, macroscopic androgenic embryos of 0.5–2 mm in size can be collected with fine forceps ca. 20 days after microspore culture initiation, and further processed for plantlet regeneration.

3.4 Plantlet Regeneration and Transfer to Soil

1. Plate embryos onto NBM Maltose solid medium (5 embryos per dish) and incubate at 28 °C in darkness for 10–20 days (depending on the embryo's original size).

2. Transfer embryos onto medium for root development (5 embryos per dish) and incubate at 28 °C with diffused light for 7–15 days (so that roots develop).

3. Transfer rooted embryos onto medium for plantlet development and incubate at 28 °C in light conditions for 7–15 days (until plantlet development).

4. Transfer plantlets to soil, using standard conditions for ex vitro acclimatization and for maize young plant culture.

5. It is possible to assess the ploidy and homozygosity status of recovered plants by flow cytometry and genotyping (see Chapters 6 and 8). Putative DH plantlets will develop into fertile DH plants which can be self-fertilized (see **Note 18**).

4   Notes

1. Similar dispersing tools of other brands or blenders may be used. In that case, adjustment of spikelet pulverizing time/speed through trial and error may be necessary to obtain the best compromise between total microspore yield and microspore viability rate.

2. Typically, Ø 45 mm stainless steel Saulas test sieves (http://www.saulas.fr) are used. Yet, either nylon or stainless steel sieves of other manufacturers or homemade devices with the same mesh may be used.

3. Cell culture treatment renders polystyrene hydrophilic. This permits to culture small volumes of microspore suspension.

4. High sugar media are somewhat viscous. Viscosity can affect pH electrode reaction time. It is thus advisable to adjust pH slowly, or to use a special purpose electrode. It is also advisable to check pH with non-bleeding pH-indicator strips, like Merck MQuant® strips.

5. Alexander staining solution contains toxic and regulated chemicals. A ready-made solution is available from Morphisto (https://www.morphisto.de/en/shop/detail/d/   ALEXANDER-Färbelösung//10924/). Also, an alternative to Alexander solution with no phenol and no chloral hydrate has been proposed [34], which may permit assessment of maize microspore stage.

6. Exact morphological stage for tassel sampling may depend on genotype or mother plant growing conditions.

7. Optimal duration of pretreatment may depend on the genotype or mother plant growing conditions.

8. Maize tassel displays gradients in cytological development of microspores along tassel branches and between first and second flower of each spikelet.

9. If the experimenter is familiar with grass microspore/pollen development stages [35], it is also possible to observe directly anther contents in ISO_S (or ISO_M), with no staining, using bright field or phase contrast/DIC modes.

10. To retain tassel segments in the bottle when emptying it, use long forceps to block them at the bottle mouth. Alternatively, it is possible to implement rinsing steps using a 50-mL pipette.

11. If necessary, it is possible to use a small funnel to help transferring microspores into the 50-mL tube.

12. The microspore suspension must be layered onto the 30% Percoll solution very gently to minimize mixing. To this aim, it is recommended to proceed as follows [36]: gently touch the

pipette tip against the meniscus at the tube wall; then move very slightly the pipette up the tube wall, leaving a trail of wetted tube connecting the pipette tip to the meniscus; then allow the suspension to flow slowly out of the pipette, while slowly moving the pipette upwards so that a trail of wetted tube connecting to the meniscus is continuously maintained during the flow.

13. Depending on sample/pellet size, use two (or more) 30% Percoll tubes, as overloading of the gradient would result in poor resolution (i.e., poor stage separation).

14. This band comprises mid-late uninucleate and early bicellular stages, while both early uninucleate, nonvacuolated microspores and mid-late bicellular pollen grains, which display higher buoyant densities, are sedimented in the pellet.

15. It is not necessary to use a pipette at this step. Directly pouring S_{150} (or M_{157}) followed by centrifugation will result in a neat separation of viable microspores from dead ones (Fig. 1e).

16. It has been observed in some genotypes that lower sugar concentrations (60 and 90 g/L) yielded higher frequencies of embryo-like structures than a high concentration (120 g/L) [2]. Thus, it is also advisable to evaluate culture media with lower sugar (sucrose or maltose) concentration.

17. In particular, after 5 days of culture, proembryos still enclosed in the exine, can be observed. It is possible to purify them for further analysis or culture [19].

18. Indeed, in some genotypes, spontaneous chromosome doubling occurs at high rates in maize (>40%) [37], and it has been shown that this process most probably results from nuclear fusion in androgenic proembryos, 5–7 days after microspore culture initiation [21]. It is also noticeable that chromosome doubling technologies have been designed and improved since the 1990s for use on haploid plantlets obtained through *in planta* haploid induction [38, 39] (*see* Chapter 5 of this Volume). It is likely that these technologies can be also adapted to haploid plantlets issued from in vitro microspore embryogenesis.

Acknowledgments

Work on maize microspores was supported by the CIFRE system (ANRT), by INRA (grants AIP 88/4684 and 89/4684) and by the European Commission (BIO4-CT96-0275, SIME project). PV is a staff member of INRAE (formerly INRA).

References

1. Büter B (1997) *In vitro* haploid production in maize. In: Jain SM, Sopory SK, Veilleux RE (eds) In vitro haploid production in higher plants. Current plant science and biotechnology in agriculture, vol 26. Springer, Dordrecht, pp 37–71. https://doi.org/10.1007/978-94-017-1862-2_2

2. Nägeli M, Schmid J, Stamp P et al (1999) Improved formation of regenerable callus in isolated microspore culture of maize: impact of carbohydrates, plating density and time of transfer. Plant Cell Rep 19:177–184. https://doi.org/10.1007/s002990050730

3. Goralski G, Lafitte C, Bouazza L et al (2002) Influence of sugars on isolated microspore development in maize (*Zea mays* L.). Acta Biol Crac 44:203–212

4. Zheng MY, Weng Y, Sahibzada R et al (2003) Isolated microspore culture in maize (*Zea mays* L.), production of doubled-haploids *via* induced androgenesis. In: Maluszynski M, Kasha KJ, Forster BP, Szarejko I (eds) Doubled haploid production in crop plants. Springer, Dordrecht, pp 95–102. https://doi.org/10.1007/978-94-017-1293-4_15

5. Beckert M (1998) Genetic analysis of in vitro androgenetic response in maize. In: Chupeau Y, Caboche M, Henry Y (eds) Androgenesis and haploid plants. Springer-Verlag, Berlin, Heidelberg, pp 24–37

6. Barnabás B (2003) Anther culture of maize (*Zea mays* L.). In: Maluszynski M, Kasha KJ, Forster BP, Szarejko I (eds) Doubled haploid production in crop plants. Springer, Dordrecht, pp 103–108. https://doi.org/10.1007/978-94-017-1293-4_16

7. Barret P, Brinkman M, Dufour P et al (2004) Identification of candidate genes for in vitro androgenesis induction in maize. Theor Appl Genet 109:1660–1668. https://doi.org/10.1007/s00122-004-1792-8

8. Jacquier NMA, Gilles LM, Pyott DE et al (2020) Puzzling out plant reproduction by haploid induction for innovations in plant breeding. Nat Plants 6:610–619. https://doi.org/10.1038/s41477-020-0664-9

9. Seguí Simarro JM, Nuez F (2008) How microspores transform into haploid embryos: changes associated with embryogenesis induction and microspore-derived embryogenesis. Physiol Plant 134:1–12. https://doi.org/10.1111/j.1399-3054.2008.01113.x

10. Żur I, Dubas E, Krzewska M et al (2015) Current insights into hormonal regulation of microspore embryogenesis. Front Plant Sci 6:424. https://doi.org/10.3389/fpls.2015.00424

11. Testillano PS (2019) Microspore embryogenesis: targeting the determinant factors of stress-induced cell reprogramming for crop improvement. J Exp Bot 70:2965–2978. https://doi.org/10.1093/jxb/ery464

12. Ikeuchi M, Favero DS, Sakamoto Y et al (2019) Molecular mechanisms of plant regeneration. Annu Rev Plant Biol 70:377–406. https://doi.org/10.1146/annurev-arplant-050718-100434

13. Sugimoto K, Temman H, Kadokura S et al (2019) To regenerate or not to regenerate: factors that drive plant regeneration. Curr Opin Plant Biol 47:138–150. https://doi.org/10.1016/j.pbi.2018.12.002

14. Vergne P, Gaillard A, Beckert M (1991) Isolation of viable microspores and immature pollen grains from cereal inflorescences. In: Negrutiu I, Gharti-Chhetri G (eds) A laboratory guide for cellular and molecular plant biology. Birkhäuser Verlag, Basel, Boston, Berlin, pp 75–87

15. Gaillard A, Vergne P, Beckert M (1991) Optimization of maize microspore isolation and culture conditions for reliable plant regeneration. Plant Cell Rep 10:55–58. https://doi.org/10.1007/BF00236456

16. Magnard JL, Vergne P, Dumas C (1996) Complexity and genetic variability of heat shock protein expression in isolated maize microspores. Plant Physiol 111:1085–1096. https://doi.org/10.1104/pp.111.4.1085

17. Gagliardi D, Breton C, Chaboud A et al (1995) Expression of heat shock factor and heat shock protein 70 genes during maize pollen development. Plant Mol Biol 29:841–856. https://doi.org/10.1007/BF00041173

18. Gaillard A, Matthys Rochon E, Dumas C (1992) Selection of microspore derived embryogenic structures in maize related to transformation potential by microinjection. Bot Acta 105:313–318. https://doi.org/10.1111/j.1438-8677.1992.tb00304.x

19. Magnard JL, Le Deunff E, Domenech J et al (2000) Genes normally expressed in the endosperm are expressed at early stages of microspore embryogenesis in maize. Plant Mol Biol 44:559–574. https://doi.org/10.1023/A:1026521506952

20. Testillano PS, Ramirez C, Domenech J et al (2002) Early microspore maize embryos show two domains with defined features also present

in zygotic embryogenesis. Int J Dev Biol 46:1035–1047

21. Testillano PS, Georgiev S, Mogensen HL et al (2004) Spontaneous chromosome doubling results from nuclear fusion during in vitro maize induced microspore embryogenesis. Chromosoma 112:342–349. https://doi.org/10.1007/s00412-004-0279-3

22. Massonneau A, Coronado MJ, Audran A et al (2005) Multicellular structures developing during maize microspore culture express endosperm and embryo-specific genes and show different embryogenic potentialities. Eur J Cell Biol 84:663–675. https://doi.org/10.1016/j.ejcb.2005.02.002

23. Hicks G, Robert S (eds) (2014) Plant chemical genomics. Methods in molecular biology (methods and protocols), vol 1056. Humana Press, Totowa, NJ. https://doi.org/10.1007/978-1-62703-592-7

24. Ku MK, Cheng WC, Kuo LC et al (1981) Induction factors and morpho-cytological characteristics of pollen-derived plants in maize (Zea mays L.). In: Proc Symp plant tissue cult, vol 1978. Science Press, Peking, pp 35–42

25. Chu CC (1981) The N_6 medium and its application to anther culture of cereal crops. In: Proc Symp plant tissue cult, vol 1978. Science Press, Peking, pp 43–50

26. Murashige T, Skoog F (1962) A revised medium for rapid growth and bioassays with tobacco tissue cultures. Physiol Plant 15:473–497. https://doi.org/10.1111/j.1399-3054.1962.tb08052.x

27. Gamborg O, Miller R, Ojima K (1968) Nutrient requirements of suspension cultures of soybean root cells. Exp Cell Res 50:151–158. https://doi.org/10.1016/0014-4827(68)90403-5

28. Genovesi AD, Collins GB (1982) In vitro production of haploid plants of corn via anther culture. Crop Sci 22:1137–1144. https://doi.org/10.2135/cropsci1982.0011183X002200060013x

29. Matthys Rochon E, Piola F, Le Deunff E et al (1998) *In vitro* development of maize immature embryos: a tool for embryogenesis analysis. J Exp Bot 49:839–845. https://doi.org/10.1093/jxb/49.322.839

30. Mòl R, Matthys Rochon E, Dumas C (1993) In-vitro culture of fertilized embryo sacs of maize: zygotes and two-celled proembryos can develop into plants. Planta 189:213–217. https://doi.org/10.1007/BF00195079

31. Heslop Harrison J, Heslop Harrison Y (1970) Evaluation of pollen viability by enzymatically induced fluorescence; intracellular hydrolysis of fluorescein diacetate. Stain Technol 45:115–120. https://doi.org/10.3109/10520297009085351

32. Alexander MP (1969) Differential staining of aborted and non-aborted pollen. Stain Technol 44:117–122. https://doi.org/10.3109/10520296909063335

33. Vergne P, Delvallée I, Dumas C (1987) Rapid assessment of microspore and pollen development stage in wheat and maize using DAPI and membrane permeabilization. Stain Technol 62:299–304. https://doi.org/10.3109/10520298709108014

34. Peterson R, Slovin JP, Chen C (2010) A simplified method for differential staining of aborted and non-aborted pollen grains. Int J Plant Biol 1:e13. https://doi.org/10.4081/pb.2010.e13

35. Bedinger P (1992) The remarkable biology of pollen. Plant Cell 4:879–887. https://doi.org/10.1105/tpc.4.8.879

36. Hames BD (1978) Methods for preparing and fractionating gradients. In: Rickwood D (ed) Centrifugation-a practical approach. Information Retrieval, London, Washington DC, pp 47–67

37. Antoine-Michard S, Beckert M (1997) Spontaneous versus colchicine-induced chromosome doubling in maize anther culture. Plant Cell Tissue Organ Cult 48:203–207

38. Chaikam V, Molenaar W, Melchinger AE et al (2019) Doubled haploid technology for line development in maize: technical advances and prospects. Theor Appl Genet 132:3227–3243. https://doi.org/10.1007/s00122-019-03433-x

39. Chaikam V, Gowda M, Martinez L et al (2020) Improving the efficiency of colchicine-based chromosomal doubling of maize haploids. Plants 9:459. https://doi.org/10.3390/plants9040459

Chapter 16

Triticale Isolated Microspore Culture for Doubled Haploid Production

Priti Maheshwari and John D. Laurie

Abstract

Here, we describe a method of triticale isolated microspore culture for production of doubled haploid plants via androgenesis. We use this method routinely because it is highly efficient and works well on different triticale genotypes. To force microspores into becoming embryogenic, we apply a 21-day cold pretreatment. The shock of cold facilitates redirecting microspores from their predestined pollen developmental program into the androgenesis pathway. Ovaries are included in our culture methods to help with embryogenesis, and the histone deacytelase inhibitor Trichostatin A (TSA) is added to further improve androgenesis and increase our ability to recover green doubled haploid plants.

Key words Androgenesis, Embryogenesis, Doubled haploid, Microspores, Ovary co-culture, Tissue culture, Trichostatin A, Triticale

1   Introduction

Androgenesis in triticale has been successfully initiated using both anther culture and isolated microspore culture (IMC) methods. While anther culture involves culturing of intact anthers, IMC involves extracting microspores from the anthers followed by culture of the microspores in vitro. Although the triticale anther culture method is low in cost and relatively simple to perform, it is less efficient than IMC, making IMC the preferred method. Since the first reports of IMC in triticale [1], significant advancements have been made [2–19] and the method is now routinely used by companies as well as breeders in academia and government.

The discovery that microspores can be coerced into entering the embryogenesis pathway using stress treatment was a milestone in the history of doubled haploid (DH) technology development [20]. Since then, several different stress factors have been applied separately and in combinations to trigger androgenesis. The ultimate choice of stress treatment may depend on the donor plant

Jose M. Seguí-Simarro (ed.), *Doubled Haploid Technology: Volume 1: General Topics, Alliaceae, Cereals*,
Methods in Molecular Biology, vol. 2287, https://doi.org/10.1007/978-1-0716-1315-3_16,
© Springer Science+Business Media, LLC, part of Springer Nature 2021

genotype but in general, the most effective trigger in triticale is a long pretreatment at low temperature [2, 5, 10, 11, 21, 22]. Molecular studies are increasing our understanding of the events that occur during microspore embryogenesis [23] and future efforts will lead to increased knowledge and further improvement of the technology. A recent advancement has come from the use of trichostatin A (TSA), a *Streptomyces* metabolite that inhibits histone deacetylases [24]. Inclusion of TSA in IMC for DH production has also improved efficiency of the technology in cereals [25].

When faced with different genotypes, a variety of IMC methods and various other factors, it may be difficult to decide how to get started in IMC [1, 26]. These factors also make it difficult for the process to be transferred between laboratories. The method described here is meant to overcome many of these challenges and if followed step-by-step will lead to a high level of success. A prerequisite for this success is a supply of healthy donor plants, free of stress and disease, that are grown under controlled environmental conditions. This method was developed over the years by continuous modification of the report published by Eudes and Amundsen [7]. It uses cold pretreatment combined with ovary co-culture, addition of phytosulfokine-alpha (PSK-α), a five amino acid sulfated peptide, is shown to play an important role in embryogenesis [15], TSA [25], and several other modifications resulting in an efficient IMC protocol that can be successfully applied to any genotype.

2   Materials

2.1 Surface Sterilized Materials

1. Paper towels.

2. Blender cups and lids (110 mL warring blender cup).

3. Stainless steel juice strainer.

4. Cell strainers (100μm).

5. Forceps.

6. Petri dishes measuring 60 × 15 mm.

7. Beaker size 200 mL.

8. Conical centrifugation tubes sizes 15 and 50 mL.

9. Pasteur pipettes.

10. Filters (0.22μm).

2.2 Growth Chamber Materials

1. Nursery pots (2.8 L).

2. Cornell mix [27].

3. Intercept™ (0.004 g/L of soil, Imidacloprid, Bayer).

4. Tilt™ (2.5 mL/L propiconazole, Syngenta).

5. Fertilizer (20–20–20).

2.3 Stock Solutions   Prepare the following stock solutions of the basal salt mixtures, amino acids, and vitamins which will be required to make the culture media (*see* Table 1, Subheading 2.4, **Note 1**).

1. CIMC macrosalts (10×): 14.15 g/L KNO_3, 2.320 g/L $(NH_4)_2SO_4$, 2.0 g/L KH_2PO_4, 0.937 g/L $MgSO_4 \cdot 7H_2O$, and 0.830 g/L $CaCl_2 \cdot 2H_2O$. Add 900 mL deionized water to a 2 L beaker on a stir plate. $CaCl_2 \cdot 2H_2O$ should be dissolved in water before adding to the other components. Stir until salts are dissolved and then adjust the volume to 1 L. Filter sterilize and store at 4 °C.

2. CIMC microsalts (100×): 0.040 g/L KI, 0.5 g/L $MnSO_4 \cdot H_2O$, 0.5 g/L H_3BO_3, 0.5 g/L $ZnSO_4 \cdot 7H_2O$, 1.25 mL of 1 mg/mL $CoCl_2 \cdot 6H_2O$ stock, 1.25 mL of 1 mg/mL $Na_2MoO_4 \cdot 2H_2O$ stock, and 1.25 mL of 1 mg/mL $CuSO_4 \cdot 5H_2O$ stock. Filter sterilize and store at 4 °C (*see* **Note 2**).

3. Iron-EDTA (100×): 0.278 g/L $FeSO_4$, 0.372 g/L Na_2EDTA. Store in an amber bottle at 4 °C.

4. CIMC-7 organic salts (100×): 0.05 g/L pyridoxine HCl, 0.5 g/L thiamine HCl, 0.05 g/L nicotinic acid, and 0.2 g/L glycine. Store at −20 °C as 10 mL aliquots to prevent multiple free-thaw cycles.

5. CIMC-4 organic salts (100×): 0.5 g/L thiamine HCl, 0.1 g/L pyridoxine HCl, 0.1 g/L nicotinic acid, 2.5 g/L ascorbic acid, 0.1 g/L aspartic acid, 0.1 g/L citric acid, and 0.4 g/L glycine. Store at −20 °C as 10 mL aliquots to prevent multiple free-thaw cycles.

6. 2,4-Dichlorophenoxyacetic acid (2,4-D) stock solution at 1 mg/mL: Dissolve 15 mg 2,4-D in 1 mL of 95–100% ethanol in a 1.5 mL snap cap tube. Dilute to 1 mg/mL with deionized water.

7. Kinetin and 2-benzyl amino purine (BAP) stock solutions at 1 mg/mL: Dissolve 15 mg of kinetin/BAP powder in 1 mL of 1 N KOH and dilute to 1 mg/mL with deionized water. Store solution at 4 °C.

8. Phenylacetic acid (PAA) stock solution at 1 mg/mL: Dissolve 50 mg of PAA in 1 mL of 95–100% ethanol, and then dilute to 1 mg/mL with deionized water. Store solution at 4 °C.

9. Larcoll stock solution at 10 mg/mL: Dissolve 300 mg Larcoll in water and prepare 1 mL aliquots in 1.5 mL snap cap tubes. Store at −20 °C.

10. Glutathione stock solution at 1 mg/mL: Prepare 30 mL and split into 1 mL aliquots in 1.5 mL snap cap tubes. Store at −20 °C.

Table 1
Composition of the tissue culture media used for extraction and culture of isolated microspores from triticale

Constituents	CIMC extraction buffer	CIMC induction medium	CIMC-4 shoot regeneration medium	Rooting medium
KNO_3	1415	1415	1415	1415
$(NH_4)_2SO_4$	232	232	232	232
KH_2PO_4	200	200	200	200
$CaCl_2 \cdot 2H_2O$	83	83	83	83
$MgSO_4 \cdot 7H_2O$	93	93	93	93
KI	–	0.4	0.4	0.4
$MnSO_4 \cdot 4H_2O$	–	5	5	5
H_3BO_3	–	5	5	5
$ZnSO_4 \cdot 7H_2O$	–	5	5	5
$CoCl_2 \cdot 6H_2O$	–	0.0125	0.0125	0.0125
$CuSO_4 \cdot 5H_2O$	–	0.0125	25	0.0125
$Na_2MoO_4 \cdot 2H_2O$	–	0.0125	0.0125	0.0125
$FeSO_4 \cdot 7H_2O$	27.8	27.8	27.8	27.8
Na_2EDTA	37.3	37.3	37.3	37.3
MES hydrate	175	–	–	–
Thiamine-HCl	–	5	10	–
Ascorbic acid	–	–	25	–
Pyridoxine-HCl	–	0.5	1	–
Nicotinic acid	–	0.5	1	–
Aspartic acid	–	–	10	–
Citric acid	–	–	10	–
Glycine	–	2	2	–
Myo-inositol	–	300	300	–
Glutamine	–	500	500	–
Proline	–	–	500	–
Glutathione	–	1	–	–
AG-Larcoll	–	10	–	–
PSK-α	–	0.085	–	–
Cefotaxime	–	100	100	–
Spermine	–	–	1	–

(continued)

Table 1
(continued)

Constituents	CIMC extraction buffer	CIMC induction medium	CIMC-4 shoot regeneration medium	Rooting medium
Spermidine	–	–	4	–
Maltose	–	90,000	60,000	–
Mannitol	72,868	9000	–	
Sucrose	–	–	–	10,000
2,4-D	–	0.2	–	–
PAA	–	1	1	0.5
Kinetin	–	0.2	–	–
BAP	–	–	2	–
Meta-topoline	–	–	0.5	–
Purified Agar	–	–	–	6000
Gelrite	–	–	3500	–
pH	6.5	7	5.8	6

Amounts are represented in mg/L

11. Spermine and spermidine stock solutions at 1 mg/μL: Prepare spermine and spermidine stock solutions in deionized water, make aliquots, and store at −20 °C.

12. Trichostatin A (TSA) stock solution: TSA solution (Sigma # T-1952-200μL) is diluted 10× in DMSO (dimethylsulfoxide, Sigma # D2650-5X5ML) to a final concentration of 0.5 mM. In a laminar air flow hood, add 800μL of sterile DMSO to the 200μL TSA solution and mix by pipetting. To prevent free thawing, prepare workable aliquots (10–25μL) in sterile 0.5 mL snap cap tubes and store at −20 °C.

13. 0.5 mg/mL PSK-Alpha (PSK-α) stock solution: Weigh 0.5 mg PSK-α into a 2 mL centrifuge tube and add 1 mL sterile optima water to make a 0.5 mg/mL stock solution. Store at −20 °C.

14. (PSK-α) Working stock solution: Take 54.4μL of the 0.5 mg/mL stock and dilute with 1945.6μL of CIMC induction medium to a final volume of 2 mL (*see* **Note 3**). Filter sterilize using a 0.22μm syringe filter in a laminar air flow bench (hood). Prepare 250μL aliquots in sterile 0.5 mL snap cap tubes. Store at −20 °C.

2.4 Culture Media   The following culture media are prepared from the above-mentioned stock solutions.

1. CIMC extraction buffer (2 L): 200 mL 10× CIMC macro salts, 20 mL 100× Iron-EDTA, 350 mg 2-(N-morpholino) ethanesulfonic acid (MES) hydrate, and 145.736 g mannitol. Autoclave in advance, a 2 L bottle that is compatible with the filtration device. Add 1600 mL ddH$_2$O to a 2 L beaker on a stir plate and then mix all media components sequentially. Adjust pH to 6.5 with 1 N KOH. Filter sterilize and store at 4 °C (*see* **Note 4**).

2. CIMC induction medium (1 L): 100 mL 10× Macro salts, 10 mL 100× Micro salts, 10 mL 100× Iron-EDTA, 10 mL 100× CIMC-7 organic salts, 1 mg glutathione, 300 mg myo-inositol, 500 mg glutamine, 1 mg Larcoll, and 90 g maltose monohydrate, 9 g mannitol, 0.2 g 2,4-D, 0.2 g Kinetin, 1 mg phenyl acetic acid (PAA), and 100 mg cefotaxime. Make up the volume to 1 L and adjust pH to 7.0 with 1 N KOH. Filter the media into an autoclaved 1 L bottle and store at 4 °C.

3. CIMC-4 shoot regeneration media (2 L): This media is prepared at 2× concentration from the stock solutions to allow mixing 1:1 with double strength Gelrite for preparation of final semi-solidified media plates. First, prepare Part 1: 200 mL 10× macro salts, 20 mL 100× micro salts, 50 mg CuSO$_4$ · 5H$_2$O, 20 mL 100× Iron-EDTA, 20 mL 100× CIMC-4 organics salts, 600 mg myo-inositol, 1 g glutamine, 1 g proline, 120 g maltose monohydrate, 2 mg spermine, 8 mg spermidine, 1 mg meta-Topoline, 2 mg PAA, 4 mg BAP, and 200 mg cefotaxime. Make up the volume to 1 L and adjust pH to 5.8 with 1 N KOH. Filter sterilize the solution into an autoclaved 1 L bottle. Second, prepare Part 2: 7.0 g Gelrite in 1 L deionized water. Autoclave for 20 min. Third, mix part 1 with part 2: Place the flask containing 2× Gelrite (part 2) on a heated magnetic stir plate inside the laminar flow hood (*see* **Note 5**). Pour the filtered part 1 into the Gelrite with constant mixing. Dispense 20 mL into 100 mm sterile Petri dishes inside the laminar flow hood. Leave the Petri dishes open for 45 min inside the hood. Stack and store at 4 °C.

4. Rooting Medium (1 L): 100 mL 10× CIMC Macro salts, 10 mL 100× CIMC micro salts, 10 mL 100× Iron-EDTA, 0.5 mg PAA, and 10 g sucrose. Bring final volume to 1 L. Adjust pH to 6.0 with 1 N KOH. Mix 6 g Agar and autoclave. Transfer 50 mL medium to sterile Magenta boxes inside a laminar flow hood. Once solidified, store boxes at 4 °C.

5. 20% maltose solution: 20 mg maltose in 100 mL deionized water.

2.5 Disinfecting Agents

1. 70% (v/v) aqueous ethanol.

2. 10% (v/v) aqueous bleach (dilution of 5.25% sodium hypo-chlorite commercial bleach).

3. Sterile distilled water.

3 Methods

The following method for triticale isolated microspore culture has been optimized over the past several years for successful production of doubled haploid plants. The method can be directly applied to any triticale genotype (*see* **Note 6**).

3.1 Handling of Donor Plants

Triticale seeds are sown every week in nursery pots to provide a regular supply of high quality donor spikes for microspore isolation.

1. Plants are seeded, two per pot, into Cornell mix [27] and grown in environmentally controlled growth rooms (*see* **Note 7**) with a photoperiod of 18/6 h (light/dark), a light intensity of $300\mu E\ m^{-2}\ s^{-1}$, a temperature regime of 15 °C/12 °C (day/night), and with a relative humidity of about 70%.

2. Pests and disease are to be kept to a minimum using good greenhouse practices. At the 2–3 leaf stage (around 3 weeks after planting), young seedlings are treated with Intercept™ and Tilt™ for preventive control against aphids and mildew.

3. Following pesticide treatment, donor plants are fertilized every 2 weeks (20–20–20, N-P-K at 200 ppm) and maintain under controlled conditions as indicated above.

3.2 Collection of Donor Spikes

1. Tillers are harvested (the first 10 per plant) when the micro-spores are at the mid to late uni-nucleate stage (*see* **Note 8**) (Fig. 1a–c).

2. Tillers with spikes at the right stage are cut at least a couple nodes below the spike.

3. For triticale, the morphological marker indicating the correct stage for microspores is when the spike is roughly 1.5–3 cm out of the boot (Fig. 1d). The stage of the microspores can be further evaluated using acetocarmine staining (*see* **Note 9**).

3.3 Cold Pretreatment of Donor Spikes

A 21-day cold pretreatment prior to microspore isolation induces microspore transition from their normal pollen fate toward embryogenesis.

1. At the right stage, the top three-fourth of the harvested tillers (*see* Subheading 3.2) are wrapped in aluminum foil for storage under dark.

Fig. 1 Tiller collection for triticale IMC. Stages of uni-nucleate microspore development ranging from (**a**) early, (**b**) mid to, (**c**) late (vacuolated). M: Micropore. N: nuclei. (**d**) Morphological marker for the collection of triticale spikes when microspores are at the mid-late uni-nucleate stage. (**e**) Spikes are wrapped in aluminum foil and placed in a beaker containing tap water for cold storage

2. The tillers are then transferred and stored at 4 °C with their stem bases submerged in a beaker filled with tap water (Fig. 1e).

3. After 21 days, the tillers are removed and used for microspore isolation (*see* **Note 10**).

3.4 Selection of Spikes for Ovary Donors

Co-culture of microspores with ovaries is carried out to improve androgenesis. Spikes for ovaries are collected from donor triticale plants grown under controlled conditions. It is important to have a supply of ovary donor plants to be able to collect spikes on the day of microspore isolation, as ovary donors are preferably collected fresh (*see* **Note 11**).

1. Cut tillers for ovary donors when spikes are completely out of the boot, but before anthesis. At this stage, the anthers are light green and have not yet turned yellow; the ovary is swollen with a fluffy stigma (*see* **Note 12**).

3.5 Disinfection of Spikes

The cold pretreated spikes are disinfected along with the ovary donor spikes on the day of microspore isolation.

1. At 21 ± 3 days of cold pretreatment, remove the spikes from their boots.

2. Use eight spikes per microspore isolation.

3. At least four fresh donor spikes should be used for ovary co-culture per extraction.

4. Trim awns with scissors (*see* **Note 13**) (Fig. 2a).

5. Transfer the trimmed spikes to a clean 1 L bottle (Fig. 2b).

6. In the laminar flow hood under sterile conditions, disinfect the spikes by adding 500 mL of 10% commercial bleach and agitate constantly for 4 min.

7. Pour off bleach and rinse four times with about 400 mL of autoclaved distilled water for 1 min/rinse with constant agitation.

3.6 Microspore Isolation Procedure

In the laminar flow hood, place ten sheets of sterile paper towel.

1. Fill a pre-chilled autoclaved 110 mL warring blender cup with 50 mL cold CIMC extraction buffer. Put the filled blender cup on ice.

2. Using sterile forceps and scissors, remove one surface sterilized microspore donor spike at a time from the bottle. Place the spike on the right side of an opened paper towel with a fold down the middle (Fig. 2c, d).

3. Use the thumb (at the base of the spike) and forefinger (at the top of the spike) to gently stretch the spike to spread the florets allowing removal of the outer glooms (Fig. 2d).

4. Remove the florets one by one and transfer immediately to the blender cup (*see* **Note 14**) (Fig. 2e).

5. Rotate the spike 180° to remove the outer glooms and transfer the florets on the opposite side of the spike to the blender.

6. Discard the used spike, turn to close the paper towel page and repeat using a new spike on a clean towel page. Do this until all florets are collected from the eight spikes (not including ovary donors) (*see* **Note 15**).

7. Upon completion, place the lid on the blender cup.

8. Save the left over paper towels in a corner of the laminar air flow bench for use in ovary isolation.

9. Pre-chill a centrifuge containing a swinging bucket rotor adjusted to appropriate settings (5 min at $100 \times g$, 4 °C) (*see* **Note 16**).

10. Blend twice for 9 s each at low speed (18,000 rpm) in a Waring blender (*see* **Note 17**).

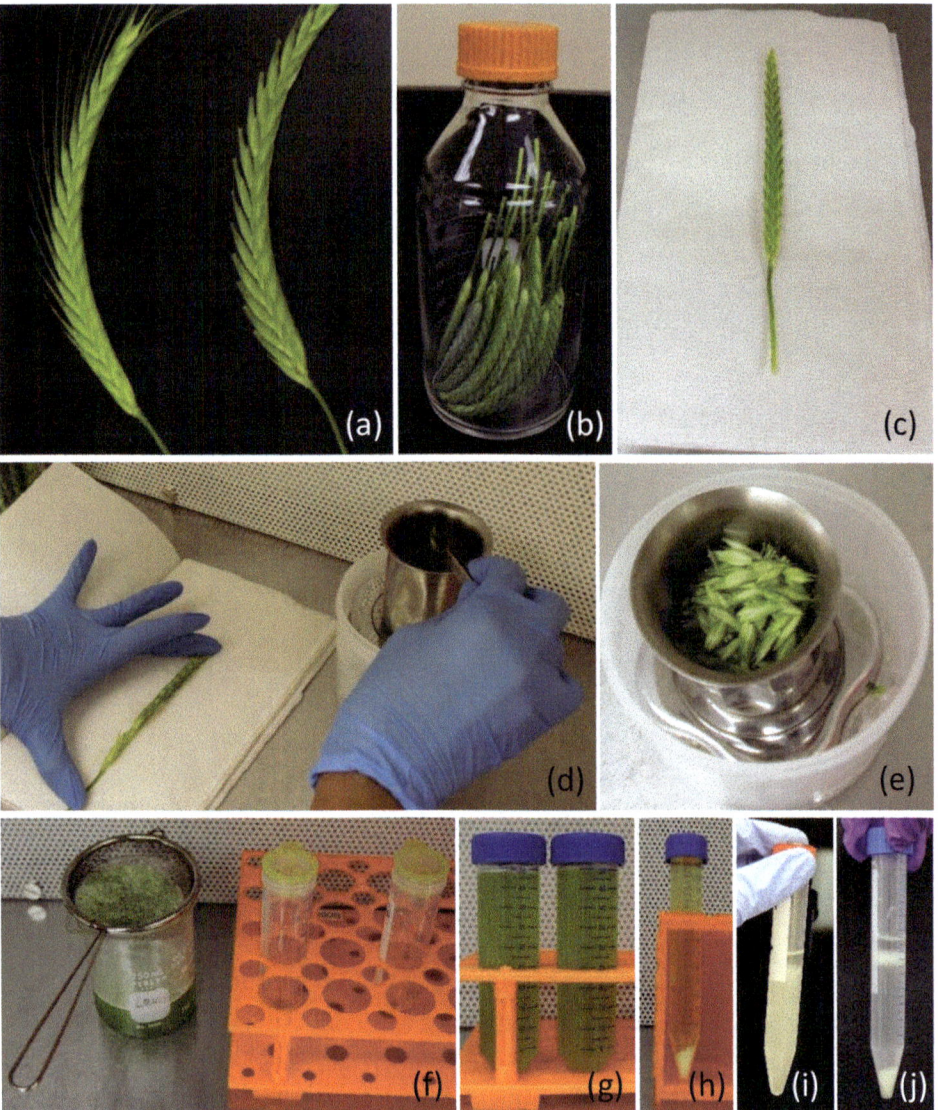

Fig. 2 Steps of the microspore isolation procedure. (**a**) Awns are trimmed from spikes. (**b**) Trimmed spikes are placed in a 1 L bottle for sterilization. (**c**) Spikes are aseptically transferred onto sterile paper towel. (**d**, **e**) Florets are transferred into a blender cup filled with cold 50 mL CIMC extraction buffer. (**f**) After blending, the slurry is filtered. (**g**) The slurry is transferred into 50 mL tubes for the first centrifugation step. (**h**) Microspores form a tighter pellet after the first spin in a 15 mL tube. (**i**) Microspores are suspended in 20% maltose and layered with 1 mL of CIMC induction medium. (**j**) After maltose density gradient centrifugation, a band of live cells forms at the interface

11. In the laminar flow hood, pour the slurry from the blender through an autoclaved stainless steel juice strainer (roughly 1 mm pore size) into a sterile pre-chilled 200 mL beaker (*see* **Note 18**; Fig. 2f).

12. Rinse the blender with an additional 50 mL cold CIMC extraction buffer and pour the rinse through the juice strainer over the crushed spikes.

13. Pour the strained slurry through 100μm sterile cell strainers into two 50 mL conical centrifugation tubes (Fig. 2f, g).

14. Centrifuge the tubes in a swinging bucket rotor for 5 min at $100 \times g$, 4 °C.

15. Pour off the supernatant and leave around 3–5 mL buffer in each tube (*see* **Note 19**). Resuspend the cell pellet in this 3–5 mL buffer. Transfer the cell pellets from both the 50 mL tubes into a single 15 mL sterile conical centrifugation tube.

16. Add CIMC induction medium to bring the volume up to 15 mL and centrifuge the tube for 5 min at $100 \times g$ @ 4 °C.

17. Pour off the supernatant (Fig. 2h) and repeat the wash step by resuspending the pellet, topping up with CIMC induction medium to 15 mL followed by centrifugation for 5 min at $100 \times g$, 4 °C.

18. Again, pour off the supernatant and resuspend the cells in roughly 6–9 mL 20% maltose solution. Carefully layer 1 mL of CIMC induction medium on top of the maltose solution to maintain a clear interface (Fig. 2i). Centrifuge the tube for 13 min at $100 \times g$, 4 °C.

19. Collect the cells that form a band at the interface (*see* **Note 20**; Fig. 2j) using a sterile glass Pasteur pipette and transfer the cells to a new sterile 15 mL conical centrifugation tube.

20. Dilute the cells to 15 mL volume with CIMC induction medium and add 3μL of the 0.5 mM stock of TSA to achieve a final TSA concentration of 1μM. Incubate the cells for 10 min at room temperature, then centrifuge for 6 min at $160 \times g$, 4 °C.

21. Remove the supernatant and in the same tube resuspend the cells in a total volume of 1400μL in CIMC induction medium.

22. Take 10μL of cell suspension and count the cell density using a hemocytometer. This estimation is then used to determine the number of dishes that will be made to contain 100,000 cells per dish.

3.7 Isolated Microspore Culture Method

1. Dilute the cells to a final volume with CIMC induction medium to a concentration of 100,000 cells for every 500μL.

2. Fill 60×15 mm sterile Petri dishes with 3.5 mL of CIMC induction medium (Fig. 3a). Add 25μL of PSK-α working stock into each Petri dish (final working concentration is 0.1μM).

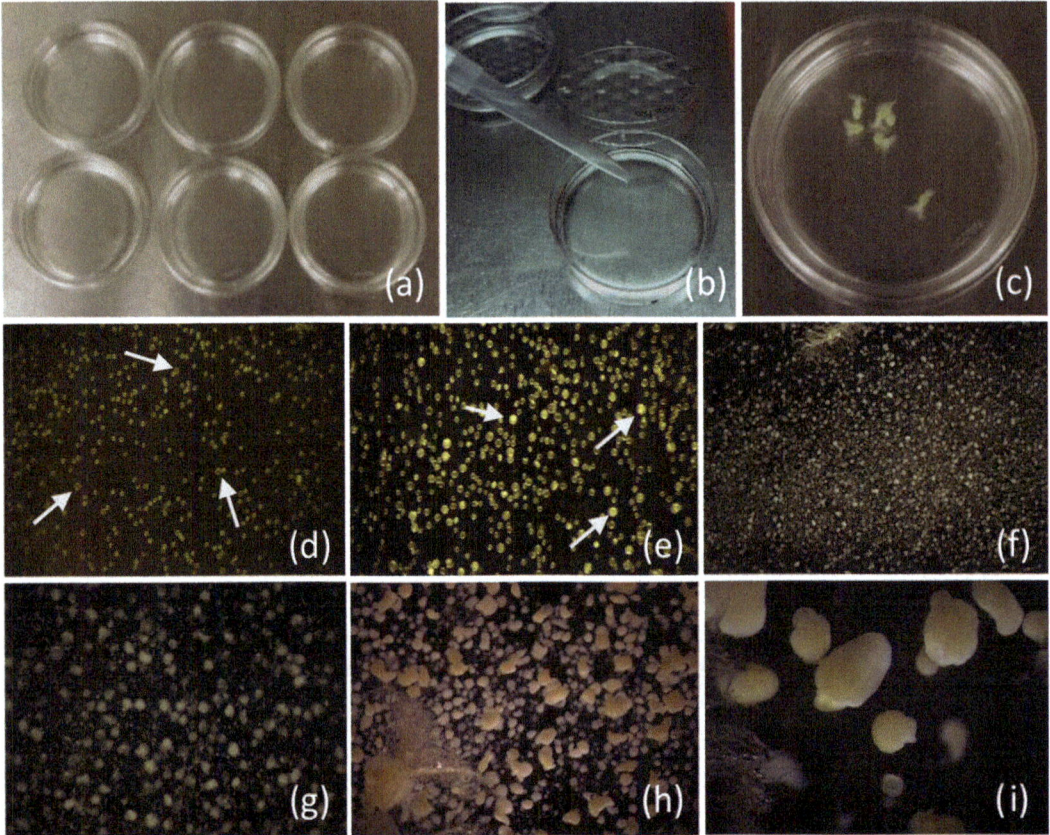

Fig. 3 Microspore culture steps. (**a**) 60 × 15 mm sterile Petri dishes are prepared with 3.5 mL of CIMC induction medium. (**b**) Microspore suspensions are pipetted into each Petri dish. (**c**) Six ovaries are added to each Petri dish. Microspore culture stages can then be observed at various time points after initiation in CIMC induction medium: (**d**) swollen floating microspores at 48 h, (**e**) multicellular structures rupturing out of exine around day 10–12, (**f**) microscopic ELS on day 15, (**g**) enlarged ELS day 18, (**h**) fully mature embryos day 25, and (**i**) microspore-derived embryos resemble zygotic embryos

3. Pipette 500μL of microspore suspension containing 100,000 cells into each dish (Fig. 3b). This will lead to a total volume of 4 mL per dish.

4. Retrieve the bottle stored at 4 °C with the sterile spikes for ovary donors. In the laminar air flow bench, take out one spike and position it on the paper towel in the same manner as spikes used for microspore isolation (*see* Subheading 3.6, **step 9**). Aseptically remove and discard the outer and inner glooms, as well as the anthers. Release the ovary by gently squeezing at the base. Transfer six ovaries to each dish containing isolated microspores (Fig. 3c).

5. Seal the dishes with parafilm and place in a closed box containing a beaker with distilled water (*see* **Note 21**) and incubate at 28 °C in the dark [1, 7].

6. Check the plates after 48 h for microbial contamination.

7. Observe the surface of the culture medium under a stereo-dissecting microscope. Round swollen microspores floating on the liquid surface after 48 h is an indication of the onset of androgenesis (Fig. 3d). Clear signs of division appear after 7 days, and between 1 and 2 weeks the microspores will sink to the bottom of the dish and exine will rupture to release multicellular structures (Fig. 3e).

8. Between days 15 and 18, several hundreds of developing embryo-like structures (ELS) can be observed (Fig. 3f). At this point in time, dishes are opened in the laminar flow hood and 3.5 mL of CIMC induction medium is added to replenish nutrients.

9. Over the next 2 weeks or so, standard embryo developmental stages will be noticeable and most of the embryoids/ELS (now macroscopic) will float to the surface (Fig. 3g–i).

3.8 Regeneration of Doubled Haploid Plantlets

1. When ELS are 1–2 mm in size (roughly 4 weeks from culture initiation; Fig. 4a), aseptically transfer them (germ side up) within a laminar flow hood (Fig. 4b) onto CIMC-4 regeneration medium plates (see **Note 22**; Fig. 4c). Not all embryos mature at the same time henceforth after the first round of transfers of ELS in CIMC-4 regeneration medium, the CIMC induction medium plates are incubated for another 2 weeks to allow development of additional embryos.

2. Incubate the plates with ELS approximately 30 cm beneath Sylvania Gro-lux wide spectrum bulbs (40 W) delivering $80\mu M \ m^{-2} \ s^{-1}$ of light (16/8 h photoperiod) at 16 °C. The ELS start to germinate between 2 and 3 weeks after transfer onto the CIMC-4 regeneration medium (Fig. 4d). There will be a mix of green and albino plants on the plates at this time.

3.9 Rooting and Soil Transfer of Plantlets

1. When the green plantlets have grown to approximately 2 cm and have least three roots of 1 cm or longer (Fig. 4e), transfer them aseptically onto rooting medium contained in Magenta boxes (see **Note 23**).

2. Maintain the plantlets in rooting medium under the same conditions as the CIMC-4 plates (Subheading 3.8, **step 2**).

3. Allow the plantlets to grow until the apex reaches the lid of the plastic container (Fig. 4f).

4. Take the plantlets out of the rooting medium (Fig. 4g) and wash the roots gently under running tap water to remove all media.

5. Transfer the plantlets to 4 × 8 cell root trainers (flats) containing moistened Cornell mix (Fig. 4h) and place the root trainer

Fig. 4 Regeneration of DH plantlets. (**a**) 1–2 mm ELS 4 weeks after culture initiation are ready for transfer to CIMC-4 shoot regeneration medium. (**b**) Fully developed embryo axes (germ) are visible on ELS. (**c**) ELS are placed on CIMC-4 shoot regeneration medium. (**d**) Upon germination, green and albino plants are observed. (**e**) Plantlets soon outgrow Petri dishes and must be transferred to rooting medium. (**f, g**) When rooted plantlets fill their containers, they are ready for soil transfer. (**h**) Finally, DH plantlets are transferred to soil in root trainers

(covered with a clear topped plastic container) in a growth cabinet initially under high humidity and 10–16 °C (*see* **Note 24**).

6. Transfer the plants from the root trainers to nursery pots when they are 8–12 in. tall (*see* **Note 25**).

4   Notes

1. Stock solutions and media are prepared using deionized water with a purity of 18.2 MΩ/cm. Tissue culture tested or analytical grade chemicals are purchased from Sigma-Aldrich and/or Phytotechnology laboratories unless otherwise stated. All macro and micro stock solutions are stored at 4 °C. Organic stocks are frozen in aliquots at −20 °C (unless indicated otherwise). Growth regulator stocks are also stored at 4 °C. Sterilization is carried out either by autoclaving at 121 °C for 15 min at 15 psi or by filtering using 0.22 μm filters, depending on the media type.

2. Prepare 50 mL solutions of 1 mg/mL $CoCl_2 \cdot 6H_2O$, 1 mg/mL $CuSO_4 \cdot 5H_2O$, and 1 mg/mL $Na_2MoO_4 \cdot 2H_2O$ and store at 4 °C. Conical centrifugation tubes, 15 and 50 mL, are convenient for storing stock solutions. Prepare 250–500 mL CIMC microsalt stock aliquots since they are used up slowly. Filter sterilization of microsalt stocks is optional.

3. The final volume of CIMC induction medium per culture dish is 4 mL (see Subheading 3.7, **step 3**) and contains 340 ng of PSK-α. By diluting 54.4 μL of the 0.5 mg/L stock solution of PSK-α with 1945.6 μL of CIMC induction medium, 25 μL of this working stock when delivered to each culture dish would correspond to 340 ng.

4. Preparing 2 L of CIMC extraction buffer in advance is desirable since it is enough for multiple microspore isolations. Using a graduated Pasteur pipette, aliquot half of the CIMC extraction buffer in sterile 50 mL conical centrifugation tubes and store at 4 °C. This avoids the entire media from being accidentally contaminated. Two 50 mL tubes with CIMC extraction buffer will be required for microspore isolation.

5. The Erlenmeyer flask containing Gelrite should be kept hot after autoclaving. Use low heat to keep Gelrite in liquid form, do not boil. Although not recommended, Gelrite solution can be stored after autoclaving in an incubator set at 70 °C.

6. Isolated microspore culture is dependent on various factors, genotype being a predominant one. This method can be applied straightaway to any triticale genotype with good success. For greater efficiency the methods may need to be optimized on a case-by-case basis.

7. Greenhouse grown plants produce a higher numbers of albinos as compared to plants grown in cabinets where the environment is more controlled [3].

8. The first eight to ten spikes gives the highest quality and number of microspores. Older spikes can be used, but the quality of the microspores and their ability to regenerate into

green plants is reduced. For optimal response, the portion of microspores at the early uni-nucleate stage should remain low. Microspores at the mid uni-nucleate stage will have a large central vacuole, with the nucleus against the cell periphery at roughly a 90° angle to the microspore pore (*see* Fig. 1b). A late uni-nucleate microspore will look similar to the mid uni-nucleate but with the nucleus against the cell periphery and at the opposite end to the microspore from the pore (*see* Fig. 1c).

9. Remove three anthers from a floret in the middle of the spike and place them on a slide. Add one drop of acetocarmine then overlay with a cover slip. Release microspores from the anthers by gently tapping the cover slip and examine using a microscope.

10. 21 days of cold stress pretreatment is ideal for the regeneration of embryos form triticale microspores [2]. However, spikes from 21 ± 3 days can be safely used without compromising the yield and quality of microspores.

11. It is ideal to collect ovary donor tillers fresh on the same day of microspore isolation. In adverse situations, tillers can be stored for use up to no more than 3 days. They can be stored in the same manner as those used for microspore isolation.

12. The quality of ovaries is crucial for a successful androgenesis. The use of under developed ovaries for co-culture will result in poor regeneration of embryoids. Pick the best spikes that have ovaries with fluffy stigmas.

13. Be careful not to trim the top of the florets as this will allow the anthers to come into contact with the bleach during subsequent sterilization and will adversely affect androgenesis.

14. Leave the florets at the top and bottom of the spike that have touched fingers while holding the spike.

15. The ovary donor spikes will be used at a later step. Leave them in the same bottle and store at 4 °C.

16. Once the anthers are blended, the clock is ticking. It is good to have the centrifuge chilled and ready to go with the appropriate settings.

17. The blender motor can be placed outside the laminar flow hood. In this situation make sure the blender lid is in place prior to removing the blender from the hood to crush the plant material.

18. It is important to strain the slurry immediately after blending. The released cell contents and shreds of plant tissue have deleterious effects on microspores.

19. To prevent cells from being lost, it is important to pour off the supernatant very slowly as the pellet of cells is loosely packed.

20. Dead cells will be in the pellet and live cells at the right stage will be in the band at the interface.

21. It is important to maintain a high humidity surrounding the culture dishes to prevent evaporation of the liquid media.

22. If left too long in the induction medium, ELS will turn to calli and will most likely not germinate into green plants. No more than 50 ELS should be transferred per plate to prevent competition for nutrients.

23. As many as 10–15 plantlets can be transferred into each Magenta box containing the rooting medium depending on the size of the plantlets.

24. It is important to maintain high humidity and lower temperatures during the first week after transfer to soil. This helps plants acclimatize to the harsher environment outside of tissue culture.

25. If colchicine treatment is required, it will be most effective and convenient to be carried out just before transferring the plantlets to nursery pots.

Acknowledgments

We thank Katherine Anderson-Bain and Jamieson Peacock for proofreading and Fengying Jiang for technical assistance.

References

1. Eudes F, Chugh A (2009) An overview of triticale doubled haploids. In: Touraev A, Forster BP, Jain SM (eds) Advances in haploid production in higher plants. Springer, Dordrecht

2. Immonen S, Robinson J (2000) Stress treatment and ficoll for improving green plant regeneration in triticale anther culture. Plant Sci 150:77–84. https://doi.org/10.1016/S0168-9452(99)00169-7

3. Li H, Devaux P (2001) Enhancement of microspore culture efficiency of recalcitrant barley genotypes. Plant Cell Rep 20:475–481. https://doi.org/10.1007/s002990100368

4. Zheng MY, Weng Y, Liu W, Konzak CF (2002) The effect of ovary conditioned medium on microspore embryogenesis in common wheat (*Triticum aestivum* L.). Plant Cell Rep 20:802–807. https://doi.org/10.1007/s00299-001-0411-2

5. Pauk J, Mihaly R, Monostori T, Puolimatka M (2003) Protocol of triticale (×Triticosecale Wittmack) microspore culture. In: Maluszynski M, Kasha KJ, Forster BP Szarejko

I (eds) Doubled haploid production in crop plants. Springer, Dordrecht

6. Oleszczuk S, Sowa S, Zimny J (2004) Direct embryogenesis and green plant regeneration from isolated microspores of hexaploid triticale (×Triticosecale Wittmack) cv. Bogo. Plant Cell Rep 22:885–893. https://doi.org/10.1007/s00299-004-0796-9

7. Eudes F, Amundsen E (2005) Isolated microspore culture of Canadian 6×triticale cultivars. Plant Cell Tissue Organ Cult 82:233–241. https://doi.org/10.1007/s11240-005-0867-9

8. Letarte J, Simion E, Miner M, Kasha KJ (2006) Arabinogalactans and arabinogalactan-proteins induce embryogenesis in wheat (*Triticum aestivum* L.) microspore culture. Plant Cell Rep 24:691–698. https://doi.org/10.1007/s00299-005-0013-5

9. Soriano M, Cistué L, Castillo AM (2008) Enhanced induction of microspore embryogenesis after n-butanol treatment in wheat (*Triticum aestivum* L.) anther culture. Plant

Cell Rep 27:805–811. https://doi.org/10.1007/s00299-007-0500-y

10. Żur I, Dubas E, Golemiec E, Szechyńska-Hebda M, Janowiak F, Wędzony M (2008) Stress-induced changes important for effective androgenic induction in isolated microspore culture of triticale (×Triticosecale Wittm.). Plant Cell Tissue Organ Cult 94(3):319–328. https://doi.org/10.1007/s11240-008-9360-6

11. Żur I, Dubas E, Golemiec E, Szechyńska-Hebda M, Gołębiowska G, Wędzony M (2009) Stress-related variation in antioxidative enzymes activity and cell metabolism efficiency associated with embryogenesis induction in isolated microspore culture of triticale (×Triticosecale Wittm.). Plant Cell Rep 28:1279–1287. https://doi.org/10.1007/s00299-009-0730-2

12. Würschum T, Tucker MR, Reif JC, Maurer HP (2012) Improved efficiency of doubled haploid generation in hexaploid triticale by in vitro chromosome doubling. BMC Plant Biol 12:109. https://doi.org/10.1186/1471-2229-12-109

13. Asif M, Eudes F, Randhawa H, Amundsen E, Yanke J, Spaner D (2013) Cefotaxime prevents microbial contamination and improves microspore embryogenesis in wheat and triticale. Plant Cell Rep 32(10):1637–1646. https://doi.org/10.1007/s00299-013-1476-4

14. Asif M, Eudes F, Goyal A, Amundsen E, Randhawa H, Spaner D (2013) Organelle antioxidants improve microspore embryogenesis in wheat and triticale. In Vitro Cell Dev Biol Plant 49:489–497. https://doi.org/10.1007/s11627-013-9514-z

15. Asif M, Eudes F, Randhawa H, Amundsen E, Spaner D (2014) Phytosulfokine alpha enhances microspore embryogenesis in both triticale and wheat. Plant Cell Tissue Organ Cult 116(1):125–130. https://doi.org/10.1007/s11240-013-0379-y

16. Asif M, Eudes F, Randhawa H, Amundsen E, Spaner D (2014) Induction medium osmolality improves microspore embryogenesis in wheat and triticale. In Vitro Cell Dev Biol Plant 50(1):121–126. https://doi.org/10.1007/s11627-013-9545-5

17. Lantos C, Bóna L, Boda K, Pauk J (2014) Comparative analysis of in vitro anther- and isolated microspore culture in hexaploid Triticale (×Triticosecale Wittmack) for androgenic parameters. Euphytica 197:27–37. https://doi.org/10.1007/s10681-013-1031-y

18. Sinha RK, Eudes F (2015) Dimethyl tyrosine conjugated peptide prevents oxidative damage and death of triticale and wheat microspores. Plant Cell Tissue Organ Cult 122(1):227–237.

https://doi.org/10.1007/s11240-015-0763-x

19. Żur I, Dubas E, Krzewska M, Zieliński K, Fodor J, Janowiak F (2019) Glutathione provides antioxidative defense and promotes microspore derived embryo development in isolated microspore cultures of triticale (× Triticosecale Wittm.). Plant Cell Rep 38:195–209. https://doi.org/10.1007/s00299-018-2362-x

20. Touraev A, Vicente O, Heberle-Bors E (1997) Initiation of microspore embryogenesis by stress. Trends Plant Sci 2:297–302. https://doi.org/10.1016/S1360-1385(97)89951-7

21. Wędzony M (2003) Protocol for anther culture in hexaploid triticale (×Triticosecale Wittm.). In: Maluszynski M, Kasha KJ, Forster BP, Szarejko I (eds) Doubled haploid production in crop plants. Springer, Dordrecht

22. Żur I, Dubas E, Krzewska M, Janowiak F, Hura K, Pociecha E, Bączek-Kwinta R, Płażek A (2014) Antioxidant activity and ROS tolerance in triticale (×Triticosecale Wittm.) anthers affect the efficiency of microspore embryogenesis. Plant Cell Tissue Organ Cult 119(1):79–94. https://doi.org/10.1007/s11240-014-0515-3

23. Żur I, Dubas E, Krzewska M, Sánchez-Díaz RA, Castillo AM, Valles MP (2014) Changes in gene expression patterns associated with microspore embryogenesis in hexaploid triticale (×Triticosecale Wittm.). Plant Cell Tissue Organ Cult 116:261–267. https://doi.org/10.1007/s11240-013-0399-7

24. Li H, Soriano M, Cordewener J, Muiño JM, Riksen T, Fukuoka H, Angenent GC, Boutilier K (2014) The histone deacetylase inhibitor trichostatin a promotes totipotency in the male gametophyte. Plant Cell 26(1):195–209. https://doi.org/10.1105/tpc.113.116491

25. Jiang F, Ryabova D, Diedhiou J, Hucl P, Randhawa H, Marillia EF, Foroud NA, Eudes F, Kathiria P (2017) Trichostatin A increases embryo and green plant regeneration in wheat. Plant Cell Rep 36(11):1701–1706. https://doi.org/10.1007/s00299-017-2183-3

26. Wędzony M, Forster BP, Żur I, Golemiec E, Szechyńska-Hebda M, Dubas E, Gołebiowska G (2009) Progress in doubled haploid technology in higher plants. In: Touraev A, Forster BP, Jain SM (eds) Advances in haploid production in higher plants. Springer, Dordrecht

27. Boodly JW, Sheldrake R Jr (1982) Cornell peat-lite mixes for commercial plant growing. An extension publication of the New York State College of Agriculture and Life Sciences, a statutory college of the state university, at Cornell University, Ithaca, New York

Chapter 17

Oat (*Avena sativa* L.) Anther Culture

Marzena Warchoł, Kinga Dziurka, Ilona Czyczyło-Mysza, Katarzyna Juzoń, Izabela Marcińska, and Edyta Skrzypek

Abstract

Production of doubled haploids (DHs) by androgenesis is a promising and convenient alternative to traditionally used breeding techniques. Low response of anther culture and strong genotype dependency in the development of embryo-like structures (ELS) was reported for oat (*Avena sativa* L.). Total homozygosity has been reached in one generation. This chapter describes a step-by-step protocol that can be useful for androgenesis studies and oat DH line production through anther culture.

Key words Androgenesis, *Avena sativa* L., Doubled haploids, Haploids, Pretreatment

1 Introduction

Androgenesis offers a rapid method of producing fully homozygous plants for many crops. This technique enables immediate evaluation of quantitative characteristics and greater selection and discrimination between genotypes within each generation. However, some economically important species are still recalcitrant to this process, e.g., oat (*Avena sativa* L.). Research on androgenesis in oat has begun in 1980s without spectacular success. The first haploid plant of oat was obtained in anther culture by Rines [1]. Later, Kiviharju and Pehu [2] reported many anther-derived embryos of *A. sativa* without successful plant regeneration. Kiviharju et al. [3] reported obtaining 30 green plants per 100 anthers using a modified procedure. Oat, as a species recalcitrant to androgenesis induction, is also strongly dependent on the genotype [4, 5]. Androgenic response in in vitro culture is determined by many factors along with their interactions among them. The most important include: genotype, condition of donor plants, microspore development stage, type of pretreatment applied, and composition of induction medium [6]. The application of suitable pretreatment factors such as low or high temperature, sucrose

Jose M. Seguí-Simarro (ed.), *Doubled Haploid Technology: Volume 1: General Topics, Alliaceae, Cereals*, Methods in Molecular Biology, vol. 2287, https://doi.org/10.1007/978-1-0716-1315-3_17,

starvation, or osmotic stress promotes responses that switch the gametophytic development of microspores toward a sporophytic one. Effective selection of triggering factors for androgenesis competence should be adapted to plant species and genotype. Success of ELS formation depends strongly on microspore development stage. The uninucleate microspore stage or early binucleate stage are the most suitable for androgenic response in cereals [7]. Pretests that establish a correlation between easily observable morphological traits are useful to identify the favorable stage of the microspore. It is possible to observe whether the tip of the panicle is inside the leaf sheath [2] or to measure the distance from the base of the flag leaf to the penultimate leaf [5]. This chapter describes in detail the effects of cultivar, pretreatment, induction medium, and morphological pretests on the efficiency of oat anther culture.

2 Materials

2.1 Equipment

1. Glasshouse under natural (solar) light and artificial lightning with sodium lamps (400 W, Philips SON-T AGRO from Philips or equivalent).

2. Hood with laminar flow of sterile air.

3. Cooling counter with a temperature of +4 °C.

4. Flow cytometer.

5. Autoclave.

6. Fume hood.

7. Air pump.

8. Magnetic stirrer.

9. pH meter.

2.2 General Materials

1. Soil and sand (3:1 v/v) mixture.

2. 3 L plant pots.

3. Forceps.

4. Spatula.

5. Filter paper.

6. Parafilm.

7. 1.5 mL conical (Eppendorf type) plastic tubes.

8. 5 mL tubes.

9. Petri dishes.

10. Magenta plant culture boxes.

11. Syringes and 25 mm filters (0.2 μm pore size) for syringes.

2.3 Solutions and Media

1. 70% ethanol solution for tiller disinfection: For 1 L, pour 700 mL 96% ethanol into 1 L cylinder. Fill with 300 mL of sterile deionized water (*see* **Note 1**) and stir with spatula.

2. 2.5% $CaCl_2$ solution for tiller disinfection: Weigh 25.0 g $CaCl_2$ and pour it to 1 L Erlenmeyer flask. Add 975 mL of deionized water and stir it using a magnetic stirrer. Filter the solution through a filter paper.

3. Acetocarmine solution: 4% acetocarmine in water.

4. Stock solutions of growth regulators at a concentration of 1 mg/mL: First, dissolve 2,4-dichlorophenoxyacetic acid (2,4-D) in a small volume of ca. 96% ethanol, then add appropriate amount of distilled water. Dissolve kinetin in 1 N NaOH before adding water (*see* **Note 2**). Aliquot dissolved growth regulators (2 mL aliquots of 2,4-D and 0.5 mL aliquots of kinetin); pour into Eppendorf tubes and store in a freezer at −20 °C.

5. Hoagland medium [8]: Prepare by mixing in water the components listed in Table 1 (*see* **Note 3**).

6. Induction medium: Oat anther culture W14 medium [9] (Table 1). To prepare 1 L of solid W14 medium, dissolve all components (except Fe-EDTA, growth regulators and vitamins) in ca. 400 mL distilled water. Prepare Fe-EDTA by dissolving 2.78 g $FeSO_4 \cdot 7H_2O$ and 3.78 g $Na_2EDTA \cdot 2H_2O$ separately in 250 mL distilled water; heating may be required. Then combine these solutions. Note that $FeSO_4 \cdot 7H_2O$ dissolves completely only after combining both solutions. Then combine Fe-EDTA with the solution of other components. After thorough mixing, adjust the volume to 980 mL with distilled water, and pH to 6.0 before adding 6 g agar. Autoclave the medium for 20 min at 121 °C and a pressure of 0.1 MPa. Then, dissolve vitamins in 17.5 mL distilled water, add 2 mL aliquots of 2,4-D and 0.5 mL aliquots of kinetin, and filter them using syringe filters into the cooled medium. Pour aseptically induction medium into 6.0 cm Petri dishes under a laminar flow hood. After cooling the medium, wrap Petri dishes with a cling film or Parafilm M® and store in the fridge.

7. Rooting medium: haploid plants rooting MS medium [10] (Table 1). Prepare 1 L of solid rooting MS medium as described above for induction medium or weigh 4.4 g of MS medium powder, 30 g of sucrose, adjust pH to 5.8 and solidify with 6% agar. Next, pour 40 mL of medium into Magenta vessels and sterilize by autoclaving for 20 min at 121 °C and a pressure of 0.1 MPa (*see* **Note 4**).

8. Acclimatization jars: 0.5 L glass jars filled to one-third volume with perlite moistened with half the original concentration of MS medium [10] and sterilized by autoclaving for 20 min at 121 °C and a pressure of 0.1 MPa.

Table 1
Composition of the different media used (in mg/L)

Compound	Medium		
	Hoagland	W14	MS
Macro salts			
$MgSO_4 \cdot 7H_2O$	520	140	180.54
KNO_3	660	2000	1900
KH_2PO_4	–	–	170
$Ca(NO_3)_2 \cdot 4H_2O$	940	–	–
NH_4NO_3	–	–	1650
$CaCl_2 \cdot 2H_2O$	–	–	332
$(NH_4)H_2PO_4$	120	380	–
K_2SO_4	–	700	–
Micro salts			
$MnSO_4 \cdot 4H_2O$	3.4	8.0	16.9
$ZnSO_4 \cdot 7H_2O$	0.2	3.0	8.6
H_3BO_3	2.8	3.0	6.2
KJ	–	0.5	0.83
$CuSO_4 \cdot 5H_2O$	0.1	0.025	0.025
$CoCl_2 \cdot 6H_2O$	–	0.025	0.025
$Na_2MoO_4 \cdot 2H_2O$	0.025	0.005	0.025
Fe-EDTA			
$Na_2EDTA \cdot 2H_2O$	37.8	37.8	37.8
$FeSO_4 \cdot 7H_2O$	27.8	27.8	27.8
Vitamins			
Myo-inositol	–	–	100
Nicotinic acid	–	0.5	0.5
Pyridoxine–HCl	–	0.5	0.5
Thiamine–HCl	–	2.0	0.1
Amino acids			
Glycine	–	2.0	2.0
Sugars			
Maltose	–	90,000	–
Sucrose	–	–	30,000

(continued)

Table 1
(continued)

Compound	Medium		
	Hoagland	W14	MS
Growth regulators			
2,4-D	–	2.0	–
Kinetin	–	0.5	–
Other			
Agar	–	6000	6000
pH	6.7	6.0	5.8

9. Plant growth mixture: soil and sand mixture (3:1 v/v) in 3 L pots.

10. Colchicine solution: 0.1% (w/v) colchicine, 4% dimethyl sulfoxide (DMSO), 0.025 g/L gibberellic acid (GA$_3$), and 20 µL/L Tween 20. To prepare 500 mL colchicine solution, using nitrile gloves and within fume hood, dissolve 0.0125 g GA$_3$ in a few drops of 96% ethanol, add 0.5 g colchicine, 480 mL distilled water, 20 mL DMSO, a drop of Tween, and mix them in a 1 L beaker.

11. Nuclei extraction buffer for flow cytometry [11]: For 1 L, prepare 0.8 L of distilled water in a beaker and add 8.00 g NaCl, 0.20 g KCl, 1.44 g Na$_2$HPO$_4$, 0.24 g of KH$_2$PO$_4$, 2.00 g EDTA, and 0.5% bovine serum albumin (BSA). Adjust pH to 7.4 with HCl and bring to 1 L with distilled water.

12. 2% Propidium iodide (PI) solution: Prepare 5 mL of distilled water in a tube and add 0.1 g PI. Store solution at 4 °C and protect from light.

3 Methods

3.1 Donor Plant Growth

1. Rinse oat seeds with a commercial fungicide and sow in pots with a plant growth mixture.

2. Keep plants in the glasshouse at 21/17 °C day/night and 16-h photoperiod.

3. Irrigate plants with tap water as required, and once a week fertilize with Hoagland liquid medium [8].

3.2 Collecting Tillers and Cold Pretreatment

1. Collect oat tillers when the panicle is enclosed within the leaf sheath (*see* **Note 5**). To correlate the stage of microspore development with tiller morphology, measure the distance from the base of the flag leaf to the penultimate panicle leaf (*see* **Notes 6** and **7**).

Fig. 1 Androgenesis in oat (*Avena sativa* L.) anther culture: (**a**) tillers of 'Chwat' covered with aluminum bags, in 1 L jars containing half the original concentrations of Hoagland liquid medium at +4 °C; (**b**) tiller of 'Chwat' in beaker prepared for disinfection; (**c**) isolated anthers and ovaries on W14 medium; (**d**) anthers with developed ELS indicated by arrow; (**e**) scanning electron microscopy of dehisced anther with ELS indicated by arrow; (**f**) ELS on W14 medium; (**g**) plants on MS medium; (**h**) plants in perlite; and (**i**) plants maturing into greenhouse conditions

2. Cut tillers and cover with aluminum bags, put in 1 L jars containing half the original concentrations of Hoagland liquid medium [8], and place in a cooling counter for 3 weeks at 4 °C in the dark (Fig. 1a; *see* **Note 8**).

3.3 Determination of Microspore Developmental Stage

1. To check the microspore developmental stage, place several anthers on a glass slide in a drop of acetocarmine solution, put on a cover slip, and gently crush them.

2. After 5 min, observe them under a light microscope. The highest androgenic response is achieved when microspores are in the middle to late uninucleate stage of development (*see* **Note 9**).

3.4 Tiller Disinfection

1. After cold pretreatment, take the tillers to a sterile laminar flow work bench.

2. Remove panicles from the leaf sheaths and place them into autoclaved 300 mL Erlenmeyer flasks (Fig. 1b). Disinfect the panicles in 70% ethanol (1 min), then in a 2.5% (w/v) solution of calcium hypochlorite (7 min), and subsequently wash three times with sterile water (*see* **Note 10**).

3.5 Anther Isolation

1. Place the panicle on a sterile piece of glass or paper (15 × 15 cm), open flowers with forceps, and pull out the anthers and five to seven ovaries.

2. Transfer carefully isolated anthers and ovaries to the induction W14 medium [9] (Fig. 1c). Treat the anthers gently, do not damage them or press too hard.

3. Place the anthers and ovaries from 1 panicle on a 6 cm diameter Petri dish. Wrap Petri dishes with Parafilm M® and place in a growth room in the dark at 28 ± 1 °C.

3.6 Regeneration of Embryogenic Structures and Rooting of Haploid Plants

1. Embryo-like structures (ELS) are observed after about 6–8 weeks under a microscope (Fig. 1d, e). All ELS should be transferred to fresh W14 medium [9] every 3 weeks (Fig. 1f). First haploid plant regeneration is observed after 2 weeks (Fig. 1g).

2. Developed haploid plants are transferred to Magenta vessels with solid rooting medium without growth regulators and maintained at 21 ± 2 °C and light intensity of $60 \mu mol\ m^{-2}\ s^{-1}$ (16/8 h light/dark).

3.7 Acclimatization of Haploid Plants

1. Haploid plants are acclimated to ex vitro conditions by transferring them to acclimatization jars (Fig. 1h).

2. Keep jars with haploid plants at 21 ± 2 °C and light intensity of $60 \mu mol\ m^{-2}\ s^{-1}$ (16/8 h light/dark).

3. Two weeks later, transfer the plants to pots with plant growth mixture and place them in the greenhouse under the conditions described at Subheading 3.1, **step 2**.

3.8 Determination of Ploidy Level

1. Collect three fresh leaf tissue samples of approximately 1 cm³ from each plant, place into a 60-mm glass Petri dish with 1 mL of nuclei extraction buffer and chop with a razor blade to release the nuclei from cells. Keep Petri dishes and nuclei extraction buffer on ice.

2. Filter the chopped tissues through a 30 μm nylon mesh filter into 5 mL tubes and add 20 μL of the 2% PI solution to stain the nuclei.

3. After 15 min of incubation on ice, mix the samples gently and measure the ploidy using a flow cytometer equipped with a blue laser (488 nm wavelength).

4. DNA content of investigated plants is determined by comparing the peak position of the nuclear 2C with a peak of the 2C diploid oat plant used as controls (*see* **Note 11**).

3.9 Chromosome Doubling and Growth of DH Plants

1. Four weeks after haploid plant acclimatization to natural conditions, remove them from the pots and wash the soil from roots with tap water.

2. Trim the roots by about 2 cm and put them into a 1 L beaker containing 0.1% colchicine solution (*see* **Note 12**).

3. Ensure that all roots and shoot meristems are immersed in the colchicine solution. Treat the plants for 7.5 h under a fume hood (*see* **Note 13**), aerating the colchicine solution with an air pump at 25 °C and 80–100 μmol m^{-2} s^{-1} light intensity.

4. Rinse the roots using tap water aerated by an air pump for 24 h.

5. Finally, transplant all plants in 3 L pots with plant growth mixture.

6. Keep double haploid (DH) plants in a glasshouse to maturity in conditions as described at Subheading 3.1, **step 2** and cover the panicles separately with cellophane, paper, or fleece bags to ensure self-fertilization (Fig. 1i).

7. Water the plants as needed and fertilize them once a week with Hoagland liquid medium [8].

4 Notes

1. The deionized water needed is prepared in advance in glass bottles autoclaved at 121 °C, 0.1 MPa for 20 min.

2. Weigh 4 g of NaOH and dissolve in 100 mL of deionized water to obtain 1 N NaOH solution, which will be used to dissolve kinetin.

3. It is best to prepare 50 L of medium solution by dissolving all salts in distilled water in an opaque container. Start fertilization when the plantlets have four leaves.

4. Magenta vessels are manufactured from polycarbonate, so that they are both autoclavable and reusable. Transparent surface allows the samples to be viewed without disturbing growth; dimensions: 77 mm × 77 mm × 97 mm.

5. This usually occurs within 5–8 weeks of plant growth in a greenhouse and depends on the genotype.

6. Young tillers, where the distance from the base of the flag leaf to the penultimate panicle leaf is from 0.0 to 4.0 cm, are favorable for androgenesis induction.

7. In our experiment, the distance from the base of the flag leaf to the penultimate leaf of the panicle was measured and four distance classes were designated: 0.0–4.0 cm, 4.1–8.0 cm, 8.1–12.0 cm, and 12.1–16.0 cm. A simple pretest establishes a correlation between easily observable morphological traits and the favorable developmental stage of microspores.

8. For 10 tillers, pour 0.5 L Hoagland liquid medium [8] into 1 L glass jar. Change the medium every 3 days.

9. The stage of microspore development is one of the critical factors for successful induction of androgenesis. In oat, the development of microspores within florets and of florets within the spike is generally asynchronous, because there are three sets of flowers in one oat spikelet, and they are of different sizes. Due to the above observations, in our laboratory we isolate green anthers from the mid part of the panicle, and we reject the anthers from the third sets of flowers.

10. Disinfection carried out in this way guarantees sterility of the anthers at the level of 96–98%. It is better to sterilize a maximum of eight panicles at a time, so as not to extend the sterilization time, which affects the viability of anthers.

11. Plant DNA content should be estimated before and after colchicine treatment. Oat plants derived from seeds with known diploid DNA content are used as controls.

12. Wear a mask when weighing colchicine. To avoid exposing coworkers to colchicine, weight it and prepare the solution under a fume hood. Wear protective gloves during the entire procedure. In our laboratory, we prepare fresh colchicine solution before each use.

13. By increasing colchicine concentration, we can shorten the duration of treatment.

Acknowledgments

The project was supported by the Ministry of Agriculture and Rural Development, grant No. HORhn–4040dec13/08 and the *Franciszek Górski* Institute of Plant Physiology, Polish Academy of Sciences.

References

1. Rines HW (1983) Oat anther culture: genotype effects on callus initiation and the production of haploid plant. Crop Sci 23:268–227

2. Kiviharju E, Pehu E (1998) The effect of cold and heat pretreatments on anther culture response of *Avena sativa* and A. sterilis. Plant Cell Tiss Org Cult 54:97–104

3. Kiviharju E, Moisander S, Laurila J (2005) Improved green plant regeneration rates from oat anther culture and the agronomic performance of some DH lines. Plant Cell Tissue Organ Cult 81:1–9

4. Kiviharju E, Moisander S, Tanhuanpää P (2017) Oat anther culture and use of DH-lines for genetic mapping. In: Gasparis S (ed) Oat. Methods in molecular biology, vol 1536. Humana Press, New York, pp 71–93

5. Warchoł M, Czyczyło-Mysza I, Marcińska I, Dziurka K, Noga A, Kapłoniak K, Pilipowicz M, Skrzypek E (2019) Factors inducing regeneration response in oat (*Avena sativa* L.) anther culture. In Vitro Cell Dev Biol Plant 55(5):595–604

6. Ferrie AMR, Irmen KI, Beattie AD, Rossnagel BG (2014) Isolated microspore culture of oat (*Avena sativa* L.) for the production of doubled haploids: effect of pre-culture and post-culture conditions. Plant Cell Tissue Organ Cult 116:89–96

7. Sood S, Dwivedi S (2015) Doubled haploid platform: an accelerated breeding approach for crop improvement. In: Bahadur B, Venkat Rajam M, Sahijram L, Krishnamurthy K (eds) Plant biology and biotechnology. Springer, New Delhi, pp 89–111

8. Hoagland DR, Arnon DI (1938) A water culture method for growing plants without soil. Circ Univ Calif, Agric Exp Stn No 347

9. Ouyang TW, Jia SE, Zhang C, Chen X, Feng G (1989) A new synthetic medium (W14) for wheat anther culture. Annual report 1987–1988. Institute of Genetics Academia Sinica, Beijing, pp 91–92

10. Murashige T, Skoog F (1962) A revised medium for rapid growth and bioassays with tobacco tissue cultures. Physiol Plant 15:473–497

11. Sambrook J, Fritschi EF, Maniatis T (1989) Molecular cloning: a laboratory manual. Cold Spring Harbor Laboratory Press, New York

Chapter 18

Oat Doubled Haploid Production Through Wide Hybridization with Maize

Edyta Skrzypek, Marzena Warchoł, Ilona Czyczyło-Mysza, Katarzyna Juzoń, Kinga Dziurka, and Izabela Marcińska

Abstract

Wide hybridization is one of the haploid-inducing techniques that can accelerate the breeding process. Obtaining new cultivars is crucial to solve the problem of the constantly growing world population and global increase in demand for food, feed and renewable energy under changing environmental conditions. Here, we present a detailed protocol for obtaining oat (*Avena sativa* L.) doubled haploids (DHs) by pollination with maize (*Zea mays* L.). After fertilization, not only oat homozygotes, but also oat × maize hybrid zygotes can be formed, and during early embryo development, maize chromosomes are preferentially eliminated, which ultimately results in haploid plant formation. This chapter describes a method to produce oat DHs by crossing oat with maize, covering all steps from crossings to haploid plant regeneration and chromosome doubling.

Key words *Avena sativa* L., Doubled haploids, Wide crossing, *Zea mays* L.

1 Introduction

Plant breeding readily applies doubled haploids due to the possibility of obtaining complete homozygosity after one generation, which significantly shortens the breeding programmes [1]. Unfortunately, oat still remains recalcitrant to haploidization [2, 3]. Previous research has shown that obtaining oat haploid plants seems to be more effective using the wide crossing method (chromosome elimination) than androgenesis. Moreover, there are no albino plants among the regenerants [4]. Numerous factors were considered to be involved in the efficiency of oat haploid production: genotype of donor plants, time between emasculation and pollination, influence of auxin on ovaries enlargement and embryo production, time between pollination and application of growth regulators and type of regeneration medium for embryo germination and plant growth [2, 5–12]. Sometimes, after fertilization, an

Jose M. Seguí-Simarro (ed.), *Doubled Haploid Technology: Volume 1: General Topics, Alliaceae, Cereals,*
Methods in Molecular Biology, vol. 2287, https://doi.org/10.1007/978-1-0716-1315-3_18,
© Springer Science+Business Media, LLC, part of Springer Nature 2021

oat × maize hybrid zygote is formed. The chromosomes of maize are preferentially eliminated during early embryo development. Nevertheless, it has been found that some of the oat plants can retain a single maize chromosome [13–17]. This chapter describes a method for oat DH production by oat × maize crossing, applying 2,4-dichlorophenoxyacetic acid for ovary enlargement, embryo conversion, plant acclimation and chromosome doubling by colchicine treatment.

2 Materials

2.1 Equipment

1. Glasshouse with artificial lighting (400 W sodium lamps Philips SON-T AGRO, or equivalent).
2. Hood with laminar flow of sterile air.
3. Flow cytometer.
4. Autoclave.
5. Fume hood.
6. Air pump.
7. Magnetic stirrer.
8. pH meter.
9. Binocular microscope.

2.2 General Materials

1. Precise surgical tweezers.
2. Fine brush.
3. Syringe with needle.
4. 25 mm filters (0.2 μm pore size) for syringes.
5. Dissecting needles.
6. Commercial fungicide.
7. Soil and sand (3:1 v/v) mixture.
8. Perlite.
9. 3 L and 6 L plant pots.
10. 5, 10 and 50 mL Falcon (or equivalent) plastic tubes.
11. Spatula.
12. Erlenmeyer flasks.
13. Glass bottles.
14. 500 mL glass jars.
15. Filter paper.
16. Sterile Petri dishes.
17. 0.3 L Magenta plant culture vessels.
18. Parafilm.
19. 30-μm nylon mesh filters.

2.3 Solutions and Media

1. Solution of 100 mg/L 2,4-dichlorophenoxyacetic acid (2,4-D): For 500 mL 2,4-D solution, weigh 500 mg of 2,4-D and pour it into a 100-mL beaker. Add 1–2 mL of 96% ethanol (*see* **Note 1**) and stir to dissolve. Gradually add water and stir it with a spatula to obtain 100 mL of a clear solution. Transfer dissolved 2,4-D to a 500-mL cylinder and add deionized water to a final volume of 500 mL. Pour the solution to 50-mL Falcon tubes, and store in a refrigerator at 4 °C.

2. 70% ethanol solution for ovary disinfection: For 0.5 L, pour 350 mL 96% ethanol into 500-mL cylinder. Fill with 150 mL of deionized water and stir with spatula.

3. 2.5% $CaCl_2$ solution for ovary disinfection: Weigh 10.0 g $CaCl_2$ and pour it to 500-mL Erlenmeyer flask. Add 390 mL water and stir it using a magnetic stirrer. Filter the solution through a filter paper.

4. Sterile water: autoclave deionized water in glass bottles at 121 °C, 0.1 MPa for 20 min.

5. Stock solutions of growth regulators at a concentration of 1 mg/mL: First, dissolve separately 1-naphthaleneacetic acid (NAA) and kinetin in a drop of 1 N NaOH, then add appropriate amount of distilled water, 0.5 mL aliquots of each growth regulator and store into Eppendorf tubes in a freezer at −20 °C.

6. Hoagland medium [18]: Prepare by mixing in water the components listed in Table 1 in an opaque container (*see* **Note 2**).

7. 190-2 regeneration medium [19] (Table 1): To prepare 1 L of solid 190-2 medium, dissolve all components (except Fe-EDTA, growth regulators and vitamins) in ca. 400 mL distilled water. Prepare Fe-EDTA by dissolving 27.8 mg $FeSO_4 \cdot 7H_2O$ and 37.8 mg $Na_2EDTA \cdot 2H_2O$ separately in 250 mL distilled water (heating may be required). Then combine these solutions. Note that $FeSO_4 \cdot 7H_2O$ dissolves completely only after combining both solutions. Then, combine Fe-EDTA with the solution of other components. After thorough mixing, adjust the volume to 980 mL with distilled water, and pH to 6.0 before adding 6 g agar. Autoclave the medium for 20 min at 121 °C and a pressure of 0.1 MPa. Then, dissolve vitamins in 19 mL distilled water, add 0.5 mL aliquots of NAA and 0.5 mL aliquots of kinetin and filter them using syringe filters into the cooled medium. Pour aseptically regeneration medium into 6.0 cm Petri dishes under a laminar flow hood. After cooling the medium, wrap Petri dishes with a cling film or Parafilm M$^®$ and store in the fridge.

8. Murashige and Skoog medium [20] (MS medium, Table 1): Prepare 1 L of MS medium as described above for 190-2 medium or weigh 4.4 g of MS medium powder (Sigma-

Table 1
Composition of the different media used (in mg/L)

| | Medium | | |
Compound	Hoagland	190-2	MS
Macro salts			
$MgSO_4 \cdot 7H_2O$	520	200	180.54
KNO_3	660	1000	1900
KH_2PO_4	–	300	170
$Ca(NO_3)_2 \cdot 4H_2O$	940	100	–
KCl	–	40	–
NH_4NO_3	–	–	1650
$CaCl_2 \cdot 2H_2O$	–	–	332
$NH_4H_2PO_4$	120	–	–
$(NH_4)_2SO_4$	–	200	–
Micro salts			
$MnSO_4 \cdot 4H_2O$	3.4	8.0	16.9
$ZnSO_4 \cdot 7H_2O$	0.2	3.0	8.6
H_3BO_3	2.8	3.0	6.2
KJ	–	0.5	0.83
$CuSO_4 \cdot 5H_2O$	0.1	–	0.025
$CoCl_2 \cdot 6H_2O$	–	–	0.025
$Na_2MoO_4 \cdot 2H_2O$	0.025	–	0.025
Fe-EDTA			
$Na_2EDTA \cdot 2H_2O$	37.8	37.8	37.8
$FeSO_4 \cdot 7H_2O$	27.8	27.8	27.8
Vitamins			
Myo-inositol	–	100	100
Nicotinic acid	–	0.5	0.5
Pyridoxine-HCl	–	0.5	0.5
Thiamine-HCl	–	1.0	0.1
Amino acids			
Glycine	–	2.0	2.0
Sugars			
Maltose	–	90,000	–

(continued)

Table 1
(continued)

Compound	Medium		
	Hoagland	190-2	MS
Sucrose	–	–	30,000
Growth regulators			
NAA	–	0.5	–
Kinetin	–	0.5	–
Other			
Agar	–	6000	6000
pH	6.7	6.0	5.8

Aldrich, product no 5519), 30 g of sucrose, adjust pH to 5.8 and solidify with 6% agar. Pour 40 mL of medium into Magenta vessels and autoclave at 121 °C, 0.1 MPa for 20 min.

9. 0.1% Colchicine solution: For 1 L of colchicine solution, weigh 1 g of colchicine under a fume hood, wearing a mask and gloves, dilute in 1 L of deionized water and add 40 g dimethyl sulphoxide (DMSO), 0.025 g gibberellic acid (GA_3) (dissolved in a few drops of 96% ethanol) and a drop of Tween 20.

10. Phosphate-buffered saline (PBS) buffer [21]: For 1 L of buffer, weight 8.00 g NaCl, 0.20 g KCl, 1.44 g Na_2HPO_4, 0.24 g of KH_2PO_4, 2.00 g EDTA, 5 g bovine serum albumin (BSA), dilute in deionized water and adjust pH to 7.4 with HCl.

11. Solution of 2% propidium iodide (PI): Dilute 0.2 g PI in 10 mL of distilled water in Falcon tube, protect it from light and store at 4 °C.

3   Methods

3.1   Donor Plant Growth

1. Rinse oat and maize seeds with a commercial fungicide and sow in pots in the following volume: 3 L for oat and 6 L for maize, filled with a mixture of soil and sand (3:1 v/v).

2. Sow the seeds of sweet maize cultivar 2 weeks before sowing oat. Repeat maize sowing three times every 2 weeks (*see* **Note 3**).

3. Keep plants in the glasshouse: maize at 28/23 °C day/night, oat at 21/17 °C day/night and 16-h photoperiod. Plants should be additionally illuminated on cloudy days.

4. Irrigate plants with tap water as required, and once a week fertilize with liquid Hoagland medium [18].

3.2 Panicles Emasculation and Pollination

1. For emasculation, select oat plants when the florets start to appear in the panicle from the leaf sheath, before anthesis.

2. Pull out the panicle from the leaf sheet. Using precise tweezers, remove the florets from the upper and bottom part of the panicle and leave the middle part when the florets do not have maturated anthers (green colour).

3. Discard all secondary florets and leave only the oldest one (primary florets) in the same phase of development.

4. Using precise tweezers, emasculate the primary florets from the anthers (Fig. 1a) and cover the panicles with fleece bags, protecting the florets against self-pollination.

5. Two days after emasculation, pollinate the florets with fresh maize pollen (*see* **Note 4**) using a fine soft brush (Fig. 1b); ovaries' stigmas should be feathery (Fig. 1c) and again cover the panicles with fleece bags.

6. Two days after pollination, apply a drop of 2,4-D solution on the floret pistils using a syringe with needle (Fig. 1d) and again cover the panicles with fleece bags (Fig. 1e).

3.3 Disinfection of Enlarged Ovaries and Isolation of Haploid Embryos

1. Three weeks after pollination, cut the panicles and remove enlarged ovaries from the panicles manually (Fig. 1f).

2. Then transfer them to 25-mL sterile beakers (ovaries from one panicle to one beaker) under a laminar air flow chamber (Fig. 1g). Disinfect ovary surfaces with 70% ethanol (1 min), 2.5% calcium hypochlorite (7 min), and subsequently wash three times with sterile deionized water.

3.4 Haploid Embryos Isolation

1. Transfer ovaries from the beakers to sterile Petri dishes.

2. Using a binocular microscope and two dissecting needles, gently open the ovaries and remove haploid embryos (Fig. 1h).

3. Transfer isolated haploid embryos (Fig. 1i, j) to 60-mm Petri dishes (one embryo per dish) containing 190-2 medium [19]. Keep haploid embryos for germination in a growth chamber at 21 ± 2 °C and a light intensity of 60μmol m^{-2} s^{-1} (16/8 h light/dark).

3.5 Regeneration of Haploid Embryos and Acclimation of Haploid Plants

1. Approximately 2 weeks later, haploid embryos should germinate (Fig. 1k, l) and develop into green plantlets. Transfer plantlets to the Magenta vessels containing 40 mL of MS medium [20] solidified with 0.6% agar (Fig. 1m) and maintain them for 2 weeks in the same condition as isolated haploid embryos.

2. Transfer haploid plants to 500 mL glass jars filled with perlite (sterilized by autoclaving at 121 °C, 0.1 MPa for 20 min) moistened with Hoagland medium [18] for acclimation and maintain it for 2 weeks (Fig. 1n).

Fig. 1 Oat × maize wide crossing: (**a**) emasculation using precise tweezers; (**b**) pollination with maize pollen using fine soft brush; (**c**) ovaries with feathery stigmas; (**d**) application of 2,4-D solution on the floret pistils using a syringe with needle; (**e**) panicles covered with glassine bags; (**f**) enlarged ovaries removed manually

3. Subsequently, transplant haploid plants to pots filled with the mixture of soil and sand and keep in the glasshouse for 4 weeks.

3.6 Ploidy-Level Determination

1. Collect three fresh leaf tissue samples of approximately 1 cm^3 from each plant, place in a 60-mm glass Petri dish with 1 mL of PBS buffer and chop with a razor blade to release nuclei from cells. Keep Petri dish and PBS buffer on ice.

2. Filter the chopped tissues through a 30-μm nylon mesh filter into 5-mL tubes and add 20μL of a 2% PI solution to stain the nuclei.

3. After 15 min of incubation on ice, mix the samples gently and measure the ploidy using a flow cytometer equipped with a blue laser (at 488 nm).

4. DNA content of investigated plants is determined by comparing the peak position of the nuclear 2C with a peak of the 2C diploid oat plant (*see* **Note 5**).

3.7 Chromosome Doubling

1. For chromosome doubling, remove the developed haploid plants from soil, wash the roots with running water and trim them ca. 2 cm.

2. Immerse roots and shoot meristems in an aerated by air pump colchicine solution for 7.5 h under a fume hood. Colchicine treatment should be carried out at 25 °C and 80–100μmol m^{-2} s^{-1} light intensity.

3. Then, wash the roots with tap water aerated by an air pump for 24 h and transplant the plants to 3-L pots with soil and sand mixture (Fig. 1o).

3.8 Growth of DH Plants

1. Keep DH plants to maturity in a glasshouse in conditions at 21/17 °C day/night and 16-h photoperiod (Fig. 1p).

2. Water the plants as needed and fertilize them once a week with Hoagland medium [18].

4 Notes

1. 2,4-D is difficult to dissolve in water; thus, it is better to dissolve it first in several mL of 96% ethanol, but not more that 10% of total volume, and subsequently add the required amount of water.

Fig. 1 (continued) from the panicle; (**g**) ovaries in beakers prepared for disinfection; (**h**) haploid embryo formed in caryopsis; (**i**) haploid embryo isolated from caryopsis; (**j**) scanning electron microscopy of haploid embryo; (**k, l**) germinated haploid embryos on 190-2 medium; (**m**) developed haploid plants on MS medium; (**n**) acclimation of haploid plant in perlite; (**o**) plants in soil after colchicine treatment; (**p**) maturation of DH plants in the greenhouse

2. The macroelements dissolve relatively easy when they are first diluted separately in a small amount of water, then pooled together and topped up with water to the required volume.

3. Sowing maize at 2-week intervals provides access to fresh pollen for several weeks.

4. We have found that it is better to collect fresh pollen in 15-min intervals. First, in the morning, shake off the pollen from maize produced during the previous day and night. Collect fresh pollen from the maize panicle to a glass Petri dish.

5. Plant DNA content should be estimated before and after colchicine treatment. Oat plants derived from seeds with known diploid DNA content are used as controls.

Acknowledgements

The work was supported by the National Centre for Research and Development, grant No. 12002904/2008 and no PBS3/B8/17/2015, and the Ministry of Agriculture and Rural Development, grant no HORhn-801-1/13.

References

1. Dziurka K, Dziurka M, Warchoł M, Czyczyło-Mysza I, Marcińska I, Noga A, Kapłoniak K, Skrzypek E (2019) Endogenous phytohormone profile during oat (*Avena sativa* L.) haploid embryo development. In Vitro Cell Dev Plant 55:221–229

2. Marcińska I, Nowakowska A, Skrzypek E, Czyczyło-Mysza I (2013) Production of double haploids in oat (*Avena sativa* L.) by pollination with maize (*Zea mays* L.). Cent Eur J Biol 8(3):306–313

3. Rines HW (2003) Oat haploids from wide hybridization. In: Maluszynski M, Kasha KJ, Forster BP, Szarejko I (eds) Doubled haploid production in crop plants. Kluwer Academic Publishers, Dordrecht

4. Rines HW, Dahleen LS (1990) Haploid oat plants produced by application of maize pollen to emasculated oat florets. Crop Sci 30:1073–1078

5. Ishii T, Tanaka H, Eltayeb AE, Tsujimoto H (2013) Wide hybridization between oat and pearl millet belonging to different subfamilies of Poaceae. Plant Reprod 26:25–32

6. Nowakowska A, Skrzypek E, Marcińska I, Czyczyło-Mysza I, Dziurka K, Juzoń K, Cyganek K, Warchoł M (2015) Application of chosen factors in the wide crossing method for the production of oat doubled haploids. Open Life Sci 10:112–118

7. Sidhu PK, Howes NK, Aung T, Zwer PK, Davies A (2006) Factors affecting oat haploid production following oat x maize hybridization. Plant Breed 125:243–247

8. Matzk F (1996) Hybrid crosses between oat and Andropogoneae or Paniceae species. Crop Sci 36:17–21

9. Noga A, Skrzypek E, Warchoł M, Czyczyło-Mysza I, Dziurka K, Marcińska I, Juzoń K, Warzecha T, Sutkowska A, Nita Z, Werwińska K (2016) Conversion of oat (*Avena sativa* L.) haploid embryos into plants in relation to embryo developmental stage and regeneration media. In Vitro Cell Dev Plant 52:590–597

10. Skrzypek E, Warchoł M, Czyczyło-Mysza I, Marcińska I, Nowakowska A, Dziurka K, Juzoń K, Noga A (2016) The effect of light intensity on the production of oat (*Avena sativa* L.) doubled haploids through oat x maize crosses. Cer Res Comm 44(3):490–500

11. Warchoł M, Skrzypek E, Nowakowska A, Marcińska I, Czyczyło-Mysza I, Dziurka K, Juzoń K, Cyganek K (2016) The effect of auxin and genotype on the production of *Avena sativa* L. doubled haploid lines. Plant Growth Regul 78:155–165

12. Warchoł M, Czyczyło-Mysza I, Marcińska I, Dziurka K, Noga A, Skrzypek E (2018) The effect of genotype, media composition, pH and sugar concentrations on oat (*Avena sativa* L.) doubled haploid production through oat × maize crosses. Acta Physiol Plant 40:93

13. Liera-Lizarazu O, Rines HW, Philips RL (1996) Cytological and molecular characterization of oat x maize partial hybrids. Theor Apel Genet 93:123–135

14. Okagaki RJ, Kynast RG, Livingston SM, Russell CD, Rines HW, Phillips RL (2001) Mapping maize sequences to chromosomes using oat-maize chromosome addition materials. Plant Physiol 125:1228–1235

15. Philips RL, Rines HW (2009) Genetic analyses with oat-maize addition and radiation hybrid lines. In: Bennetzen JL, Hake S (eds) Maize handbook, vol. II: genetics and genomics. Springer, The Netherlands, pp 523–538

16. Kynast RG, Davies DW, Philips RL, Rines HW (2012) Gamete formation via meiotic nuclear restitution generates fertile amphiploid F1 (oat × maize) plants. Sex Plant Reprod 25:111–122

17. Skrzypek E, Warzecha T, Noga A, Warchoł M, Czyczyło-Mysza I, Dziurka K, Marcińska I, Kapłoniak K, Sutkowska A, Nita Z, Werwińska K, Idziak-Helmcke D, Rojek M, Hosiawa-Barańska M (2018) Complex characterization of oat (*Avena sativa* L.) lines obtained by wide crossing with maize (*Zea mays* L.). PeerJ 6:e5107

18. Hoagland DR, Arnon DI (1938) A water culture method for growing plants without soil. Circ. Univ. Calif., Agric Exp Stn, No. 347

19. Zhuang JJ, Xu J (1983) Increasing differentiation frequencies in wheat pollen callus. In: Hu H, Vega MR (eds) Cell and tissue culture techniques for cereal crop improvement. Science Press, Beijing, p 431

20. Murashige T, Skoog F (1962) A revised medium for rapid growth and bioassays with tobacco tissue cultures. Physiol Plant 15:473–497

21. Sambrook J, Fritschi EF, Maniatis T (1989) Molecular cloning: a laboratory manual. Cold Spring Harbor Laboratory Press, New York

Chapter 19

Anther Culture and Chromosome Doubling in Mediterranean Japonica Rice

Isidre d'Hooghvorst, Irene Ferreres, and Salvador Nogués

Abstract

Anther culture is the most used technique to produce doubled haploid lines in rice. This technique is well developed in a wide range of indica rice genotypes. However, in japonica type, and more specifically, the Mediterranean japonica, the protocols are yet to be optimized. Japonica and indica have different androgenic response, as well as different induction and regeneration rates, albinism ratios and chromosome doubling competence. The step-by-step anther culture protocol presented in this chapter allows to regenerate doubled haploid rice plantlets from anther microspores in 8 months. We also include an in vitro chromosome doubling protocol to induce doubled haploids from haploid plantlets by immersion in a colchicine solution. This chromosome doubling protocol complements the anther culture by taking advantage of the regenerated haploid plantlets.

Key words Androgenesis, Anther culture, Chromosome, Doubling, Colchicine, Doubled haploid, Rice

1 Introduction

Doubled haploid (DH) lines in rice are mainly obtained through androgenesis by means of anther culture. Niizeki & Oono were the first to report this technique in rice [1]. The culture of rice anthers in specific medium causes the induction of calli from haploid microspores, which are deviated from its original gametophytic pathway towards a sporophytic pathway [2]. Androgenesis in rice implies originally a two-step process. First, callus induction and then regeneration of spontaneous green DH plantlets from the embryogenic callus. Regeneration of spontaneous DH plants in rice occurs at a rate of around 30–40% [3]. Nevertheless, to increase the efficiency of androgenesis, a third step of chromosome doubling can be coupled to take advantage of the green haploid plantlets regenerated [4].

Jose M. Seguí-Simarro (ed.), *Doubled Haploid Technology: Volume 1: General Topics, Alliaceae, Cereals,*
Methods in Molecular Biology, vol. 2287, https://doi.org/10.1007/978-1-0716-1315-3_19,
© Springer Science+Business Media, LLC, part of Springer Nature 2021

Androgenesis in rice has many limiting factors that reduce the efficiency of the process. Endogenous factors such as the genotype are key in androgenesis. In general, the main issues that determine anther culture efficiency in rice are the high genotype dependence, low callus induction, low plantlet regeneration, high ratio of albino plantlets, low frequency of DHs and high frequency of haploid plantlets [5]. In particular, indica rice genotypes have a success rate lower than the japonica ones due to early necrosis, poor callus induction and a high ratio of regenerated albino plantlets [6, 7].

Many improvements have been achieved to overcome those exogenous limiting factors and achieve a minimum level of efficiency at each step of the process. Selection of the panicles at booting stage and their cold pretreatment are two important factors that increase the number of rice anthers at mid- to late-uninucleate pollen stages, where microspore response to anther culture is better [8]. Culture medium is a crucial factor in anther culture, and basic media such as N_6 [9] and MS [10] are commonly used and supplemented with a wide range of growth regulators to control the differentiation process [4, 11, 12]. The carbon source, usually sucrose, is also necessary due to its osmotic and nutritional properties [13]. It has also been proved that the addition of sorbitol enhances the effectiveness of the regeneration medium [14]. Other medium additives such as colchicine may increase calli induction, reduce the number of albino plantlets regenerated and increase the number of regenerated DH lines [4, 15]. Besides, colchicine treatment of regenerated haploid plantlets has a chromosome doubling efficiency up to 42% [16].

In this chapter, we present a protocol for anther culture of Mediterranean japonica rice, which includes the production of DH lines from rice cultivars, hybrids or inbred lines, increasing the efficiency in each step. The protocol has been applied to obtain DH lines for commercial use.

2   Materials

2.1 Plant Material

We used different Mediterranean japonica rice genotypes such as stabilized commercial lines and F2, F3 and F4 hybrids, provided by the Càmara Arrossera del Montsià SCCL.

2.2 General Labware

1. Sterile paper.
2. Sterile forceps and scalpel.
3. Razor blades.
4. Parafilm®.
5. Plastic tubes of 50 mL.
6. Sterile Petri dishes 90 mm.

7. Flow cytometer tubes.

8. 75 μm nylon filter.

9. 0.22 μm filters.

10. Substrate pots of 4 L.

11. Steel bead of 3 mm of diameter.

2.3 Equipment

1. Laminar flow hood.

2. Flow cytometer Gallios™ Flow Cytometer.

3. Tissue lyser.

4. Orbital shaker.

5. Growth chamber at 25 °C, light intensity of 50–70 μmol m^{-2} s^{-1} and a 16/8 h day/night photoperiod.

6. Greenhouse.

2.4 Solutions, Culture Media and Rice Substrate

1. 70% ethanol (v/v).

2. Sterile distilled water autoclaved at 121 °C for 20 min.

3. 10% sodium hypochlorite with 30 drops/L Tween 20.

4. Chromosome doubling solution: 500 mg/L colchicine supplemented with 4 drops/L Tween 20, filtered with a 0.22 μm filter.

5. Colchicine-supplemented induction medium: Chu N6 Basal Salt Mixture (Table 1) [17], supplemented with 1 g/L casein enzymatic hydrolysate, 250 mg/L L-proline, 0.5 g/L 2-(*N*-morpholino) ethanesulfonic acid monohydrate (MES monohydrate), 30 g/L sucrose, 1 mg/L 2,4-D, 1 mg/L kinetin, 150 mg/L colchicine, 3 g/L Gelrite® and pH adjusted to 5.8.

6. Colchicine-free induction medium: Chu N6 Basal Salt Mixture (Table 1) [17], supplemented with 1 g/L casein enzymatic hydrolysate, 250 mg/L L-proline, 1 mg/L 2,4-D, 1 mg/L kinetin, 0.5 g/L MES, 30 g/L sucrose, 3 g/L Gelrite® and pH adjusted to 5.8.

7. Plantlet regeneration medium: Chu N6 Basal Salt Mixture (Table 1) [17], supplemented with 1 g/L casein hydrolysate, 250 mg/L L-proline, 1 mg/L naphthaleneacetic acid, 2 mg/L kinetin, 0.5 g/L MES, 30 mg/L sucrose, 3 g/L Gelrite® and pH was adjusted to 5.8.

8. Plantlet growth medium: Murashige and Skoog Basal Salt Mixture (Table 1) [10], supplemented with 30 g/L sucrose, 0.5 g/L MES, 2 g/L Gelrite and pH was adjusted to 5.8.

9. Cold lysis buffer: 0.1 M citric acid and 0.5% Triton® X-100 dissolved in distilled water.

Table 1
Composition of Chu N6 and MS basal salt mixture

Component (mg/L)	Chu N6 [17]	MS [10]
Ammonium nitrate	–	400
Ammonium sulphate	463	–
Boric acid	1.6	6.2
Calcium chloride anhydrous	125.33	332.2
Cobalt chloride · 6H$_2$O	–	0.025
Cupric sulphate · 5H$_2$O	–	0.025
Na$_2$-EDTA	37.25	37.26
Ferrous sulphate · 7H$_2$O	27.85	27.8
Magnesium sulphate	90.37	180.7
Manganese sulphate · H$_2$O	3.33	16.9
Molybdic acid (sodium salt) · 2H$_2$O	–	0.25
Potassium iodide	0.8	0.83
Potassium nitrate	2830	1900
Potassium phosphate monobasic	400	170
Zinc sulphate · 7H$_2$O	1.5	8.6

10. Propidium iodide (PI) stain solution: 0.25 mM Na$_2$HPO$_4$, 9 M PI and 10 mL 10× stock (100 mM sodium citrate, 250 mM sodium sulphate) and made up to 100 mL.

11. Rice substrate: peat moss and vermiculite (2:1 v/v), supplemented with 1 g/L of controlled-release fertilizer mix per substrate and 1 g/L of CaCO$_3$ per peat litter to adjust the substrate pH to ~6.

3   Methods

3.1   Donor Plant Growth Conditions

1. Sow and grow donor rice plants during the summer in the greenhouse at 25–40 °C and 80–100% relative humidity (Fig. 1a).

2. Sow 10–15 seeds each week for 1 month to ensure tiller collection during a period of 2 months after 6–8 weeks from sowing.

3.2   Tiller Collection and Cold Pretreatment

1. Early in the morning, collect tillers at booting stage (*see* **Note 1**; Fig. 1b).

2. Surface disinfect the tillers with 70% ethanol for 1 min, then rinse twice with distilled water.

Fig. 1 Rice anther culture steps: (**a**) rice donor plants growing in the greenhouse; (**b**) tillers rice plant at booting stage; (**c**) cold pretreatment of tillers; (**d**) panicle with spikelets at the optimum stage containing anthers; (**e**) anthers sown in induction medium with induced callus; (**f**) callus growing on regeneration medium; and, (**g**) callus regenerating green and albino plantlets

3. Keep the tillers in a polystyrene bag for cold pretreatment for 9 days at 10 °C in darkness (Fig. 1c).

3.3   In Vitro Anther Culture

1. Dissect tillers to obtain the panicles in the laminar flow hood.

2. Surface disinfect panicles in 70% ethanol for 1 min and rinse twice with sterile distilled water. Then, disinfect with sodium hypochlorite for 3 min and finally wash three times with sterile distilled water.

3. Place the panicles over sterile paper, excise the spikelet (Fig. 1d) to obtain the anthers and place anthers in Petri dishes (*see* **Note 2**) with colchicine-supplemented induction medium (Fig. 1e; *see* **Note 3**). Seal dishes with Parafilm and incubate them at room temperature in the darkness.

4. After 24 h, transfer the anthers to colchicine-free induction medium in sterile conditions. Seal dishes with Parafilm and incubate them at room temperature in darkness.

5. After 3–6 weeks, microspore-derived calli of 1–2 mm diameter should emerge from anthers (Fig. 1e).

6. Transfer the callus to plantlet regeneration medium under sterile conditions and discard the anthers that induced each callus (*see* **Note 4**). Seal dishes with Parafilm and keep them in a growth chamber. Transfer the callus to fresh plantlet regeneration medium each 4 weeks (Fig. 1f).

7. When plantlets regenerate from callus (Fig. 1g) and are 3 cm long or more, transfer into tubes with plantlet growth medium. Transfer the plantlets to fresh plantlet growth medium every 4 weeks.

8. Eliminate black-coloured (necrosed) callus or after 10 weeks from the first transfer to regeneration medium.

3.4 Ploidy-Level Determination

1. Excise young leaves from in vitro plantlets and place them in 2-mL tubes containing a steel bead.

2. Add 300 μL of cold lysis buffer and cool tubes at −20 °C for 10 min.

3. Shake the samples in the tissue lyser at 25 Hz for 48 s (*see* **Note 5**) and filter through a 75-μm nylon filter collecting the cell suspension in the flow cytometer tubes.

4. Add 150 μL of propidium iodide stain solution.

5. Keep tubes on ice for 1 h in darkness before flow cytometry analysis. Count a minimum of 5000 cells per sample (Fig. 2).

3.5 Chromosome Doubling of Haploid Plantlets

1. Take the in vitro plantlets determined as haploids (*see* **Note 6**) to the laminar flow hood (Fig. 2a).

2. Trim plantlet stem and roots up to 3 cm long, place them in 50 mL tubes an add the colchicine solution to a final volume of 45 mL (*see* **Note 7**).

3. Incubate the 50 mL tubes in an orbital shaker for 5 h at 120 rpm at room temperature.

4. Once incubation time elapses, rinse plants with water to eliminate the excess of colchicine solution.

5. Transfer the plantlets to plantlet growth medium until they develop 10–15 cm in size and excise young leaves to perform ploidy-level determination (*see* **Note 8**).

3.6 Acclimation and Seed Recovery

1. Acclimate plantlets to individual pots with soil in a growth chamber. Only confirmed DH plantlets are used for this stage of acclimatized. Mixoploid plants may be acclimatized too, in order to check out the fertility. Haploid plantlets are eliminated.

2. After 2 weeks of acclimatization, transfer the plants to the greenhouse to recover DH seed (*see* **Note 9**).

3. Rice spikelet is bagged before the anthesis for self-fecundation.

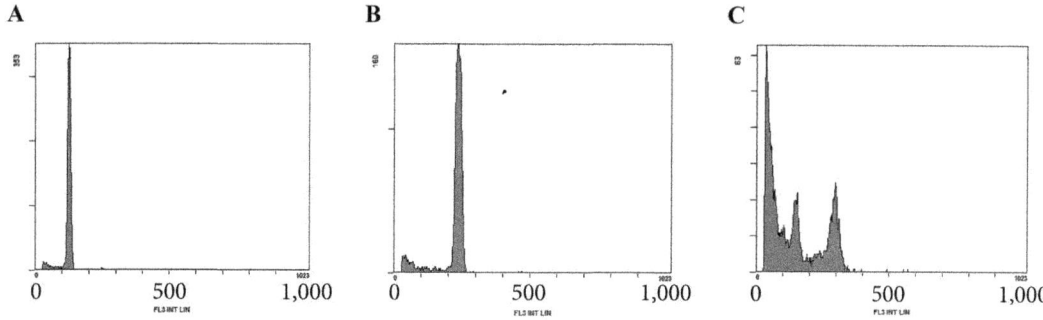

Fig. 2 Flow cytometry histograms of regenerated plants. (**a**) Flow cytometry histogram of a haploid plant showing a peak at channel number 150; (**b**) flow cytometry histogram of a diploid plant showing a peak at channel number 300; and (**c**) flow cytometry histogram of a mixoploid plant showing peaks at channels numbers 150 and 300

4. After 2–3 weeks since anthesis, DH seeds are collected and ready for further use.

4 Notes

1. The booting stage of tillers is defined as the stage when the distance from the flag leaf to the auricle of the penultimate leaf is between 5 and 12 cm. This is the optimal developmental stage for induction of androgenesis due to the high proportion of microspores at mid- and late uninuclear stage. Microspores at those stages have higher induction and DH regeneration rates. Cytological examination of the microspore stage can be carried out for each genotype to ensure the optimal stage of booting.

2. Panicles are placed on sterile paper and the spikelet is dissected under sterile conditions. Spikelet tips are first held with forceps and then the basal part is cut in the lower third level, so the anther filaments are cut at the same time. Then, the spikelet tip is placed above the petri dish containing induction medium and anthers are released by hitting the forceps against the dish edges.

3. Microspores cultured in colchicine-supplemented medium have higher rates of callus induction and DH plant regeneration.

4. Callus develop fast and can be disaggregated during the process. To avoid counting of unwanted clonal calli, usually only one callus is considered per anther.

5. Cold lysis buffer should be green after trituration due to the cell leaves in suspension. Trituration may be adjusted depending on the toughness of the leaf, which depends on the

genotype. To ensure the trituration, fold the leaves to make it a ball before trituration. The chopping trituration method can be used with the same lysis buffer too.

6. Haploid plantlets can be cultured in MS medium supplemented with hormones to induce a higher sprouting of shoots and therefore, increase the number of haploid clones used for chromosome duplication. Usually, 50–70% of the androgenetic plantlets produced are DH, including those arising directly DH, and those needing an additional colchicine treatment.

7. The colchicine treatment induces chromosome doubling in about 40% of the androgenetic haploid plantlets. Other antimitotic compounds have been assayed, as oryzalin and trifluralin, to reduce the hazardous exposition to colchicine in routine lab work. Nevertheless, worse chromosome doubling results have been detected when using oryzalin or trifluralin.

8. Ploidy analysis of rice plants usually shows only one clear peak in flow cytometry histograms, except for mixoploid individuals, which present two. Rice plants do not usually show a visible G2 peak as for the majority of species.

9. Spontaneous DH or induced DH plants can present partial sterility in some spikelet. Sterile plants or spikelets are detected according to the phenotypical characteristics of flowers, which have smaller size and do not set grain after pollination.

References

1. Niizeki H, Oono K (1968) Induction of haploid rice plant from anther culture. Proc Jpn Acad 44:554–557

2. Seguí-Simarro JM (2010) Androgenesis revisited. Bot Rev 76(3):377–404

3. Rukmini M, Rao GJN, Rao RN (2013) Effect of cold pretreatment and phytohormones on anther culture efficiency of two Indica rice (*Oryza sativa* L). Exp Biol Agr Sci 2:69–76

4. Hooghvorst I, Ramos-Fuentes E, López-Cristofannini C, Ortega M, Vidal R, Serrat X, Nogués S (2018) Antimitotic and hormone effects on green double haploid plant production through anther culture of Mediterranean japonica rice. Plant Cell Tissue Organ Cult 134(2):205–215

5. Lentini Z, Roca WM, Martinez CP (1997) Cultivo de anteras de arroz en el desarrollo de germoplasma, vol 293. CIAT

6. He T, Yang Y, Tu SB, Yu MQ, Li XF (2006) Selection of interspecific hybrids for anther culture of indica rice. Plant Cell Tissue Organ Cult 86(2):271–277

7. Chen C-C, Tsay H-S, Huang C-R (1991) Factors affecting androgenesis in rice (*Oryza sativa* L.). In: Bajaj YPS (ed) Rice. Springer, Berlin, pp 193–215

8. Kaushal L, Balachandran SM, Ulaganathan K, Shenoy V (2014) Effect of culture media on improving anther culture response of rice (*Oryza sativa* L.). Int J Agric Innov Res 3(1):218–224

9. Chu CC (1978) The N6 medium and its application to anther culture of cereal crops. Symp. Plant Tiss. Cult. Science Press, Beijing

10. Murashige T, Skoog F (1962) A revised medium for rapid growth and bioassays with tobacco tissue cultures. Physiol Plant 15(3):473–497

11. Ball ST, Zhou H, Konzak CF (1993) Influence of 2, 4-D, IAA, and duration of callus induction in anther cultures of spring wheat. Plant Sci 90(2):195–200

12. Trejo-Tapia G, Maldonado Amaya U, Salcedo Morales G, De Jesús Sánchez A, Martínez Bonfil B, Rodríguez-Monroy M et al (2002) The effects of cold-pretreatment, auxins and

carbon source on anther culture of rice. Plant Cell Tissue Organ Cult 71(1):41–46

13. Powell W (1988) The influence of genotype and temperature pre-treatment on anther culture response in barley (*Hordeum vulgare* L.). Plant Cell Tissue Organ Cult 12(3):291–297

14. Yoshida KT, Fujii S, Sakata M, Takeda G (1994) Control of organogenesis and embryogenesis in rice calli. Jpn J Breed 44(4):355–360

15. Alemanno L, Guiderdoni E (1994) Increased doubled haploid plant regeneration from rice (*Oryza sativa* L.) anthers cultured on colchicine-supplemented media. Plant Cell Rep 13(8):432–436

16. Hooghvorst I, Ribas P, Nogués S (2020) Chromosome doubling of androgenic haploid plantlets of rice (*Oryza sativa*) using antimitotic compounds. Plant Breed 139 (4):754–761. https://doi.org/10.1111/pbr.12824

17. Chu C-C (1975) Establishment of an efficient medium for anther culture of rice through comparative experiments on the nitrogen sources. Sci Sinica 18:659–668

Chapter 20

High-Throughput Doubled Haploid Production for Indica Rice Breeding

Swapan K. Tripathy

Abstract

Anther culture is an important biotechnological tool for quick recovery of fixed breeding lines with unique gene combinations that might otherwise disappear in the course of an extended series of segregating generations in conventional breeding methods in rice. The haploid microspores in culture or the resultant haploid plants are converted to doubled haploids (homozygotes). Variation in doubled haploid lines from F_1 hybrids is due to the recovery of rare gene combinations by single round of recombination following meiosis. Androgenesis in rice is largely species- and genotype-specific. *O. glaberrima* responds better to anther culture than *O. sativa*; and japonica sub-group is more responsive to microspore embryogenesis than indica types. The author provides a detailed protocol of the anther culture technique for doubled haploid production in indica rice hybrids amenable for genetic improvement.

Key words Anther culture, Androgenesis, Doubled haploids, Genetic improvement, Indica rice, Plantlet regeneration

1 Introduction

Anther culture for haploid plant production was demonstrated in rice in 1968 [1], just 2 years after the accidental discovery of such novel technique in *Datura* [2]. In rice, anther culture is highly species- and genotype-specific [3]. *Oryza glaberrima* L. responds better to callus induction and plantlet regeneration than *Oryza sativa* genotypes. Asian cultivated rice (*O. sativa*, ssp. indica) is more recalcitrant to anther culture than japonica types (more than 20-fold response than indica), which limits its practical application in rice breeding [4]. The major hindrance lies with limited morphogenic potential of anther-derived calli due to chromosome number and structure variation, gene mutation and epigenetic changes (DNA methylation) and higher percentage of regenerated albino plants due to changes in chloroplast DNA. However, a number of researchers got differential responses and it took a

Jose M. Seguí-Simarro (ed.), *Doubled Haploid Technology: Volume 1: General Topics, Alliaceae, Cereals*,
Methods in Molecular Biology, vol. 2287, https://doi.org/10.1007/978-1-0716-1315-3_20,
© Springer Science+Business Media, LLC, part of Springer Nature 2021

long way to standardize a protocol for effective androgenic response in indica rice. Each single anther contains thousands and thousands of pollen grains. Anthers exposed to various kinds of stresses (cold, heat, osmotic stress, sugar starvation, gamma irradiation and chemical treatment) prior to in vitro culture are reported to induce androgenesis [5] and inhibit callus induction from somatic origin (anther wall and tapetum). The type and duration of pretreatments vary with the species and variety of rice. However, cold pretreatment is more effective and commonly used by different researchers. The anther culture technique involves plating of anthers containing microspores at the late uni-nucleate (vacuolated microspores) to early binucleate stage of development into a suitable medium under aseptic condition to induce callus followed by green plantlet formation. The anther-derived haploids with deleterious genes that usually interfere with survivability and adaptability, are perished through the 'haploid sieves'. The resulting haploids in culture are then converted to doubled haploid (DH) status by colchicine treatment (chromosome doubling) to fix homozygosity in just a single generation. Thus, large numbers of pure lines produced through bypassing the normal fertilization are then grown to select the best superior line(s). This shortens the breeding cycle, saves time, space and labour. Thus, the doubled haploid (DH) breeding technique is a potential tool for quick success in developing new varieties (within 5–6 years) of rice with novel gene combinations from anthers of F_1 (first filial generation) hybrids without waiting for several selfing generations for recovery of pure breeding lines. Presently, more than 40 rice varieties have been developed using double haploid breeding. Some of the double haploid lines derived from diverse parents are reported to be equal or exceed the heterotic level of the hybrid [6]. Besides, anther culture of BC_1 (first back cross generation) plants is also employed for quick introgression of specific trait(s) from elite donors. DH technique helps in fixing a transgene while simultaneously removing unwanted selectable marker gene [7]. Some of the DHs may turn out to have potential breeding implications, for example, biotic and abiotic stress tolerance, improved agronomic traits and yield potential, and better nutrient and cooking quality traits. Such elite DHs may be amenable for development of heterotic hybrids. Besides, the DH lines serve as fixed mapping population for construction of genetic linkage map, tagging of valuable genes/quantitative trait loci (QTLs) using marker-trait association studies [8], marker-aided selection [9], use in functional genomics [10] and in vitro mutant selection [11].

2 Materials

2.1 Plant Materials

Diverse putative F_1 plants of *Oryza sativa*, indica type.

2.2 Equipment

1. Water bath.
2. Horizontal electrophoresis system.
3. Micropipette.
4. Thermocycler.
5. Gel documentation system.
6. Compound microscope.
7. Phase-contrast microscope.
8. Flow cytometer.
9. BOD incubator with oxygen supply.
10. Gamma ray chamber.
11. Electronic balance.
12. Double-door refrigerator.
13. pH meter.
14. Water purification system.
15. Distillation unit.
16. Autoclave.
17. Laminar flow hood.
18. Environmentally controlled incubators.
19. Air conditioners.
20. Temperature and relative humidity (RH) recorder.
21. Timer for fluorescent tubes to maintain photoperiod.
22. Photographic camera with photography hood for documentation.

2.3 Infrastructures

1. Inoculation room with double partition entry doors having hydraulics facility.
2. Culture room with double door facility, steel racks with fluorescent tube light fittings.
3. Green house (with controlled light, temperature, RH and sprinkler irrigation facility for acclimatization).
4. Net house (for plant establishment).

2.4 Glassware

1. Culture tubes (25 × 150 mm, borosilicate).
2. Petri dishes (100 mm × 15 mm, Borosilicate).
3. Flasks (100 mL, 250 mL and 500 mL capacity, borosilicate).

4. Beakers (100 mL, 250 mL and 1000 mL).

5. Reagent bottles (100 mL, 200 mL, 500 mL and 1000 mL).

6. Wide mouth screw cap bottle (70 × 100 mm).

7. Coplin jar.

8. Glass slides and cover slips.

9. Watch glass.

2.5 General Labware

1. Surgical dissection box containing scissor, forceps and needle (stainless steel, 110 mm length each).

2. Materials for plant labelling: plastic clips, butter paper bags, threads, aluminium labels.

3. Mortar and pestle.

4. Eppendorf tubes.

5. Pipette tips.

6. Whatmann paper.

7. Sterilized non-absorbent cotton (defatted, 350 gm roll).

8. Gloves.

9. Miscellaneous: cello tap and aluminium foil.

10. Culture tube racks(holds 40, 25 mm diameter culture tubes).

11. Face mask and laboratory gown.

12. Gas lamp, Bunsen lighter or equivalent.

13. Stainless steel forceps (serrated, 200 mm length) and scalpel (180 mm length) for in vitro culture.

14. Water-proof permanent marker pen.

15. Plastic pots (3″ diameter).

16. Sterilized pot mixture (peat moss: perlite 2:1).

17. Earthen pots.

2.6 Chemicals and Reagents

1. DNA extraction buffer: 100 mM Tris–HCl, pH 8.0, 20 mM ethylenediamine-tetraacetic acid (EDTA), 2% Cetyltrimethy-lammonium bromide (CTAB), 1.4 M Sodium chloride (NaCl), 0.4% β-mercaptoethanol (β-ME) and 2% polyvinyl pyrrolidone (PVP).

2. Liquid nitrogen.

3. Molecular grade water.

4. 0.1 M sodium acetate

5. Ultra pure iso-propanol.

6. Phenol:chloroform:isoamyl alcohol (25:24:1).

7. TE buffer (Tris 10 mM and EDTA 1 mM).

8. RNase A formulation: 10 mM Tris, 15 mM NaCl and 10 mg RNase A for a final volume of 1 mL. Prepare RNase-A, incubate at 37 °C for 1 h to inactivate any DNase contaminants if any and store at −20 °C in aliquots.

9. PCR mix: 1× reaction buffer (10 mM Tris–HCl, pH 9.0, 1.5 mM MgCl$_2$, 50 mM KCl, 0.01% gelatine); 2.5 mM each of deoxynucleotide phosphates (dNTPs); 10 ng of each gene-specific microsatellite (SSR) primer pair, 20 ng of genomic DNA and Taq polymerase −1 unit).

10. Agarose Gel reagents: Agarose (2.5%), 1× TBE buffer (89 mM Tris, pH 8.3, boric acid 89 mM and 10 mM EDTA), ethidium bromide (0.1%); gel loading dye.

11. Ethrel (4000 ppm) and 0.4 M mannitol for chemical pretreatment of anthers.

12. 0.1% HgCl$_2$ (for surface sterilization of central spikelets)

13. Ethanol (98% pure).

14. 70% Ethyl alcohol

15. Aceto-alcohol fixative: 1:3 glacial acetic acid:ethyl alcohol.

16. 8-hydroxyquinoline

17. Carnoy's fixative: 6:3:1 absolute ethanol: chloroform: glacial acetic acid.

18. Enzyme mixture: 6% cellulase and 2% pectinase, prepared in 0.01 M citrate buffer.

19. Distilled water.

20. Acetocarmine stain (1%) for cytological study.

2.7 Culture Media and Solutions

1. Sugar-free MS medium: add the different components of the medium as detailed in Table 1, but excluding sucrose.

2. Half-strength MS basal liquid medium: add the different components of the medium as detailed in Table 1, but at half the concentration and without agar.

3. CIM medium (callus induction medium): Add the different components of the medium as detailed in Table 1. Supplement with 1.5 mg/L 2,4-dichloro-phenoxyacetic acid (2,4-D)and 0.5 mg/L kinetin (Kn) for somatic embryogenic callus induction within 3–4 weeks.

4. RM Medium (regeneration medium): Add the different components of the medium as detailed in Table 1. Supplement with 2 mg/L 6-benzylaminopurine (BAP) and 0.5 mg/L α-naphthalene acetic acid (NAA) for green plantlet regeneration.

5. Colchicine 0.2–0.5 g/L.

Table 1
Composition (mg/L) of minimal media for anther culture in indica rice

Components	MS	CIM	RM
NH_4NO_3	1650.0	1650.0	1650
KNO_3	1900.0	2250.0	1900
$MgSO_4 \cdot 7H_2O$	370.0	370.0	370
$MnSO_4 \cdot 4H_2O$	22.3	22.3	22.3
$ZnSO_4 \cdot 7H_2O$	8.6	8.6	8.6
$CuSO_4 \cdot 5H_2O$	0.025	0.025	0.025
$(NH4)_2SO_4$	–	–	232
$CaCl_2 \cdot 2H_2O$	440.0	440	400
KI	0.83	0.83	0.83
$CoCl_2 \cdot 6H_2O$	0.025	0.025	0.025
$KH_2PO_4 \cdot 7H_2O$	170.0	170.0	170.0
H_3BO_3	6.2	6.2	6.2
$Na_2MoO_4 \cdot 2H_2O$	0.25	0.25	0.25
$FeSO_4 \cdot 7H_2O$	27.85	77.8	55.7
$Na_2EDTA \cdot 2H_2O$	37.25	104.3	74.5
Myo-inositol	100	100	100
Nicotinic acid	0.5	0.5	0.5
Pyridoxine–HCl	0.5	0.5	0.5
Thiamine–HCl	0.1	0.5	0.25
Glycine	2.0	2.0	2.0
Glutamine	–	500	500
Tryptophan	–	100	100
Cysteine	–	40	40
Casein hydrolysate	–	500	500
Adenosine sulphate	–	–	200
Proline	–	200	500
Agar	8000	8000	8000
Sucrose	30,000	60,000	60,000
pH	5.8	5.8	5.8

3 Methods

3.1 Development of Crosses

Rice is a highly self-pollinated crop. DH lines must be developed from anther culture of F_1 hybrids to produce new true breeding lines with unique gene combinations. For this, select appropriate parents to achieve the success.

3.1.1 Hybridization

1. Selection of Parents: Choose parents that have been maintained for genetic purity (*see* **Note 1**). In each cross combination, include at least one parent with acceptable agronomic features e.g., maturity duration, plant type, tillering ability, panicle features. Include a parent with high yield potential and that has wide adaptability. Select the proven donors for target traits, for example, disease resistance, pest resistance and tolerance to abiotic stresses. Besides, choose parents focusing on grain quality and nutritional content.

2. Hybridization: Use healthy and disease-free plants for hybridization. Use as female parent the genotype-carrying recessive allele(s) for the contrasting trait(s) to discard selfed progeny, if any, in the F_1 generation due to uncontrolled selfing (*see* **Note 2**).

3. Emasculation: In the previous evening, clip the top one one-third and bottom one-third portions of the panicle (of mother tiller) of desired female parent by using scissor leaving the middle spikelets. Remove anthers by splitting the lemma and palea carefully keeping the lower portion of the spikelet intact. Alternatively, cut each spikelet slantly (Clipping method) and remove the anthers by using forceps. Cover the panicle with butter paper bag after emasculation in the evening.

4. Pollination: Next day morning (before 8.00 am), open the paper bags and place ripened bright yellow anthers from selected male parent over the stigma of emasculated spikelets (at least one anther per spikelet), using forcep to effect pollination. In case of clipping method, cut the paper bags at top portion and insert 2–3 bloomed panicles from the male parent in an inverted position into the butter paper bag and turn in both ways. After ensuring dusting of pollens over the emasculated spikelets, close the top portion of the butter paper bag by a plastic clip.

5. Harvesting of Seeds: Following effective fertilization, harvest the seeds from the female parent with proper labelling at physiological maturity. Allow sun drying of the seeds and store in a desiccator.

3.1.2 Raising F₁ Plants

1. Break the dormancy (if required) of clean, healthy seeds of different crosses in the oven at 50 °C for 3–5 days and place these for germination in Petri dishes lined with moistened filter paper for 12–24 h.

2. Sow the germinated seeds in seed boxes and add water to wet the soil till 21 days to allow normal growth of the F_1 seedlings.

3. At day 21, transplant the seedlings in well-drained field.

4. Apply at least one prophylactic spray of pesticides followed by additional spray(s) if needed to protect the plants from insect and diseases.

3.1.3 Test of Hybridity

1. Phenotyping: Observe contrasting trait(s) between parents that inherit to F_1 plants. The true F_1 plants must reveal dominant traits (*see* **Note 2**, Fig. 1). Discard plants carrying recessive traits. Save and label individual F_1 plants carefully for test of hybridity by genotyping.

2. Genotyping: In the absence of contrasting simple inherited traits between the parents, use polymorphic SSR markers to confirm successful crosses (*see* **Note 3**). Keep the F_1 plants that show co-dominance (presence of both alleles of SSR marker).

3.2 Anther Culture for DH Production

The success of anther culture depends upon exogenous and endogenous factors [12].

3.2.1 Preparatory Stage: Stage I

1. Genotype: Be sure that the parents in the cross combination of indica rice respond to anther culture (*see* **Note 4**).

2. Physiological status of donor plant: Germinate seeds in Petri dishes and then transfer them to field to raise F_1 seedlings. Transplant 25-day-old seedlings, one seedling/hill in the field as anthers from field-grown plants respond better than pot-grown plants with intensive care (*see* **Note 5**). Target mother tiller for collection of boot (*see* **Note 6**).

3. Collection of boots: Choose boots at the appropriate stage (*see* **Note 6**), which corresponds to the late uni-nucleate to early bi-nucleate stage of pollen development in anthers of the central spikelets [13]. Wrap boots with tissue paper followed by aluminium foil, pack in an ice box and carry them to the laboratory.

4. Developmental stage of pollen: Fix the cytological stages of central spikelets of few collected boots in acetic acid: ethanol (1:3) mixture for 24 h, followed by transfer to 70% ethanol and store in refrigerator. Stain the pollen grains with 1% acetocarmine stain solution and check the microspore developmental stage under the microscope (*see* **Note 7**).

5. Pretreatment of anther: Wash the boots thoroughly with tap water to remove surface dirt, remove the flag leaf blade, spray the boots with 70% ethanol and then wrap the boots with

Fig. 1 Phenotyping of parents and F₁ in a cross "Khandagiri" (K) × "Dular" (D). N22 and Sahbhagidhan are drought-tolerant checks; Images (**a–e**) show parental polymorphism (**a**: tall vs semi-dwarf, **b**: drought tolerance vs drought sensitive, **c**: deep root vs short fibrous root system, **d**: black pigmented stigma vs greenish white stigma, **e**: presence vs absence of epiculous pigmentation in cv. Dular and cv. Khandagiri, respectively). F₁ shows dominant traits present in cv. Dular

aluminium foil. Incubate boots at 4 °C for a week in darkness (*see* **Note 8**). Alternatively, other treatments may be applied (*see* **Note 9**).

3.2.2 Initiation of Anther Culture: Stage II

1. Surface sterilization of spikelets: After the cold treatment, remove the leaf blades, disinfect the spikelets from the middle portion of the panicles with aqueous mercuric chloride (0.1%, w/v) for 1 min under aseptic conditions. Wash the disinfected spikelets five times with sterile distilled water and soak excess water using sterile Whatmann paper.

2. Plating of anthers (primary culture): Hold individual spikelets at the apex by flame-sterilized forceps and cut slantly at basal 3/4th position (just below the anther) using surgical scissor to detach anther filaments. Release 30–40 anthers from selected spikelets of the same panicle on CIM medium under laminar air flow cabinet just by tapping the forcep at the brink of the culture vessel. Special care should be taken to ensure that the anthers are not injured in any way.

3. Incubate the culture vessels in culture room under dark at 25 ± 1 °C.

4. Avoid inoculation of too many anthers per culture vessel (*see* **Note 10**).

Fig. 2 Callus induction (**a**), sub-culture of calli (**b**) and green plantlet regeneration (**c** and **d**) from anther culture of indica rice cross (Khandagiri × Dular)

5. Carry out all operations under aseptic conditions of laminar airflow cabinet.

6. After completion of work, always label the cultures with names, code and date of work.

3.2.3 Callus Induction and Proliferation: Stage III

1. Observe callus induction 4 weeks after anther plating (Fig. 2a) and assess callus induction frequency (CIF%) as percentage of anthers plated.

2. Transfer calli induced in primary culture to CIM with 1.0 mg/L 2,4-D + 0.5 mg/L Kn for callus proliferation (Fig. 2b; *see* **Note 11**).

3.2.4 Green Plantlet Regeneration: Stage IV

1. Avoid long-term series of sub-culturing (*see* **Note 12**).

2. Transfer callus clumps of approximately 0.5 cm diameter to RM medium containing 2.0 mg/L BAP and 0.5 mg/L NAA for green plantlet regeneration (Fig. 2c, d). Record the number of plants regenerated per responding callus clump.

3. Consider embryogenic calli for plant regeneration (*see* **Note 13**).

4. When roots are slow growing, transfer the plantlets to MS medium without phytohormones for root initiation and better root growth.

5. For microtillering, transfer the regenerated shootlets to MS medium containing 4 mg/L Kn [14]. However, the plants derived from multiple shoots of a single callus show higher phenotypic similarity than those derived from different calli [15].

Fig. 3 Anther culture-derived putative doubled haploid production in rice. (**a**) Rice F₁ plant as source of explants, (**b**) Boots, (**c**) Spikelets, (**d**) Anther, (**e**) late uninucleate (upper) and early binucleate (lower) microspore, (**f**) Callus induction from anthers, (**g**) Regeneration of plantlets, (**h**) Plant establishment in pot mixture and (**i**) Field-grown putative DH plant at panicle stage

3.2.5 Production of DHs:
Stage V

1. Omit colchicine treatment for chromosome doubling in rice unless desired for increased DH production [16] (*see* **Note 14**, Fig. 3).

2. Be sure for requirement of colchicine treatment, as the response to colchicine treatment varies from genotype to genotype (*see* **Note 15**).

3. If needed, incubate anther culture-derived calli for 24–48 h in callus proliferation medium with 0.2–0.5 g/L colchicine followed by transfer to colchicine-free regeneration medium to induce as high as 65–70% viable DH plantlets [17] (*see* **Note 16**).

3.2.6 Acclimatization
and Plant Establishment:
Stage VI

1. Individualize the plantlets derived from each callus clump.

2. Initially, transfer the plantlets to half-strength MS basal liquid medium for 1 week in culture room.

3. In the follow-up step, transfer the plantlets to pots filled with mixture of peat moss: perlite 2:1 (v/v) and grow those in green house for 15 days.

Fig. 4 Plant establishment of regenerants in pot culture

4. Transfer the plantlets to partial shade to acclimatize with the external environment for 2–3 days. About 60% plantlets may survive.

5. Transplant the plants to earthen pots (Fig. 4) or well-drained and puddled main field.

3.2.7 Test for Homozygosity and Recovery of DHs: Stage VII

1. Check the plant morphology, fertility status, excessive vigour and any physiological abnormality of regenerants to discard haploids, aneuploids, higher order euploids and hyperploids.

2. Carry out progeny test for agronomically normal regenerants (after self-pollination) for test of homozygosity to confirm DH status (*see* **Note 17**). Other direct and indirect approaches are available, though each has its own limitations:

    (a) Chloroplast count: Count the number of chloroplast in guard cells and determine the ploidy level (*see* **Note 18**).

    (b) Stomata length: Observe stomata length under a light microscope. In haploids, the stomata through leaf 1 to leaf 8 stage of seedlings is smaller as compared to normal parents and DH plantlets.

    (c) Chromosome count: Count the number of chromosomes in cytological preparations of root tip cells at pro-metaphase [18]. It seems to be a direct approach (Fig. 5), but it may not eliminate heterozygous diploids possibly resulted from anther wall and tapetum cells in culture (*see* **Note 19**).

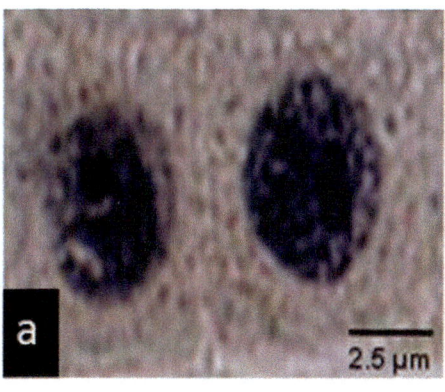
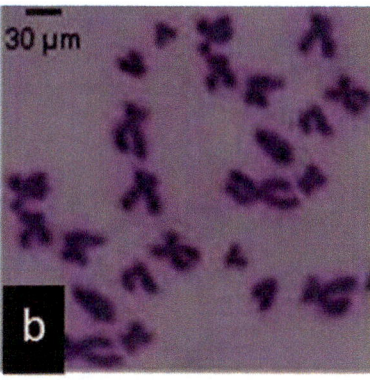

Fig. 5 Cytological preparation of doubled haploids. (**a**) Spontaneous chromosome doubling in cells of culture (two haploid chromosomal sets appear to undergo fusion), (**b**) Prometaphase plate of root tips of a doubled haploid regenerant with 24 distinct chromosomes

(d) Nucleolus count: Count the number of nucleoli in cytological preparations of root tip cells at late telophase to early prophase. The number of nucleoli per cell increases with the ploidy level. Haploids, diploids (here DHs), triploids and tetraploids contain 1, 2, 3 and 4 nucleolus, respectively [19].

(e) C-value: Measure the C-value (nuclear DNA content/cell) using flow cytometry technique [20]. C-value varies with ploidy level. Haploids, diploids (here DHs), triploids, tetraploids are designated as 1C, 2C, 3C and 4C, respectively.

3. Chromosome duplication: identify haploids by flow cytometry, treat these with colchicine (0.2%) for chromosome doubling and then transfer to a greenhouse for acclimatization.

4. After acclimatization, carry out flow cytometry again to check double haploid status of in vitro-derived plantlets [21].

5. Genotyping: Amplify genomic DNA of regenerants using SSR markers (Fig. 6) to identify DHs and to eliminate any heterozygous diploid plants possibly resulted from heterozygous somatic cells, for example, anther wall and tapetum cells in culture.

4   Notes

1. The parents in each cross combination must be genetically diverse based on available genetic variation.

2. Female parents in each cross combination should carry recessive allele of the contrasting trait(s) to discard any selfed progeny resulted in F_1 generation. The true F_1 plants must reveal

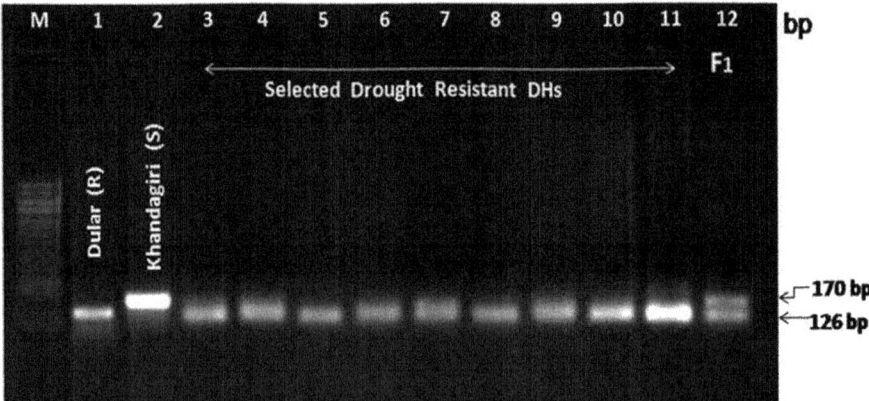

Fig. 6 Molecular profiling of parents, F₁ and selected drought-resistant doubled haploids of a cross Khandagiri × Dullar using gene-specific SSR primer RM 8085 (on chromosome 1). M: 100 bp Mol. Wt. marker, Lane 1: Dular (drought resistant), Lane 2: Khandagiri (drought sensitive), Lane 3–11: DHL 1, DHL 4, DHL 8, DHL 24, DHL 30, DHL 41, DHL 43, DHL 48 and DHL 53) and Lane 12: F₁ of the cross Khandagiri × Dular

dominant traits. For example, F₁ of a cross Dular × Khandagiri revealed dominant traits present in Dular, e.g., tallness, black pigmented stigma and epiculous pigmentation (in matured grains) as compared to semi-dwarf plant type, greenish white stigma and absence of epiculous pigmentation revealed by Khandagiri (Fig. 1).

3. Use co-dominant molecular marker for test of hybridity (of F₁ plants) which shows parental polymorphism. For this, isolate genomic DNA from parents and F₁ plants using CTAB method [22]. Remove contaminating RNAs by using DNase free RNase-A (20μg/mL of DNA extract). For RNase treatment, take 5μL of RNase-A stock solution for each 1 mL of DNA extract and incubate at 37 °C for 1 h to degrade contaminating RNAs. Quantify genomic DNA using UV-VIS Nanodrop-2000 spectrophotometer at 260 nm. Dilute each genomic DNA sample using TE buffer to a working concentration of 10 ng/μL for PCR analysis. Prime and amplify each individual genomic DNA sample (parents and F₁ plants) using gene-specific SSR primers in a reaction volume 25μL. For this, carry out DNA amplification in a thermocycler at appropriate thermal regime till loading to the agarose gel. Load the amplified products in 2.5% agarose gel (depending on the size of the amplicon) containing 0.5 mg/mL of ethidium bromide and electrophoresed at a constant voltage (50 V). Scan the gel(s) by gel documentation system for detection of gene/QTL specific alleles. Determine the size of amplicons by comparing the 100 bp lambda DNA ladder with known size fragments in base pairs (bp). Retain the F₁ plants that show co-dominance (presence of both alleles of SSR marker).

4. Callus induction and plantlet regeneration have a genetic basis and as such, anther culture is highly species- and genotype-specific. F_1 hybrids are more responsive to anther culture than their parents [15]. Alternatively, combine the high-yielding indica rice with high anther culture-responding japonica genotype to improve anther culture response.

5. Skip foliar spray of urea in the field as plants with much vegetative growth are less responsive to androgenesis. Adjust sowing and planting date to coincide panicle initiation with a day length of more than 12 h and day/night temperatures of 34 °C/25 °C. Allow few intermittent mild moisture stresses at the vegetative stage for better anther culture response.

6. Collect boots from primary tillers of F_1 at around 8.00 am, when the anthers occupy one-third to one-half of the spikelet length or when the auricle distance between the flag leaf and penultimate leaf reaches 5–8 cm.

7. The stages suitable to be switched to embryogenesis are microspores at the late uni-nucleate stage (vacuolate microspores) to early binucleate pollen grains.

8. Cold pretreatment of boots (4 °C for a week in darkness) induces and ensures continuance of sporophytic mode of development of microspores [23], induces embryogenesis [24], accelerates the frequency of spontaneous development of DH plants [25], eliminates weak or non-viable microspores in culture [26], delays anther wall senescence, prevents pollen abortion and increases symmetric division of pollen grains and release of necessary substances (cold shock proteins and amino acids) for androgenesis [27].

9. As alternatives to the 4 °C incubation, other treatments are also possible: (1) apply a hot pretreatment by placing the boots in incubator at 30 °C for 3–4 h [28], (2) spray Ethrel (4000 ppm) at the panicle initiation stage (just before meiosis in PMCs) [29], (3) apply 4000 ppm Ethrel pretreatment to boots for 48 h at 10 °C in BOD incubator [30], (4) treat anthers in 0.4 M mannitol for 3–4 days [31], (5) incubate anthers in sugar-free medium for just 2 days at the beginning of culture [32], or (6) administer low dose (20 kR) of gamma ray treatment [33] to improve anther culture response.

10. Inoculation of too many anthers per culture vessel increases the risk of mixing calli originated from different segregating microspores.

11. Avoid the use of 2,4-D concentrations higher than 2 mg/L in callus induction medium in order to prevent a reduction in plantlet regeneration. Allow maximum one sub-culture in CIM medium with reduced concentration of 2, 4-D (1.0 mg/L) and the same amount of Kn (0.5 mg/L as in primary culture) for callus proliferation with more frequency of embryoids.

12. Shorten the culture period to reduce gametoclonal variation that hinders green plant regeneration. Sexual hybridization of rice sub-groups (japonica/indica) may be adopted to improve green plant regeneration from anther-derived callus in indica rice.

13. Embryo formation gives rise to plantlets with simultaneous defined shoots and roots. To avoid mixoploidy, better use embryogenic calli, since organogenic calli may lead to chimeric plant regeneration.

14. In rice, as high as 60% of haploids in culture undergo spontaneous chromosome doubling (endoreduplication and nuclear fusion) [34, 35]. Rout et al. [36] recovered even 81.10% fertile diploid plants based on morpho-agronomic characters among 186 green plants from indica rice hybrid, CRHR32. The author recovered 129 plants (38.5%) out of 335 anther culture-derived plantlets as spontaneous double haploid (DH) status, each with 12 pairs of chromosomes (Fig. 5). These DH plants maintained normal growth without any abnormality and set seeds upon flowering. In the follow-up generation, progenies within each of the 129 DH lines exhibited uniformity in morpho-agronomic traits. Therefore, the use of colchicine may not be necessary unless desired for increased DH production [16].

15. Treatment with high concentrations of colchicine may lead to polyploidy. If needed, use colchicine treatment at early stages of androgenesis to increase the recovery of DHs and to avoid risk of chimerism due to mixoploidy.

16. Alternatively, roots of very young haploid plantlets may be immersed in a filter-sterilized solution of colchicine (0.4%) for 2–4 days followed by their transfer to normal MS basal liquid medium for further growth. However, there is risk of mixoploidy resulting chimeric plants and low seed set.

17. Anther-derived plants have different ploidy levels. The test of homozygosity of regenerants mainly relies on progeny testing after self-pollination to confirm DH status. For example, progenies within each of the 129 regenerants out of 335 plants recovered by the author were homozygous (true breeding) and homogeneous (highly uniform). The rest of the plantlets either did not survive during plant establishment in the field or showed abnormal growth without flowering (sessile).

18. Normally, plants with 5–8, 10–15 and 18–24 chloroplasts/cell are designated as monoploids, diploids (here DHs) and tetraploids, respectively.

19. For chromosome counting, pretreat the meristematic root tips (2–3 mm) with 0.002 M 8-hydroxyquinoline for 2 h at 25 °C to spread and arrest the metaphase chromosomes followed by

fixing the material with Carnoy's fixative for at least 2 days. Prior to mitotic slide preparation, digest the root tips with enzymatic mixture of 6% cellulose and 2% pectinase (prepared in 0.01 M citrate buffer) in an Eppendorf tube at 37 °C for 60 min. Using water bath. Wash the root tips with 0.01 M citrate buffer ($3\times$, 10 min each) followed by wash with distilled water ($2\times$, 5 min each) to remove the enzymes. Take one root tip on an ethanol-cleaned glass slide, remove excess water using blotting paper and then add a drop of aceto alcohol fixative. Squash thoroughly by a needle, add another 1–2 drops of fixative and spread the content over the slide and leave it for an overnight. The next day, stain the chromosomes by adding a drop of 1% acetocarmine stain to the air dried cells, cover with a cover slip, warm the slide on a spirit flame for just 2–3 s. Apply light pressure by thumb, remove excess stain using blotting paper and count chromosomes of cells at prometaphase stage under a phase-contrast microscope.

Acknowledgements

The author is grateful to all researchers for their valuable contributions included in this pursuit.

References

1. Niizeki H, Oono K (1968) Induction of haploid rice plant from anther culture. Proc Jap Acad 44:554–557

2. Guha S, Maheswari SC (1966) Cell division and differentiation of embryos in the pollen grains of datura *in vitro*. Nature 212:97–98

3. Tripathy Swapan K, Swain D, Mishra D, Prusty AM, Behera SK, Tripathy P, Chakma B (2018) Elucidation of the genetic basis of anther culture response and its breeding perspective in rice. Eur J Biotech Biosci 6(2):26–30

4. Grewal D, Manito C, Bartolome V (2011) Doubled haploids generated through anther culture from crosses of elite indica and japonica cultivars and/or lines of rice: large-scale production, agronomic performance, and molecular characterization. Crop Sci Soc Am 51 (6):2544–2553. https://doi.org/10.2135/cropsci2011.04.0236

5. Shariatpanahi ME, Bal U, Heberle-Bors E, Touraev A (2006) Stresses applied for the re-programming of plant microspores towards *in vitro* embryogenesis. Physiol Plant 127:519–534

6. Goncharova YK, Vereshchagina SA, Gontcharov SV (2020) Nutrient media for double haploid production in anther culture of rice hybrids. Plant Cell Biotech MolBiol 20 (23–24):1215–1223

7. Kapusi E, Hensel G, Coronado MJ, Broeders S, Marthe C, Otto I (2013) The elimination of a selectable marker gene in the doubled haploid progeny of co-transformed barley plants. Plant Mol Biol 81:149–160

8. Fan Y, Shabala S, Ma Y, Xu R, Zhou M (2015) Using QTL mapping to investigate the relationships between abiotic stress tolerance (drought and salinity) and agronomic and physiological traits. BMC Genomics 16:43

9. Tripathy Swapan K, Maharana M, Dash M, Bastia DN (2018) Marker aided selection of double haploids for drought tolerance in upland rice. In: 3rd ARRW Int. symposium—2018 on "Frontiers of rice research for improving productivity, profitability and climate resilience". Feb 6–9, 2018, NRRI, Cuttack, India, p 159

10. Jiang GH, He YQ,·Xu CG, Li XH, Zhang Q (2004) The genetic basis of stay-green in rice analyzed in a population of doubled haploid lines derived from an indica x japonica cross. Theor Appl Genet 108:688–698

11. Chaleff RS, Stolarz A (1982) The development of anther culture as a system for in vitro mutant selection. In: Rice tissue culture planning conference. International Rice Research Institute, Los Banos, pp 63–74

12. Tripathy Swapan K, Lenka D, Prusti AM, Mishra D, Swain D, Behera SK (2019) Anther culture in rice: progress and breeding perspective. Appl Biol Res 21(2):87–104

13. Datta SK, Wenzel G (1998) Single microspore derived embryogenesis and plant formation in barley (Hordeum vulgare). Arch Zeucht 18:125–131

14. Roy B, Mandal A (2011) Profuse microtillering of androgenic plantlets of elite indica rice variety IR 72. Asian J Biotechnol 3 (2):165–176

15. Tripathy Swapan K, Swain D, Mohapatra PM, Prusti AM, Sahoo B, Panda S, Dash M, Chakma B, Behera SK (2019) Exploring factors affecting anther culture in rice (Oryza sativa L.). J Appl Biol Biotech 7(2):87–92

16. Tripathy Swapan K (2018) Anther culture for double haploid breeding in rice—a way forward. Rice Genomics Genet 9(1):1–6

17. Premvaranon P, Vearasilp S, Thanapornpoonpong S, Karlade D, Gorinstein S (2011) In vitro studies to produce double haploid in Indica hybrid rice. Biologia 66 (6):1074–1081. https://doi.org/10.2478/s11756-011-0129-8

18. Kurata N, Omura T (1978) Karyotic analysis in rice. I. A new method for identifying all chromosome pairs. Jpn J Genet 53(4):251–255

19. Chawla HS (2004) Anther culture (Chapter 14). In: Chawla S (ed) Plant biotechnology: laboratory manual for plant biotechnology. Special Indian Edition. Oxford & IBH publishing Co. Pvt. Ltd., New Delhi, pp 1003–1112

20. Ochatt S, Pech C, Grewal R, Conreux C, Lulsdorf M, Jacas L (2009) Abiotic stress enhances androgenesis from isolated microspores of some legume species (Fabaceae). J Plant Physiol 166(15):1314–1328

21. Ribeiro CB, Pereira FC, .Filho LN, Rezende BA, Dias KOG, Braz GT, Ruy MC, Silva MB, Cenzi G, Techio VH, Souza JC (2018) Haploid identification using tropicalized haploid inducer progenies in maize. Crop Breed Appl Biotechnol 18. 16–23 https://doi.org/10.1590/1984-70332018v18n1a3

22. Doyle JJ, Doyle JL (1990) Isolation of DNA from plant tissue. Focus 12:13–15

23. Matsushima T, Kikuchi S, Takaiwa F, Oono K (1988) Regeneration of plants by pollen culture in rice (Oryza sativa L.). Plant Tissue Cult Lett 5:78–81

24. Chen CC, Tsay HS, Huang CR (1991) Factors affecting androgenesis in rice (Oryza sativa L.). In: Bajaj YPS (ed) Rice biotechnology in agriculture and forestry, vol 14. Springer, Berlin, pp 193–211

25. Amssa M, De Buyser J, Henry Y (1980) Origin of diploid plants obtained by in vitro culture of anthers of young wheat (Triticum aestivum L.). Natl Acad Sci Costa Rica 290:1095–1097

26. Lenka N, Reddy GM (1994) Role of media and plant growth regulators in callusing and plant regeneration from anthers of indica rice. Proc Indian Natl Sci Acad 60:87–92

27. Kiviharju E, Pehu E (1998) The effect cold and heat pretreatments on anther culture response of Avenasativa and A. sterilis. Plant Cell Tiss Organ Cult 54:97–104

28. Ferrie AMR, Keller WA (1995) Development of methodology and applications of doubled haploids in Brassica rapa. In: Proceedings of the 9th international rapeseed congress. Cambridge, UK, pp 807–809

29. Chawla HS (2010) In vitro production of haploids. In: Introduction to plant biotechnology, 3rd edn. Oxford & IBH Pub Co Pvt Ltd, New Delhi, p 95

30. Wang RF, Zuo QZ, Zheng SW, Tiang WZ (1979) Induction of plantlets from isolated pollen culture in rice (Oryza sativa L.). Acta Genet Sin 6:7

31. Raina SK, Irfan ST (1998) High-frequency embryogenesis and plantlet regeneration from isolated microspores of indica rice. Plant Cell Rep 17:957–962

32. Ogawa T, Fukuwa H, Ohkawa Y (1995) Plant regeneration through direct culture of isolated pollen grains in rice. Breed Sci 45:301–307

33. Mkuya MS, Si HM, Liu WZ, Sun ZX (2005) Effect of 137Cs gamma rays to panicles on rice anther culture. Rice Sci 12(4):299–302

34. Germana MA (2011) Anther culture for haploid and doubled haploid production. Plant Cell Tissue Organ Cult 104:283–300

35. Kasha KJ, Hu TC, Oro R, Simion E, Shim YS (2001) Nuclear fusion leads to chromosome doubling during mannitol pretreatment of barley (Hordium vulgare L.) microspores. J Expt Bot 52(359):1227–1238. https://doi.org/10.1093/jxb/52.359.1227

36. Rout P, Naik N, Ngangkham U, Verma RL, Katara JL, Singh ON, Samantaray S (2016) Doubled haploids generated through anther culture from an elite long duration rice hybrid, CRHR32: method optimization and molecular characterization. Plant Biotechnol 33 (3):177–186

INDEX

Jose M. Seguí-Simarro (ed.), *Doubled Haploid Technology: Volume 1: General Topics, Alliaceae, Cereals*, Methods in Molecular Biology, vol. 2287, https://doi.org/10.1007/978-1-0716-1315-3, © Springer Science+Business Media, LLC, part of Springer Nature 2021

Printed by Books on Demand, Germany